Elementary Physics

Second Edition

ELEMENTARY

PRENTICE-HALL, INC., Englewood Cliffs, New Jersey

PHYSICS

F. W. VAN NAME, JR.
Late Professor and Chairman
Department of Physics
Pratt Institute

DAVID FLORY
Associate Professor
Chairman, Department of Physics
Fairleigh Dickinson University

Library of Congress Cataloging in Publication Data

Van Name, F. W.
Elementary physics.

1. Physics. I. Flory, David, joint author. II. Title.
QC23.V27-1974 530 73-12289
ISBN 0-13-259515-X

PRENTICE-HALL PHYSICS SERIES
CONSULTING EDITORS, Francis M. Pipkin and George Snow

ELEMENTARY PHYSICS, Second Edition

F. W. Van Name, Jr. and David Flory

10 9 8 7 6 5 4 3 2 1

Printed in the United States of America

Prentice-Hall International, Inc., London
Prentice-Hall of Australia, Pty. Ltd., Sydney
Prentice-Hall of Canada, Ltd., Toronto
Prentice-Hall of India Private Limited, New Delhi
Prentice-Hall of Japan, Tokyo

Contents

10

Wave Motion 163

11

Light 189

Part Four
Modern Physics

12

Relativity 227

13

Quantum Mechanics and the Atom 248

Preface

This book presents the central ideas of physics, classical and modern, without using any mathematics more advanced than simple algebra. There is no trigonometry and vectors are treated graphically. It is especially designed for the one-semester or one-quarter course usually offered to non-science majors. The topics covered and their organization are, for the most part, traditional.

The introductory chapter discusses the nature of physics, the scientific method, and previews the forms taken by the laws of physics. This is followed by three chapters on mechanics, three on thermodynamics, four on electrodynamics and waves and three on modern physics. An appendix briefly summarizes some elementary mathematics. Simple numerical problems and discussion questions are given at the end of each chapter. The principal ideas of each chapter are applied to everyday situations, so that the student's interest will be aroused as he is able to relate physics to the world in which he lives. In addition, recent advances and applications are discussed where appropriate to give the student a feeling for physics as a dynamic, expanding field of science.

In preparing this revision of *Elementary Physics,* I have tried to modernize and revise the original without losing the informal, relaxed presentation of Professor Van Name. Wherever pedagogically feasible, topics have been presented in a manner consistent with contemporary thought on the subject. I have indicated the intrinsic limitations of each physical theory,

its realm of applicability and whether it will require subsequent modification. When discussing the various physical laws I have also attempted to distinguish carefully between the fundamental and the phenomenological.

The process of modernizing the book in accord with these guidelines involved substantial revision of the textual material. The section on scientific method has been rewritten. I have defined mass in terms of inelastic collisions, a definition that is independent of the concepts of acceleration and force. This leads directly into a derivation of conservation of momentum based only on symmetry and simple intuition. The conservation laws have been woven into the entire text and are emphasized as the most fundamental statements made in physics. Permanent magnetism is discussed in terms of circulating currents rather than pole strength. The treatment of special relativity has been expanded to an entire chapter, and the key effects of time dilation and length contraction are derived using light clocks. The twin "paradox" and its recent experimental varification by Hafele and Keating are discussed. The Bohr model is discussed as an intuitively useful but fundamentally incorrect intermediate step between classical orbits and the correct quantum mechanical picture of quantized states. Additional material on elementary particles has been provided. The SU(3) classification scheme and the quark model are briefly discussed.

In addition to the textual changes the book has been redesigned to increase its pedagogical effectiveness. The number of chapters has been increased to fourteen and the amount of material covered in each has been reduced. The number of illustrations has been substantially increased. They have all been redrawn and color has been added to improve their visual impact.

It is my hope that this book will continue to provide a vehicle whereby the non-science student with limited interest and background in mathematics can gain exposure to the world of physics, see the ways in which physics relates to and explains his everyday experience, and sense the excitement inherent in trying to confront nature and discover the laws that govern it.

D. F.

Elementary Physics

1 Introduction

1.1 THE NATURE OF PHYSICS

Traditionally, physics has been defined as the study of the properties of inanimate materials and their interactions with the rest of the universe. Today, this description is too restrictive. The boundaries between the various natural sciences (physics, chemistry, and biology) that seemed so clear in the 19th century have virtually disappeared.

A better approach would be to define physics in terms of the role that it plays in modern science. Physics provides the laws and principles on which the rest of modern science is built. It attempts to explain and correlate the inherent physical properties of the universe in terms of a few basic theories, which can then be used to explain other phenomena in physics and in the rest of science. The tools and experimental techniques of physics are now used in chemistry, biology, medicine, geology, astronomy, and all the other areas of scientific research. New specialties have arisen, such as biophysics, geophysics, astrophysics, and physical chemistry. Scientists in each of these disciplines are applying physics in areas that traditionally had not been associated with it.

Our definition of physics uses the word "property" in its most general sense to include almost all physical aspects of any situation. For instance, the physicist may be interested in a body's weight, the effect of gravity on it, its temperature, its electrical state, and many other properties.

On the other hand, he may want to know the structure and composition of a body, beginning with the theory of atoms and molecules and proceeding to the large-scale aspects of the body. Or he may deal with stars, planets, and satellites, so that he would ultimately be concerned with the structure, development, and origin of the universe. Perhaps he studies the properties of waves, such as light and radio waves. We could give many more examples, but these indicate typical branches of physics. In this book we shall take up the central ideas of physics without spending much time on applications or detailed areas that are found in courses in chemistry, engineering, and advanced physics.

1.2 SCIENTIFIC METHOD

The term *scientific method* refers to the principles and processes that guide scientific investigation. To many nonscientists, this much-abused term is a magical method by which scientists discover *truth,* with only the proper application of the method being necessary to achieve the desired result. In reality, the scientific method merely provides the framework within which scientists try to discover the laws governing the observable world. It also regulates the manner in which their inquiries are conducted.

In the early part of the 17th century, men like Galileo first began to make careful and systematic observations of natural phenomenon and to try explaining them. Galileo made many careful measurements of moving bodies. From these measurements, he was able to formulate rules for predicting motion. He also observed the planets and became immersed in the controversies surrounding the various models for the solar system. During the 2000 years before Galileo, scientists had shaped theories about the working of the physical universe, but their investigations had been motivated by philosophical or theological considerations and had relatively little connection with experimental observations. Galileo and some of his contemporaries introduced the concept that a theory was valid only insofar as it predicted and explained the observations on which it was based, i.e., empirically. This idea was a radical departure from previous thought; thus, it marked the beginning of modern scientific inquiry. The close connection between experimental observations and theory is the essence of scientific method. In the rest of this section we shall discuss this relationship and some of its implications.

The first step in constructing a physical theory is *observation,* which means making measurements of the properties of a system. These measurements can range from finding the dimensions of an object or the time at

which an event occurred to the highly complex indirect measurements that tell us the properties of elementary particles or stars. All of the properties ascribed to a given system must be based on some measurement. For example, consider an experiment to determine the color of an object. For a large object, such as a plastic block, this experiment is relatively simple. You illuminate the block with white light and look at it. If the block appears red, you can then say it is red. This experiment could be repeated by several different people; as long as one of them is not color-blind, they will all agree. On a more sophisticated level, you could measure the wavelength of the light reflected by the block, and you would find that the predominant wavelength present would correspond to that of red light.* Notice, however, that we have described a specific experiment for determining color. Had you used a pure blue light to illuminate the block, the outcome of your measurement would have been different, as can be attested by anyone who has seen a red car turn "brown" under mercury vapor highway lights. Any statement about a property of a system must be accompanied by a method for verifying that statement. If no such method exists, then the statement is outside the range of science.

An important characteristic of measurement is that its accuracy is always limited. This limitation comes in part from inaccuracies inherent in any measuring apparatus and in part from fundamental limitations set by quantum mechanics, which we shall study in Chap. 13. If you were given a stainless steel cube and a ruler, it might appear that the cube measured 2 in. on a side. However, you could not be absolutely certain, because an ordinary ruler is not accurate to more than about $\frac{1}{32}$ inch. A skilled machinist might use a micrometer and obtain a measurement good to $\frac{1}{1000}$ in. A testing laboratory might be able to measure the cube to an accuracy of $\frac{1}{1,000,000}$ in. However, the testing laboratory, and perhaps even the machinist, might find that, after all, they did not really have a cube but a slightly irregular rectangular object. For Euclid, perfect cubes existed in the philosophical sense because he could imagine one. For a scientist, an object such as a cube is defined only to the extent that he can measure its dimensions and only to the accuracy of his measurements. Today, a perfect cube would exist simply as an idealization, because the inherent limitation on the accuracy of any measurement makes the physical existence of such an object impossible to verify experimentally. Thus, for any observation or prediction in science to be meaningful, it must be qualified by some indication of its

*In this chapter we sometimes refer to concepts and terms that you may not be familiar with yet. In this case, substitute some other concept that you are familiar with and try to complete the argument. Perhaps, when you have finished the book, you can return and reread this chapter with the new perspective of what you have learned.

accuracy. One of the major reasons for evolution in science has been the increased accuracy of experimental observations, which has forced the development of more precise theory.

The development of a *theory* to explain a given body of data may come in many ways. Sometimes years of painstaking data analysis is needed while the scientist searches for patterns and regularities in his data; rarely, a sudden flash of creative genius is sufficient. Sometimes the scientist works inductively, finding regularities, fitting equations to them, and then seeing if those equations will explain other observations. Sometimes he works deductively, starting from other known laws and trying to apply them to his problem in a logical fashion. After a theory has been formulated, it must be tested. It must predict the observations; and if it does not, it must be discarded and another found. In mathematics, the only criterion for truth is logical consistency. In science, however, the situation is different. For a scientific theory to be valid, it must be experimentally verified. It must make specific numerical predictions that can be confirmed by measurement.

Two important limitations on any physical theory should be noted. First, all theories have some limits on the range of phenomena that they describe. For example, classical Newtonian mechanics, which we shall study in the next three chapters, describes and predicts correctly the dynamics of motion on the earth. However, for speeds in excess of about 10,000 miles per second, it is not accurate and must be modified by special relativity. We do not say that the theory is wrong (it is accurate to better than one part in a billion for ordinary motion) but only that in certain extreme situations it must be modified and extended.

The second limitation is that all theories involve some type of approximation or idealization. Approximation plays an important role in science because most real problems are too complicated to analyze exactly. For instance, in describing projectile motion, we begin by neglecting air resistance and the curvature of the earth. With this modification, the equations are relatively simple and can be solved. However, the predictions of the theory will not agree exactly with experiment. Again, we do not mean that the theory is wrong, only incomplete. Once the simplified problem is solved, then the changes produced by the neglected factors can be found, a more precise theory formulated, and better agreement between theory and experiment obtained. This process of refinement is never complete. There are always some approximations that must be made, and any theory always has some limit on the accuracy to be expected from its predictions.

In the development of science there is a continual tug-of-war between theory and experiment. Sometimes an experimental breakthrough will occur, producing a large amount of new data requiring years of theoret-

ical work to understand and explain. Sometimes a theoretical development will give predictions that are beyond the ability of the experimentalist to test. Then the theoretician must wait for new techniques to be developed before his theory can be tested. But always theory grows out of experiment and in turn is verified by experiment. Since carefully taken data is rarely incorrect, the theories that explain old experiments are not discarded in this process of development. Instead they are extended and refined, or unnecessary concepts are removed or replaced by better ones. Sometimes several different theories are combined into one more general one. Only very rarely is an old theory discarded as wrong. In this century, Newtonian mechanics has been modified by special relativity for extremely high-speed motion and by quantum mechanics for very small distances. However, both of these theories reduce to Newton's equations when applied to those situations for which Newtonian mechanics were originally intended.

After a theory has survived many years of experimental investigation, it is frequently accorded the status of a *law.* The usage of this term is rather inconsistent. For example, Hooke's Law, which describes the behavior of springs, is valid only under certain circumstances, but it is very useful within its range of application, so it is called a law. At the other extreme are the Laws of Conservation of Energy and Momentum, which are so powerful and general that no exceptions to them have ever been observed. They are believed to hold absolutely. Any putative theory that disagreed with them would be discarded immediately as incompatible with observation. Another class of theories provides a formalism within which specific laws and physical systems can be investigated. Examples of this type are classical (Newtonian) mechanics, relativity theory, and quantum mechanics. Quantum mechanics is presently believed by most scientists to provide a formalism sufficiently flexible and general to encompass all present theory, although the task of doing so probably will never be completed.

We have seen how science explains its observations of the world in terms of theories that survive the test of experimental verification. Some theories it explains in terms of other more fundamental theories. Modern science can predict the behavior of the physical world with tremendous accuracy and confidence. There is one question, however, it can never answer: the question "why?" Phenomena can be explained in terms of theories and theories explained in terms of other theories, but ultimately the scientist is always reduced to saying that this phenomenon is what he observed. For instance, the Law of Conservation of Momentum can be derived from a very few simple self-evident statements about the world. These statements are so obvious that no one would question them, yet we

cannot say *why* they are true. This question must be answered by teleology, not science.

1.3 THE LAWS OF PHYSICS

In this section we shall discuss in rather general terms some of the forms taken by the laws of physics. We shall also introduce some concepts and notation that will prove useful in later chapters.

The simplest statement that can be made about a physical quantity is that it does not change. Laws of this sort are called *conservation laws.* Although at first glance the concept of something not changing may not seem particularly useful, conservation laws are perhaps the most profound and fundamental statements in physics. As an example of a conservation law, consider combining 2 grams (gm) of hydrogen with 16 gm of oxygen and igniting the mixture. The result, after combustion is completed, will always be exactly 18 gm of water. (Why this particular combination works must be answered by a chemistry course.) This experiment is an example of mass conservation, a law that holds with tremendous accuracy except in the relativistic region, where it must be combined with conservation of energy (as we shall see in Chap. 12). Conservation laws are particularly useful because they give us partial information about any system, even one so complex that a complete solution is impossible. Suppose that you know the initial energy of a system and you can keep track of just how much energy enters or leaves the system as it undergoes some change. Even if the detailed evolution of the system is unknown, the final energy can be found from the Law of Conservation of Energy.

Most laws of physics describe how observable quantities change. Newton's Laws of Motion predict how the position of an object will change in time. His Law of Universal Gravitation tells how the gravitational force between two objects changes when their distance or mass is altered. Since change is so important, we denote the change in a quantity by a special symbol, Δ, a capital Greek delta. This symbol can be read as "the change in. . . ." For instance, suppose that you enter a highway at mileage marker 67 mi and exit at marker 101 mi. The distance that you traveled is 34 mi. If we use an x to denote distance, then we say $\Delta x = 34$ mi, to be read as "the change in distance equals 34 miles." The distance traveled along the highway could have been calculated in another way. Perhaps the odometer of the car read 26,248 mi as you entered the highway. Then as you left, it would have read 26,282 mi, giving $\Delta x = 34$ mi again. A last possibility is that the car be equipped with a trip odometer that would be set at zero as you entered the highway. This odometer would then have given Δx directly.

In each of these examples, Δx was calculated in some appropriate distance scale by subtracting the initial value of the car's position from the final value. We always find Δx in this manner. In general, let some dynamical quantity A have the value A_{initial} at the beginning of some interval and A_{final} at the end of the interval. We then define ΔA, the change in A, by the equation

$$\Delta A = A_{\text{final}} - A_{\text{initial}} \tag{1.1}$$

This equation can be rewritten by adding A_{initial} to both sides, giving

$$A_{\text{final}} = A_{\text{initial}} + \Delta A \tag{1.2}$$

In words, ΔA is a quantity that we add to the initial value of A to give the final value. In the definition of ΔA [Eq. (1.1)], if A increases, then ΔA is positive; and if A decreases, then ΔA is negative. For example, consider a person with an initial weight W of 187 lb who goes on a diet and finishes with a final weight of 175 lb. Using Eq. (1.1), we find that

$$\Delta W = W_{\text{final}} - W_{\text{initial}} = 175 - 187 = -12 \text{ lb}$$

A negative ΔW represents a loss in weight.

Another important physical quantity that we often describe with our Δ notation is time. The measurement of time intervals is very important in physics, because we are always interested in how fast a quantity changes. The calculation of a time interval Δt is the same as the calculation of the change in distance or weight. The initial time is subtracted from the final time, giving Δt. Notice that it does not matter whether we use standard time, daylight time or Greenwich Mean Time for t_{initial} and t_{final}. As long as our clock is accurate, Δt will be the same. It is often convenient to make t_{initial} equal zero by starting the measuring clock as the process under study begins; then we can read Δt directly from the clock.

You will find this notation useful because many laws of physics are written in the form

$$\Delta V = k\,\Delta t \tag{1.3}$$

where V and t are some observable quantity and k is a constant. The law represented by Eq. (1.3) says, in words, that the change in V is *directly proportional* to the change in t. As an example of direct proportion, suppose that you are filling a 100-gal oil tank at a constant rate. After 1 min, you

see that 5 gal of oil have flowed into the tank. You then know that after 2 min, 10 gal will have been added; after 3 min, 15 gal; and so on. You could then calculate that it would take 20 min to fill the tank. The reasoning is very simple. For each time interval of 1 min, the amount of oil in the tank increases by 5 gal; the change in the amount of oil is directly proportional to the time elapsed. To put this calculation in the form of an equation, we write Eq. (1.3) as

$$\Delta V = 5\,\Delta t \tag{1.4}$$

where V stands for the volume of oil in gallons and t the time in minutes. We have evaluated the constant k given in Eq. (1.3) by solving for k:

$$k = \frac{\Delta V}{\Delta t} \tag{1.5}$$

Then when $\Delta t = 1$ min and $\Delta V = 5$ gal, $k = 5$ gal/min. In general, if we know that Eq. (1.3) holds, and we know ΔV for some Δt, then the constant k can be evaluated by using Eq. (1.5).

With Eq. (1.4), you could also find the volume of oil added for any given time interval. If you started with an empty tank, so that $V_{\text{initial}} = 0$, and you wished to fill the tank, $V_{\text{final}} = 100$, then Eq. (1.4) gives

$$\Delta V = V_{\text{final}} - V_{\text{initial}} = 100 - 0 = 5\,\Delta t$$

This equation then can be solved for the time needed to fill the tank:

$$\Delta t = \frac{100}{5} = 20 \text{ min}$$

On the other hand, if the tank had started with 20 gal, then the ΔV needed to fill it would have been only 80 gal, and only 16 min would have been needed.

Many other physical laws take the form of Eq. (1.3). For instance, when an object is heated, it generally expands. As we shall see in Chap. 6, the change in its length is directly proportional to the change in the temperature. Numerous other examples abound; you will meet many of them in the remainder of this book.

DISCUSSION QUESTIONS

1. Discuss divine creation and evolution in terms of their value as scientific theories. Do they have predictive power? Can they be tested?

2. Give several examples of conservation laws that operate in ordinary life. Discuss their usefulness.

3. The earth's axis of rotation is tipped at an angle of 23.5 deg to the axis of the solar system and presently points at Polaris, the North Star. Over a period of some 26,000 years, the earth's axis will circle the axis of the solar system at this angle; thus, 26,000 years from now Polaris will again be the polestar. Discuss the limitations in accuracy of a navigation system based on Polaris. How should this system be modified so that it will work 13,000 years from now?

4. Give examples of quantities that change at a constant rate and thus satisfy simple proportion laws.

5. Galileo found that when an object rolls down an inclined plane, it acquires velocity at a constant rate. Will the distance that it travels be proportional to time? If not, why?

Part One
Mechanics

2 Kinematics

In the next three chapters we shall study how and why things move. This subdivision of physics is called *mechanics*. Mechanics is divided into two main parts. The first, *kinematics,* deals with the description of motion. The second, *dynamics,* deals with the causes of motion. We begin by learning to describe motion.

2.1 STANDARD UNITS

Physics is a quantitative science. To describe things quantitatively, we must make measurements. Although the information that a town is "not far away" may be helpful, we would much rather be told that it is about 5 miles away, or better yet, that it is 5.3 mi away. Thus, the first step in describing or measuring a quantity is to set up for each quantity a standard unit that can be discussed objectively. For instance, the standard unit of the monetary system in the United States is the dollar. Like all other units of measure, the dollar has multiples, such as 5 and 10 dollars, and submultiples, such as the dime ($0.10) and the cent ($0.01). Note that the dollar is defined quite arbitrarily and that the related units are chosen for their convenient

magnitudes, which are often related to the standard unit by powers of 10. The same method is followed in science.

Until the end of the 18th century, each country had its own set of units, which usually had no relation to units used in other countries. A unit of length, for instance, the inch, was equal to the length of a part of the king's finger. Another unit of length was the fathom, which was the distance between a man's fingertips when his arms were outstretched. Furthermore, relations among the units were not simple. English-speaking countries, for example, still use these equivalents: 1 yard = 3 feet = 36 inches.

At the time of the French Revolution (1789), an attempt was made to introduce rational and permanent standards as well as to derive the related units by various powers of 10. Called the *metric system,* the standards were gradually adopted by scientists and are now used throughout most of the world. This system is the one that we shall primarily follow in this book. For your convenience, however, references will be made to the British system of units, which is commonly used in the English-speaking countries.

The standard unit of length established by the French is called the *meter,* from Greek, Latin, and Anglo-Saxon words meaning "measure." Originally, the meter was supposed to be 1/40,000,000 of the earth's circumference measured from a great circle passing through Paris. Thus, the earth itself could always be used to redetermine the length of the meter. According to the best knowledge of the earth's size available at the time, the length of the meter was determined as the distance between two scratches on a platinum-iridium bar. When later measurements of the size of the earth showed that the original distance was slightly incorrect, the meter was redefined as being *exactly* the distance marked on the bar. The original meter bar is kept just outside Paris at Sevres, France. Other laboratories, such as the U.S. Bureau of Standards, have copies of this meter bar; they are compared with the original bar from time to time.

Recently, the art of making precision measurements had advanced to such a degree that the scratches on the platinum-iridium bar at Sèvres were not accurate enough to satisfy scientists. The meter was redefined as a certain multiple of the wavelength of a red light emitted by krypton. As a result, scientists are now able to measure distances to better than one part in a million. However, we shall not be making such precise statements in this book, so this sophistication will not concern us further.

Because the meter is a bit over 3 ft in length, it is not a convenient size for some measurements. In the metric system, multiples and submultiples of the meter are derived by using various powers of 10. Prefixes taken from the Greek designate multiples, as shown in the following table:

deka	10	1 dekameter = 10 meters
hecto	100	1 hectometer = 100 meters
kilo	1000	1 kilometer = 1000 meters
mega	10^6	1 megameter = 10^6 meters
giga	10^9	1 gigameter = 10^9 meters

Similarly, prefixes derived from the Latin indicate submultiples:

deci	$\frac{1}{10}$	1 decimeter = $\frac{1}{10}$ meter
centi	$\frac{1}{100}$	1 centimeter = $\frac{1}{100}$ meter
milli	$\frac{1}{1000}$	1 millimeter = $\frac{1}{1000}$ meter
micro	10^{-6}	1 micron = 10^{-6} meter
nano	10^{-9}	1 nanometer = 10^{-9} meter
pico	10^{-12}	1 picometer = 10^{-12} meter

The only common units of length are the kilometer, meter, centimeter, millimeter, and micron. However, the same prefixes are used to designate multiples and submultiples of other scientific quantities, so they appear in all areas of physics.

We can convert between the metric and British systems of units by using the following equivalency table, which is also shown in Fig. 2.1.

1 inch (in.) = 2.54 centimeter (cm)
1 foot (ft) = 30.5 cm
1 meter (m) = 39.37 in.

The first of these equivalents is exact, because it is the legal definition of the inch adopted by Congress. The other two relationships are sufficiently accurate for your use in this book.

One frequent problem is the conversion from one unit to another of the same type but of different size. Suppose that we wish to convert 8 ft to its equivalent value in inches. Consider the fraction 12 in./1 ft. The numerator and denominator are equal, so the value of the fraction is 1. Thus, we can multiply by this fraction without changing the physical length, but we do change the unit in which it is measured. We can write:

$$8 \text{ ft} = 8 \text{ ft} \times \frac{12 \text{ in.}}{1 \text{ ft}} = 8 \times 12 \text{ in.} = 96 \text{ in.}$$

We cancel feet in the numerator against feet in the denominator, leaving the inch as the remaining unit. The remainder of the problem consists of multiplying 8 and 12 in order to get the numerical part of the result.

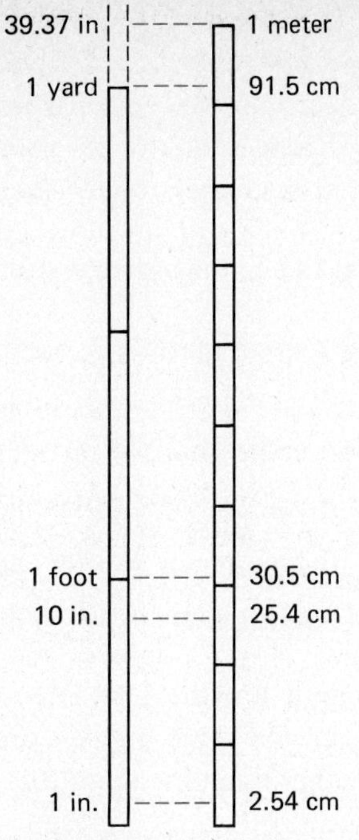

FIGURE 2.1

The relation between the English and metric systems is illustrated by a meter stick and a yard stick side by side.

EXAMPLE As a somewhat more difficult example, let us convert 60 miles (mi) per hour (hr), or 60 mph, into feet (ft) per second (sec). We write

$$60 \text{ mph} = \frac{60 \text{ mi}}{1 \text{ hr}} \times \frac{5280 \text{ ft}}{1 \text{ mi}} \times \frac{1 \text{ hr}}{60 \text{ min}} \times \frac{1 \text{ min}}{60 \text{ sec}}$$

$$= \frac{60 \times 5280}{60 \times 60} \text{ ft/sec} = 88 \text{ ft/sec}$$

In this case, we cancel miles against miles, hours against hours, and minutes against minutes. The remaining units are ft/sec. The numerical part of the result is $(60 \times 5280)/(60 \times 60)$, which equals 88.

In order to describe motion, we must measure time as well as distance. The standard unit of time used in science is the second. Fortunately, the second and its familiar multiples, the minute (60 sec), the hour ($60 \times 60 = 3600$ sec), the day ($24 \times 3600 = 86{,}400$ sec), and the year (about 3.16×10^7 seconds) are in universal use, so there is no problem of conversion. The second was originally defined in terms of astronomical measurements. In the simplest approach, 1 second equaled 1/86,400 of an average day. However, the actual length of a day is not constant. The earth never

goes around the sun in exactly the same way nor does it spin on its axis in a precisely regular manner. Consequently, it became difficult to define astronomical time sufficiently accurately to serve the needs of precision scientific work. In 1967, scientists agreed to define the second as a certain multiple of the vibration time of a cesium atom. This standard, called *atomic time,* is the one now used in scientific work. But for our purposes, the second is still $\frac{1}{60}$ min and can be measured with sufficient accuracy by any good watch.

Physics deals with numbers that range from the very big to the very small. The time intervals that we are able to perceive with our unaided senses range from a fraction of a second to a lifetime or, in numbers, from a few hundredths of a second to a few billion. Scientists, aided by their tools, have extended this range tremendously. Figure 2.2 shows some representative time intervals, beginning with a time of 0.000,000,000,000,000, 000,000,02 sec and going up past 200,000,000,000,000,000 sec. Writing these numbers with so many extra zeros is clearly very cumbersome. Therefore, scientists use what is called the *power-of-10-notation.* In this system powers of 10 are factored from each number so that the numerical part left over is between 1 and 10. This notation is explained in detail in the appendix on exponential notation. For instance, since 1 billion has 9 factors of 10 in it, we write 1,000,000,000 as 10^9. Two billion sec, the average life expectancy of a man in the United States, would be written as 2×10^9 sec, which is how it appears in Fig. 2.2. The very small number that we wrote in Fig. 2.2 is actually 2 divided by 10^{23}, and we write it as 2×10^{-23} sec. This number indicates the time for light, or a very energetic elementary

6×10^{17} Seconds	Age of universe
2×10^{17}	Age of earth
2×10^{13}	Time of man on earth
2×10^{11}	Age of human civilization
2×10^{9}	Human lifetime
3×10^{7}	One year
9×10^{4}	One day
5×10^{2}	Light travels from sun to earth
8×10^{-1}	Time between heartbeats
1×10^{-5}	Duration of stroke flash
3×10^{-9}	Light travels one meter
2×10^{-16}	Half-life of charged pion
4×10^{-19}	Light crosses atom
2×10^{-23}	Light crosses small nucleus

FIGURE 2.2
The order of magnitude of various time intervals ranging from the longest to the shortest. The representative intervals are approximate.

2×10^{26} Meters	"Radius" of Universe
2×10^{22}	Distance to nearest galaxy
4×10^{16}	Distance to nearest star
2×10^{11}	Distance to sun
7×10^{8}	Radius of sun
6×10^{6}	Radius of earth
1×10^{4}	Highest mountain
1×10^{2}	Football field
2×10^{0}	Man
1×10^{-4}	Thickness of paper
1×10^{-8}	Small virus
5×10^{-11}	Diameter of atom
1×10^{-15}	"Radius" of proton

FIGURE 2.3
The order of magnitude of some distances observable in the physical world.

particle, to cross a small nucleus. It is also the approximate lifetime of some extremely unstable "particles" studied by elementary-particle physicists. The very large number in Fig. 2.2 is written as 2×10^{17} sec, about 5 billion years, and is the approximate age of the earth. The largest time shown in Fig. 2.2 is 6×10^{17} sec, which is the age of the universe according to some of the current theories about cosmological evolution. Figure 2.3 shows a group of representative distances encountered in physics. These distances range from 10^{-15} m to 10^{26} m. The former, 10^{-15} m, is the "radius" of a proton, the smallest object with a discernible size. The latter, 10^{26} m, is the distance to the "edge" of the universe, that is, the distance to the farthest observable objects. This distance is also about the distance that light could have traveled since the "birth" of the universe. Note that the ratio of the largest distance to the smallest distance and the ratio of the largest time to the shortest time are both about 10^{40}, a rather large range of numbers.

2.2 SPEED AND ACCELERATION

We all have an idea of what is meant by the word "speed." For instance, we say that we are driving at 40 miles per hour (mph) or that our average speed during a trip was 35 mph. Similarly, when we speed up a car, we have a feeling of acceleration, which we associate with the change in speed. In this section we shall make these ideas quantitative as we study motion in a straight line. In Sec. 2.4, we shall consider motion in a curved path.

Suppose that we travel a distance of Δx in a straight line during

a time Δt, as shown in Fig. 2.4. Then we define our average speed $\bar{v}$ during this time interval by the equation

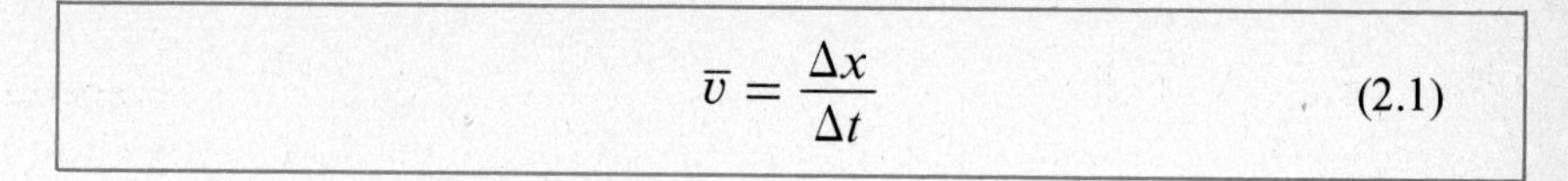

$$\bar{v} = \frac{\Delta x}{\Delta t} \tag{2.1}$$

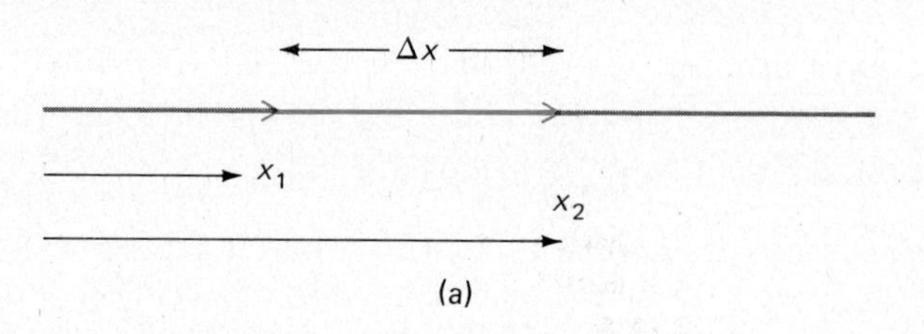

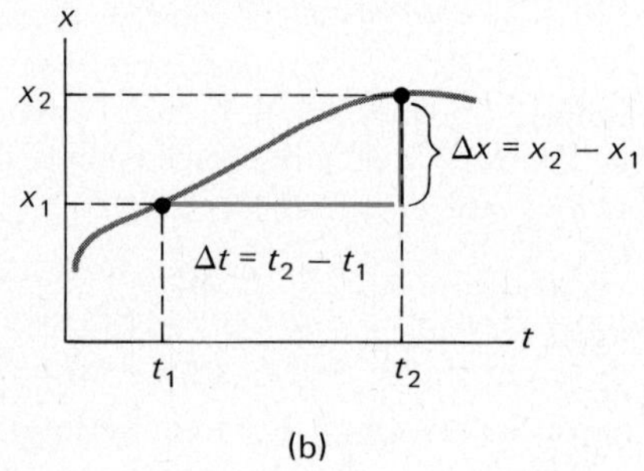

FIGURE 2.4
(a) The colored arrow shows motion from point x_1 to point x_2, a change in distance of $\Delta x = x_2 - x_1$. The time elapsed is $\Delta t = t_2 - t_1$, where t_2 and t_1 are the final and initial times. (b) The same quantities shown on a graph of distance versus time.

In words, we would say that the average speed is the distance traveled (change in distance $= \Delta x$) divided by the time it took to travel that distance (time elapsed $= \Delta t$). Note that it is customary to indicate an *average* value of a quantity by putting a line over its symbol. This is the reason that the average speed was written as $\bar{v}$.

Clearly, the unit of speed can be any unit of distance divided by any unit of time. Because we have decided to use the meter and the second as our units of time and distance, our fundamental unit of speed will be meters per second (m/sec), although other units, such as centimeter/second, feet/second, and miles/hour (mph), are often used.

We can rewrite Eq. (2.1) algebraically in order to solve for either the distance covered during a certain time interval or the time required to cover a specified distance at a given average speed. Thus, we can obtain

the following two equations:

$$x = \bar{v}t \tag{2.2}$$

$$t = \frac{x}{\bar{v}} \tag{2.3}$$

We have dropped the Δ from x and t, but you should remember that x is the change in distance and t the change in time. These equations are simply different ways of writing Eq. (2.1).

EXAMPLE Suppose that you run 200 m in 20 sec. Your average speed is then given as $\bar{v} = 200\ \text{m}/20\ \text{sec} = 10\ \text{m/sec}$. Suppose now that we would like to compute the distance that you would run at this same average speed during 4 min. Because 4 min is the same as 240 sec, from Eq. (2.2) we see that the distance would be $d = 10\ \text{m/sec} \times 240\ \text{sec} = 2400\ \text{m}$, which is a great deal more than a mile. Similarly, the time for you to cover 1500 m (the metric mile) could be found by using Eq. (2.3) in which $t = 1500/10 = 150\ \text{sec} = 2{:}30\ \text{min}$, which is more than a minute below the world record for this foot race.

In many cases, the average speed calculated from Eq. (2.1) will be different at different times. For example, when you drive a car, your speed is zero when you are stopped by a traffic light and will have other values as you start up. After you reach cruising speed (say, 60 mph on a highway), your speed will be constant. What meaning can we give to the word "speed" during the time that you are accelerating? If we calculate the average speed for a very short time interval, one short enough that the speed does not change very much, then the average calculated will be sufficiently close to the actual speed for us to say that they are the same. Thus, we define the *instantaneous speed* at a point as the average speed over a small distance near the point. Equation (2.1) becomes

$$v = \frac{\Delta x}{\Delta t} \tag{2.4}$$

when Δx and Δt are very small. The size of Δx and Δt depends on the accuracy desired. For instance, in the case of a freight train pulling out of a station, measuring the distance Δx that the train covered in a time $\Delta t = 1$ sec might be sufficiently accurate. For a drag race, however, we might have to use a Δt as small as 10^{-2} sec to get the same accuracy. The speedometer on your car indicates the average speed of your car over a short distance, which is almost exactly your instantaneous speed.

Suppose that your speed changes by an amount $\Delta v = v_2 - v_1$ during a time Δt, as shown in Fig. 2.5. We then say that you have accelerated

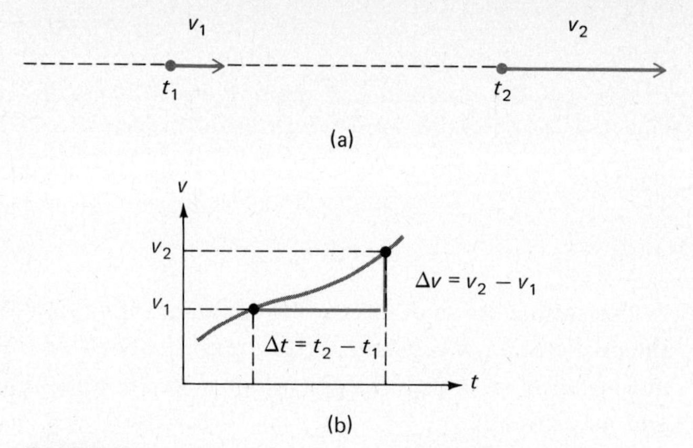

FIGURE 2.5
(a) An object moving from point 1 to point 2 and increasing its speed. The length of the colored arrows is proportional to the speed at the two points. (b) The speed versus time graph for (a).

during this time interval, and we define your *average acceleration* $\bar{a}$ during this time interval by the equation

$$\bar{a} = \frac{\Delta v}{\Delta t} \tag{2.5}$$

If your speed is varying rapidly, we must use a small time interval to calculate your actual or *instantaneous acceleration* near a point. All examples in this book deal with cases of constant acceleration, so that the average and instantaneous accelerations will be the same. From now on, we shall ignore the distinction.

From the definition of acceleration, Eq. (2.5), we see that any unit of speed divided by a unit of time will equal a unit of acceleration. In our examples we measure speed in meters per second and time in seconds, so that the unit of acceleration is meters per second squared or m/sec^2. Other units are possible, but we shall avoid them.

EXAMPLE Suppose that your speed increased from 30 m/sec to 100 m/sec during 5 sec. Then your acceleration would be

$$\bar{a} = \frac{100 - 30}{5} = 14 \text{ m/sec}^2$$

If your speed dropped from 40 m/sec to 10 m/sec during 6 sec, your acceleration would then be

$$\bar{a} = \frac{10 - 40}{6} = -5 \text{ m/sec}^2$$

In this case, the acceleration is negative, which is usually called a *deceleration*. Generally, however, the word "acceleration" means either an increase or decrease in speed.

2.3 LINEAR MOTION

In this section we shall discuss the equations of motion for situations where the acceleration is constant. With these equations we can find the location and speed of an object at a given time if we know its initial position, speed, and acceleration.

We call the initial speed of an object v_0. We always start our clocks when the object is at its starting position; therefore, $\Delta t = t - 0 = t$, and $\Delta v = v - v_0$. If we multiply both sides of Eq. (2.5) by t, we obtain

$$at = v - v_0$$

From this equation, we see that v, the speed after time t, is given by the equation

$$v = v_0 + at \tag{2.6}$$

Figure 2.6 shows graphically how the speed changes when the acceleration is constant. Because the speed changes uniformly with time, we see from

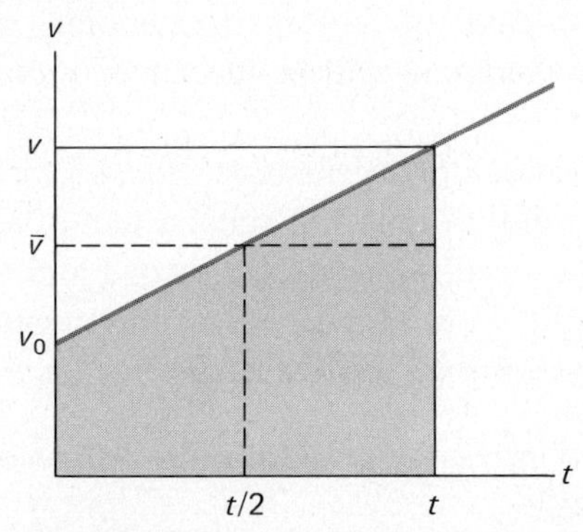

FIGURE 2.6
A graph of speed versus time for constant acceleration. The area of the colored region and the area under the dotted line are both equal to the distance traveled.

Fig. 2.6 that the average speed over the time interval t is simply the average of initial and final speeds. Thus, we can say that the average speed during the time t is

$$\bar{v} = \frac{v + v_0}{2} \qquad (2.7)$$

As a general rule, therefore, for motion with constant acceleration, the average speed over any time interval is just the average of the initial and final speeds. Now we can find the distance traveled during the time t. According to Eq. (2.2), the distance x is

$$x = \bar{v}t \qquad (2.8)$$

If we use Eqs. (2.6), (2.7), and (2.8), we can solve any problem involving motion with constant acceleration. Depending on the given information, we must begin with one of these three equations and solve in turn for the unknown quantities.

Equation (2.8) has an interesting geometrical interpretation. If we refer to Fig. 2.6, we see that $\bar{v}t$ is just the area inside the rectangle whose width is t and height $\bar{v}$. This area, however, is exactly equal to the area of the shaded region of the graph, the area *under* the curve representing the speed at each instant of time. Although a proof is outside the scope of this book, this rule holds true for any type of motion, whether the acceleration is constant or not. To find the distance traveled during a certain time interval, all we have to do is to calculate the area under the graph of speed versus time. Thus, if we know the speed of an object at any time, the only limitation on our knowledge of the distance that the object has traveled is our ability to calculate areas.

EXAMPLE Suppose that an object with an initial speed of 10 m/sec is subjected to an acceleration of 3 m/sec^2 for a time of 5 sec. The problem is to find its speed at the end of the time interval and the distance covered during the time interval. In this case, we can say that $v_0 = 10$, $a = 3$, and $t = 5$. Using Eq. (2.6), we find for the final velocity:

$$v_1 = 10 + 3 \times 5 = 25 \text{ m/sec}$$

Although the average velocity was not asked for, we must use Eq. (2.7) to find it; thus,

$$\bar{v} = \frac{10 + 25}{2} = 17.5 \text{ m/sec}$$

Now from Eq. (2.8) we find for the distance covered:

$$x = 17.5 \times 5 = 87.5 \text{ m}$$

You should note two features of this example. First, it is usually worthwhile to identify the given information with the standard letters used in the equations, such as v_0, a, and t. Second, in many cases, you must compute quantities that are not asked for, as was the case here when we computed the value of the average velocity.

Because of its frequency in everyday life, motion caused by gravity is an important example of motion in a straight line with constant acceleration. Scientists have found experimentally that any object, regardless of its weight, always falls with a constant acceleration of 9.80 m/sec^2 or 32.2 ft/sec^2 under the influence of gravity when air resistance is neglected. This acceleration of a freely falling body is called the *acceleration of gravity* and is usually denoted by g. The value of g is not really a constant. There are small variations in its value at different places on the earth's surface because of variations in the density of the earth's crust. Its value also decreases farther away from the earth's surface. The reasons for these variations will become clear in Chap. 4 when we study gravitation. We shall take $g = 10$ m/sec^2 or $g = 32$ ft/sec^2 as our values for the acceleration of gravity. These numbers will simplify calculations and are accurate to within 5 per cent, which is sufficient for our purposes.

EXAMPLE Consider the following problem, which is amusing but not really typical. A stone is released near the top of a building and requires 8 sec to fall to the pavement below, as shown in Fig. 2.7. What is the address of the building? This is an extreme example of a problem in which there is no apparent relation between the facts given

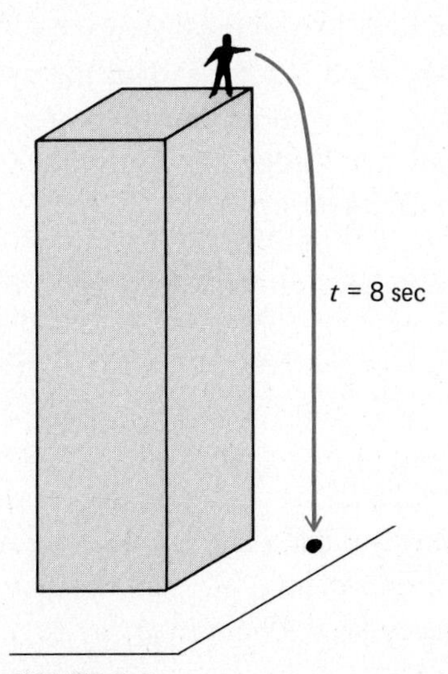

FIGURE 2.7
A stone is dropped from the top of a skyscraper. It hits the ground 8 sec later.

and the result required. In this case, we must use the facts given to calculate whatever quantities we can, in the hope that eventually the way to the answer will become clear. This problem concerns motion with constant acceleration (due to gravity, so that $a = 32$ ft/sec^2. Also, $v_1 = 0$. Using Eq. (2.6), we calculate that

$$v_2 = 32t = 256 \text{ ft/sec}$$

Equation (2.7) gives us

$$\bar{v} = \frac{v_2}{2} = 128 \text{ ft/sec}$$

From Eq. (2.8) we have

$$d = \bar{v}t = 128 \times 8 = 1024 \text{ ft}$$

Thus, we find that the height of the building from the point at which the stone was released is 1024 ft. Only one building in the world satisfies this condition, namely, the Empire State Building, which is located at 34th St. and Fifth Ave., New York City. Here is the answer to our problem. However, we had to compute the values of quantities that were not called for in the problem until we finally got the required information.

2.4 VELOCITY AND ACCELERATION USING VECTORS

Obviously, not all motion takes place in a straight line. In this section we shall consider motion in a plane where the path of an object may be curved. You will learn a more general definition of kinematical quantities as well as study vectors.

Imagine that you are in a car traveling on a highway, and you wish to describe your motion. You would certainly give your speed, perhaps 60 mph. However, this figure does not completely specify your motion. In addition, you would want to tell which direction you were moving in, say north. In other words, a complete description of your motion would involve giving both the magnitude of your speed and the direction in which you are moving. We call this your *velocity*.

Velocity is an example of a physical quantity that is determined by a magnitude and a direction. Such quantities are called *vector quantities* and are described by mathematical objects called *vectors*. Other examples of vector quantities are displacement, acceleration, force, and momentum, some of which we shall study later. Quantities such as speed, time, distance, temperature, and mass, which can be completely described by their magnitude and have no directional properties, are called *scalars*. Thus, speed is the scalar magnitude of the velocity vector. Ordinary algebra and arithmetic,

indeed most of mathematics, deal with scalars. But in the following paragraphs we shall describe vectors and develop the special rule for adding vectors, which is quite different from the addition of ordinary numbers or scalars.

Consider the illustration of walking from point *A* to point *B* in Fig. 2.8. To describe this motion, we draw an arrow. The length of the arrow represents the distance from *A* to *B*, and the direction of the arrow tells what direction we moved in. When we wish to describe such motion, i.e., distance moved in a certain direction, we use the word *displacement*. Displacement is a vector quantity because it requires a magnitude and a direction for its description. The arrow drawn from *A* to *B* is called a *vector*. Notice that we did not say exactly how we went from *A* to *B*. Regardless of the actual path taken, as long as we started at *A* and ended at *B*, we describe the displacement by the same vector. All vector quantities can be represented by arrows, with the arrow's length giving the magnitude of the quantity and the arrow pointing in the direction of the vector. To represent velocity as a vector, we draw an arrow whose length is proportional to our speed and which points in the direction we are moving.

FIGURE 2.8
The colored arrow represents the displacement from *A* to *B*. Its length is proportional to the distance and it points in the direction of the displacement. The curved line is a possible path taken in going from *A* to *B*.

To learn how to add vectors, again consider displacement as an example. All other vector quantities will add according to the rules for displacements. Suppose that you walk first 4 mi east from point *A* to point *B* and then 3 mi north from point *B* to point *C*. These two displacements are shown in Fig. 2.9. We have chosen a scale where 1 cm represents 1 mi, so that the vector **a,** which points from *A* to *B* is 4 cm long, and vector **b,** which points from *B* to *C* is 3 cm long.* Our net, or *resultant,* displacement is represented by vector **c,** which points from our starting point *A* to our final point *C* and is 5 cm long, the distance from *A* to *C*. Because the effect of the two displacements **a** and **b** in succession is the same as the single

*In order to distinguish vectors from scalars we will print them in bold-face type.

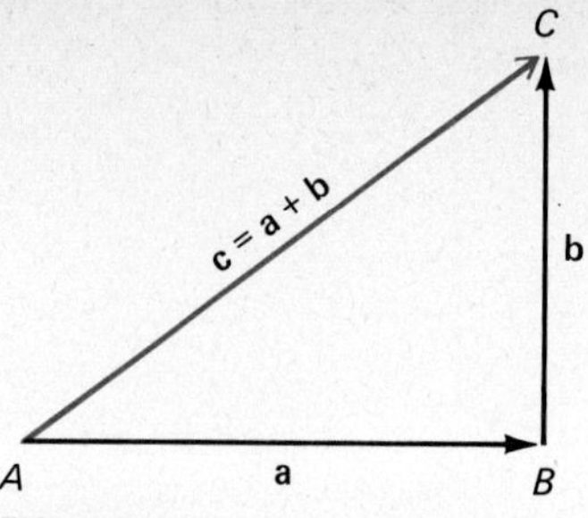

FIGURE 2.9
The colored vector **c** is the sum of **a** and **b.**

displacement **c,** we *define* **c** as the sum of **a** and **b,** i.e., **c** = **a** + **b.** Notice that if you had first walked 3 mi north and then 4 mi east, you would have ended at the same place. This movement is shown in Fig. 2.10. In this case, our first displacement would have been **b** and our second **a,** but our net displacement is still **c.** This route would be described by **b** + **a,** the sum of **b** and **a.** Clearly we have **a** + **b** = **b** + **a** = **c.** In general, the following rule is true: The order in which we add vectors does not matter. Thus, to add two vectors, we first draw one of them as an arrow and then draw the second as an arrow whose tail starts at the head of the first. The arrow from the tail of the first to the head of the second represents the vector sum. This procedure is shown in Figs. 2.9 and 2.10. The magnitude of the resultant, or vector sum, can be found by measuring the length of the resultant vector. Notice that the magnitude of the vector sum is *not* the sum of the individual magnitudes. (Can you tell when it *will* be?) For instance, in Fig. 2.9, the magnitudes of **a** and **b** are 3 and 4, respectively; but the magnitude of the resultant is 5, not 7.

Exactly the same process is used when more than two vectors are involved or when the vectors are not at right angles. We first choose a suitable scale, which will keep the diagram on the paper. We lay off the

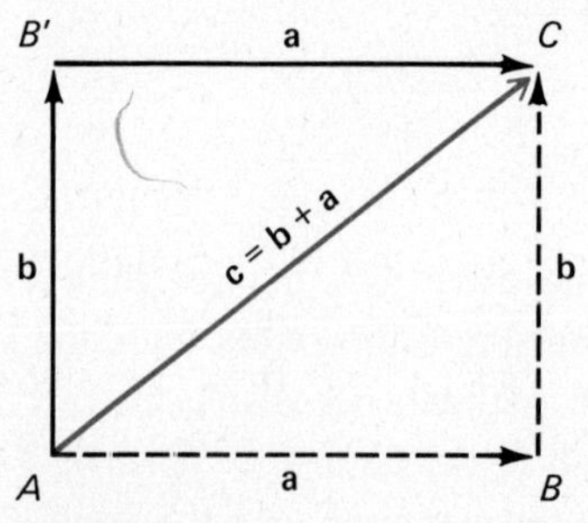

FIGURE 2.10
The vector **c** is found by calculating **b** + **a** giving the same answer as **a** + **b.**

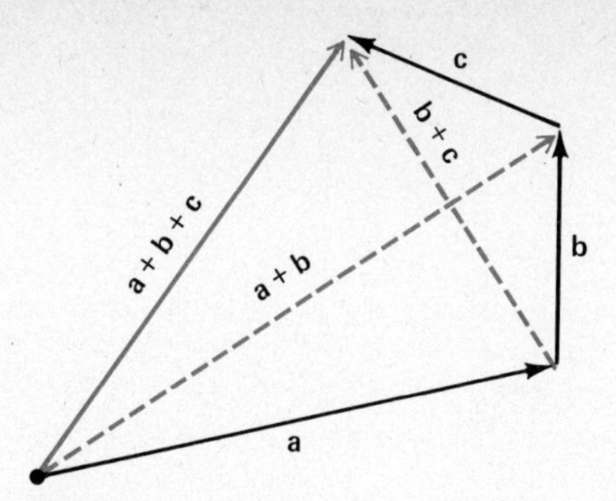

FIGURE 2.11
The sum of three vectors.

first vector to scale in its proper direction. Then we lay off the second vector to scale in its own direction, beginning at the arrowhead of the first vector. The third vector is then laid off from the arrowhead of the second vector, and so forth. The vector drawn from the beginning point of the first vector to the arrowhead of the last vector is then the sum of all these vectors. Again note that the order in which we decide to lay off the vectors has no effect on the result. An example of adding three vectors is shown in Fig. 2.11.

Having learned to add vectors, you must now learn to subtract them. But in order to subtract, you must decide what is meant by the negative of a vector. Since motion to the north is just opposite to motion to the south, it seems reasonable to define the negative of a given vector to be an arrow with the same length but pointing in the *opposite* direction. Figure 2.12

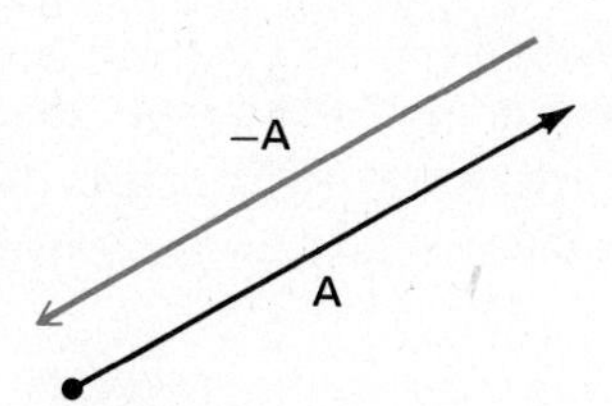

FIGURE 2.12
A vector and its negative (in color).

shows a vector **A** and its negative −**A.** Notice that $\mathbf{A} + (-\mathbf{A}) = 0$; that is, if you walk a given distance and then walk back the same distance, you will end where you started with no net displacement. Suppose **A** and **B** are two vectors, and we wish to calculate $\mathbf{C} = \mathbf{A} - \mathbf{B}$. This calculation can be written $\mathbf{C} = \mathbf{A} + (-\mathbf{B})$. So we must add the vector −**B** to **A.** In Fig. 2.13, we show **A, B,** and −**B;** Fig. 2.14 illustrates the subtraction. Thus, subtraction of vectors is virtually the same as addition.

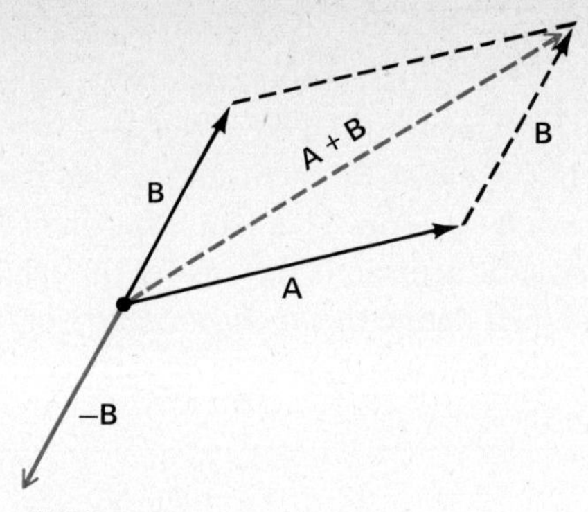

FIGURE 2.13
The vectors **A, B,** and −**B** (in color). The sum of **A** and **B** is also shown.

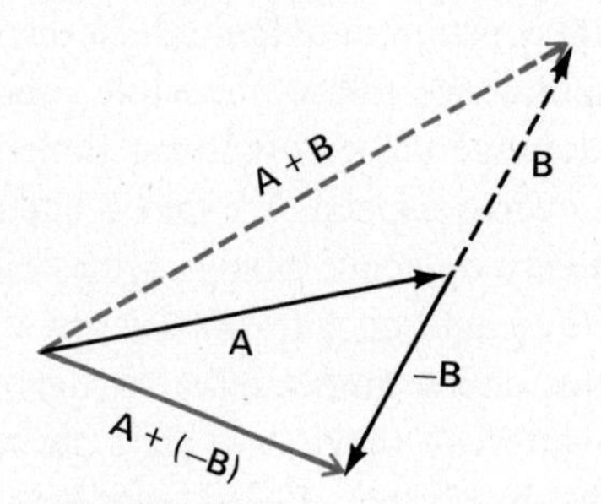

FIGURE 2.14
The vectors **A,** −**B,** and **A** + (−**B**) = **A** − **B** (in color).

EXAMPLE Suppose that a ship sails 40 mi east and then sails 50 mi northeast. We are expected to find the resultant displacement. Let us choose a scale in which 10 mi equals 1 cm. The vector diagram is then shown in Fig. 2.15. By direct measurement we find that the resultant displacement is 85 mi in a direction 26 deg north of east.

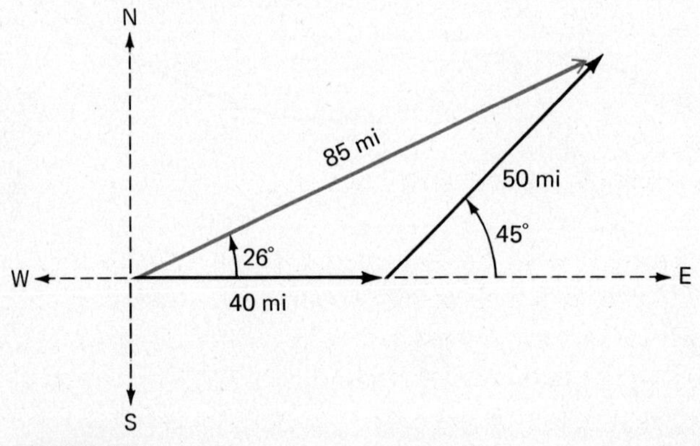

FIGURE 2.15
The colored vector is the net displacement.

We mentioned earlier that velocity is a vector quantity. Suppose that the location of a particle relative to a point taken as the origin is specified by a vector $\mathbf{r}_1$ at time t_1 and by a vector $\mathbf{r}_2$ at a time $t_2 = t_1 + \Delta t$, as shown in Fig. 2.16. The displacement during this time interval is then $\Delta\mathbf{r} = \mathbf{r}_2 - \mathbf{r}_1$, that is, $\mathbf{r}_2 = \mathbf{r}_1 + \Delta\mathbf{r}$. This sum is shown in the diagram. (Can you find $\Delta\mathbf{r}$ by vector subtraction? Is the answer the same as the $\Delta\mathbf{r}$ shown in the diagram?) We then define the *average velocity* of the particle during this time interval by the equation

$$\mathbf{v} = \frac{\Delta\mathbf{r}}{\Delta t} = \frac{\mathbf{r}_2 - \mathbf{r}_1}{t_2 - t_1} \tag{2.9}$$

If the velocity is not constant in magnitude *or* direction and we wish to find the velocity at a point, we follow the same procedure in relating instantaneous velocity to average velocity that was done in Sec. 2.2 for the case of instantaneous and average speed. We take a sufficiently small time interval Δt so that the velocity does not change appreciably, and then use Eq. (2.9) to calculate the average velocity during this time interval. The average velocity during this small time interval is defined as the instantaneous velocity during the interval. Obviously, if greater accuracy is desired, a smaller time interval must be chosen. If Δt is very small, then the vector $\Delta\mathbf{r}$ points in the direction of the motion of the particle. Because the instantaneous velocity is always taken over a small time interval, the vector $\mathbf{v}$ points in the direction of the motion. The magnitude of the velocity vector is the instantaneous speed of the particle.

Note that velocity, as we have defined it, is a vector, with properties

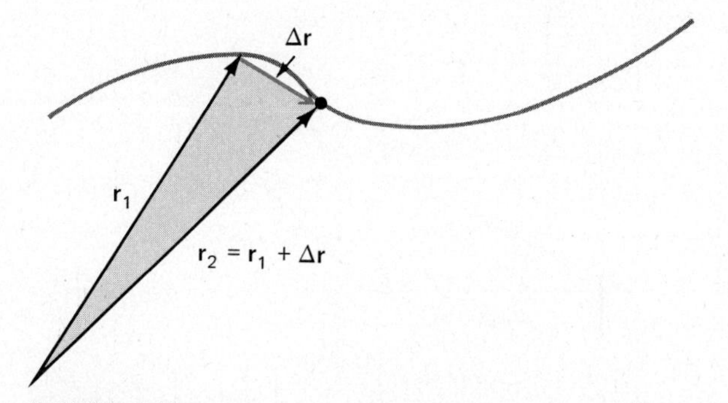

FIGURE 2.16

The position of a particle at two successive times t_1, and $t_2 = t_1 + \Delta t$. The vector sum shows that $\mathbf{r}_2 = \mathbf{r}_1 + \Delta\mathbf{r}$, where $\Delta\mathbf{r}$ is the colored arrow.

of magnitude and direction; speed, in contrast, has only the property of magnitude. Although the two words are used interchangeably in everyday life, "speed" and "velocity" have distinctly different meanings in physics. We shall run into other examples of words that ordinarily mean the same thing in common usage but have different technical meanings.

Since velocities are vectors, they are added in the same way as displacements. For instance, a man rowing a boat across a river that was flowing smoothly and slowly could drop a piece of wood into the water and measure his velocity relative to the water. The vector obtained would point in the direction in which he was rowing, perhaps perpendicular to the river bank. However, a man watching this action from the shore would see the motion and velocity of the boat differently. He would see the river carrying the boat downstream. If the velocity of the river relative to the bank is **u** (the river's velocity as seen from the shore), pointing downstream, and the velocity of the boat relative to the river is **v** (the velocity measured by the man in the boat), then the man on the shore would say that the velocity of the boat relative to the shore was

$$\mathbf{V} = \mathbf{u} + \mathbf{v} \tag{2.10}$$

Figure 2.17 shows this motion. The boat's velocity relative to the shore is in the direction of **V**, which points across and down the river.

EXAMPLE Assume that the river in the preceding discussion flowed with a speed of 3 mph and the man rowed with a speed of 4 mph relative to the water. Then his speed relative to the shore would be 5 mph. If the river were 2 mi wide and he rowed for $\frac{1}{2}$ hr, he would just make it to the other side. He would then be 1.5 mi downstream from his original position and would have traveled a total distance of 2.5 mi.

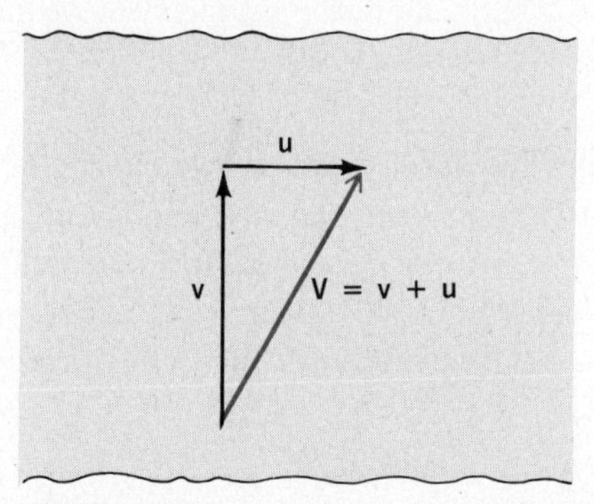

FIGURE 2.17
The velocity **V** of the boat relative to the shore (colored arrow) is the vector sum of the velocity of the boat in the water, **v**, plus the velocity of the river **u**.

Note these two main points. First, since velocity is a vector, if we wish to add velocities, we must use the rules of vector addition, *not* scalar addition. Nor is speed added in a simple way (the speed of the boat relative to the shore was 5 mph, not $3 + 4 = 7$ mph) unless the velocity vectors are all parallel, as was the case in Sec. 2.2. Second, *velocity is relative;* it depends on who is measuring it. Different observers moving relative to each other will measure different velocities for the same object. Their answers can be related by using vector addition, but they will not observe the same velocities. Imagine two spaceships approaching each other with speed v in deep interstellar space. If both ships were coasting with their drives off, they would both feel as if they were at rest (just as the man in the boat would not be able to detect any motion if he were not rowing and were in a dense fog). Then men in one ship might assert that they were at rest and the other ship was moving with a velocity v pointed towards them. However, the inhabitants of the second ship could make the same assertion, except that they would measure the velocity vector of the first ship as being in the opposite direction. If they could communicate, they might even argue about who was moving and who was at rest. The argument would be pointless; each is, in a sense, correct. Only relative motion can be measured. If one of the ships claims to be at rest, he is saying that his velocity is zero with respect to a reference system that is not available in this situation. Neither has a valid claim to be at rest, but both are correct about their relative velocities. We shall return to this type of question in Chap. 12 where we discuss relativity.

EXAMPLE Suppose that an airplane maintains a speed of 300 mph relative to the air in a northward direction, while the air itself is moving with a speed of 100 mph eastward relative to the ground. We wish to find the velocity of the airplane relative to the ground. If we use the scale of 1 cm equaling 50 mph, the vector diagram for this situation is shown in Fig. 2.18. By direct measurement we find that the airplane has a velocity of 316 mph relative to the ground in a direction 20 deg east of north.

Earlier we defined acceleration as the change in speed per unit of time for motion in a straight line. Because velocity has properties of both magnitude and direction, we can generalize and state that a change in either of these properties represents an acceleration. Thus, when the velocity changes from $\mathbf{v}_1$ to $\mathbf{v}_2$ in a time interval Δt we define the *average acceleration* $\mathbf{a}$ by the equation

$$\mathbf{a} = \frac{\Delta \mathbf{v}}{\Delta t} \tag{2.11}$$

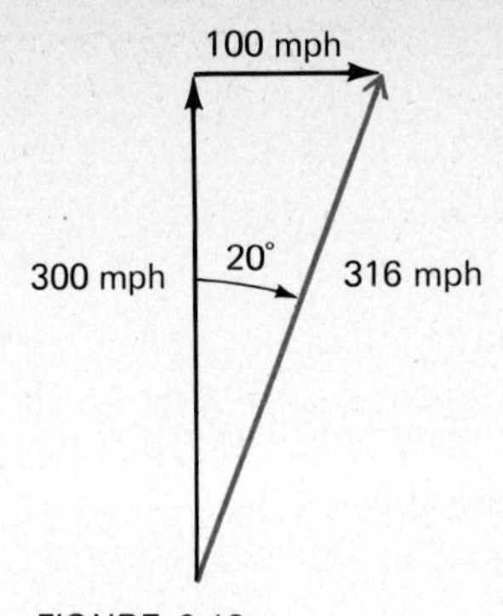

FIGURE 2.18
The colored arrow is the speed of the airplane relative to the ground.

where $\Delta \mathbf{v} = \mathbf{v}_2 - \mathbf{v}_1$. If the acceleration is variable, we must choose a very small time interval if we wish to calculate the instantaneous acceleration. Similarly, the greater the accuracy we require, the smaller the time interval must be.

In Secs. 2.2 and 2.3 we discussed motion in a straight line. In such a case, only the magnitude of the velocity vector changes. Equations (2.9) and (2.11) reduce to Eqs. (2.4) and (2.5) if the motion occurs in only one direction. In Sec. 4.2 we shall treat another situation, in which the magnitude of the velocity remains constant while its direction changes. Problems where both properties of the velocity change simultaneously are taken up in more advanced books on mechanics.

DISCUSSION QUESTIONS

1. A particle is thrown vertically upward. What is its acceleration at the peak of its path? What is its acceleration when it returns to its starting point?

2. Do the speed and the acceleration of a particle have to be in the same direction?

3. What units of length are used by astronomers in expressing distances within the solar system and distances to stars?

4. Accelerations experienced by airplane pilots during radical maneuvers are often expressed as $5g$, $8g$, and so forth. What is meant by such a statement?

5. Since the acceleration of free fall is quite large, measurements on freely falling bodies are difficult to make. Discuss how Galileo was able to observe much smaller accelerations by studying balls rolling down an inclined plane.

6. A body is known to be accelerated. Can you say with certainty anything about the magnitude and direction of the body's velocity?

7. Is it possible for a particle to have a velocity in one direction while experiencing an acceleration in another direction?

8. A boy in a railroad car moving with constant speed in a straight line throws a ball vertically upward. Does the ball land in front of him, in his hand, or behind him?

PROBLEMS

1. What is a millikilogram? (*Ans.:* 1 gm)

2. Compute the number of minutes in 1 microcentury.

3. Convert 100 km/hr into miles per hour. (*Ans.:* 62.1 mph)

4. Compute the average speed in feet per second for a man to run the 100-yard dash in 9.0 sec.

5. A particle is accelerated from rest to a speed of 20 m/sec during a time of 5 sec. Compute the acceleration of the particle and the distance it covers. (*Ans.:* $a = 4$ m/sec^2; $d = 50$ m)

6. A particle falls from rest with an acceleration of 10 m/sec^2. Assume that it falls for 5 sec. Compute its speed and the distance fallen.

7. A particle moving with a speed of 50 m/sec is brought to rest in a distance of 20 m. Compute the deceleration of the particle and the time before it is brought to a stop. (*Ans.:* $a = -62.5$ m/sec^2; $t = 0.800$ sec)

8. A boat is to be steered directly across a river that runs north and south. The speed of the boat is 10 mph relative to the water, and the current in the river is 4 mph. Determine the direction in which the boat must be steered.

If the river is 2 mi wide, compute the time required for the boat to cross the river. (*Ans.:* 0.218 hr)

9. Compute the average speed of a horse that can run a mile in 2 min.

10. If the average speed of an earthworm is 2.5 cm/min, how long would the earthworm take to cover a distance of 30 cm? (*Ans.:* 12 min)

11. An automobile accelerates from rest to a speed of 60 mph in 10 sec. Compute the acceleration (assumed constant) of this automobile in feet per second squared.

12. Compute the deceleration required for an automobile traveling at a speed of 45 mph to come to rest in 6 sec. (*Ans.:* -11 ft/sec^2)

How far would the automobile travel while coming to rest?

13. A particle falls from rest with an acceleration of 32 ft/sec^2. Compute its speed at the end of 4 sec. (*Ans.:* 128 ft/sec)

Compute how far the particle has fallen during the 4-sec period.

14. A boat can travel at 6 mph relative to the water in a river in which there is a current of 2 mph. Compute how long it would take the boat to travel to a buoy 2 mi downstream and return to its starting point. (*Ans.:* $\frac{3}{4}$ hr)

15. Combine Eqs. (2.6), (2.7), and (2.8) to derive the following equation for motion with constant acceleration:

$$v_2^2 = v_1^2 + 2ax$$

16. Combine Eqs. (2.6), (2.7), and (2.8) to derive the following equation for motion with constant acceleration:

$$x = v_1 t + \tfrac{1}{2}at^2$$

3 Dynamics

In Chap. 2 we described the movement of bodies without considering why they moved or what could change their motion. In this chapter we shall discuss Newton's Laws of Motion. These three laws enable us to predict the motion of objects. Because Newton's laws are the foundation of dynamics, you should be especially attentive in studying this chapter.

3.1 MASS AND MOMENTUM

Defining mass is a difficult task. We all have some physical feeling for *inertia,* which is a qualitative property of mass. For example, the impact of a medicine ball is greater than the impact of a basketball when they are both thrown with the same speed. Although they are about the same size, the medicine ball has more inertia, or, in the language of physics, more mass.

Let us now give a quantitative meaning to the concept of mass. Imagine that you are in a laboratory in deep space, free from any large gravitational pulls and free from such complications as air resistance. If you gently release an object, it will "float" in space without moving. This phenomenon is weightlessness, a condition many of you have seen in telecasts from the space flights. If you give the object a small push, it will move off in some direction. Because there is no air resistance, it will continue

to move in the same direction with constant speed until it strikes the wall or some other object. In other words, an object's velocity will remain constant if the object is isolated (that is, if there are no forces acting on it; however, as we have not yet defined "force," we simply take "isolated" to mean that nothing from the outside is influencing the object). Recall that velocity is a vector; for it to be constant, neither the direction of motion nor the speed of the object can change. Clearly, being at rest is a special case of constant velocity.

As we shall see later, the rule for an isolated body is actually a special case of Newton's First Law. Obviously, this description cannot apply to objects on earth where an object that is released falls promptly to the ground because of gravity. This is the reason why, in our previous example, we located our laboratory in deep space, away from strong gravitational pulls.

Now let us perform an experiment in our laboratory. Our apparatus consists of two identical objects, which are equipped so that they will stick together when they collide. The experiment is performed by pushing the two objects at each other with exactly equal but opposite velocities. They then approach each other along the same line with identical speeds and collide as shown in Fig. 3.1. What will their motion be after the collision? They will stop. Because they are identical and moving with equal and opposite velocities, the collision is completely left-right symmetric. After they strike each other and stick together, the situation must still be left-right symmetric. Thus, they *must* be at rest since motion in any direction would spoil this symmetry. You may consider that symmetry arguments such as this one are superficial. However, conclusions based on symmetry are perhaps the most fundamental of the many laws of physics. All of the conservation laws in physics can ultimately be traced to some symmetry of the system.

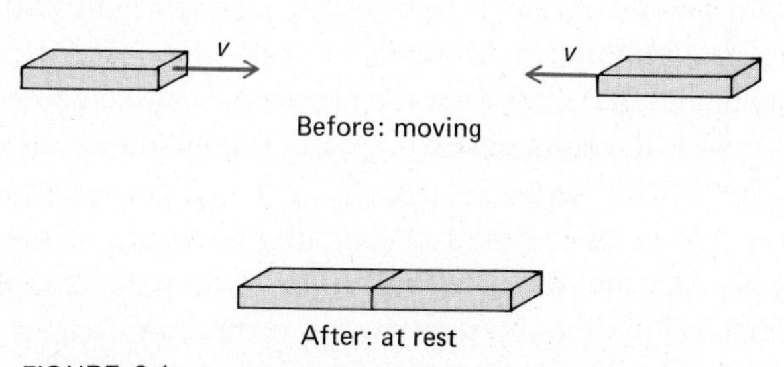

FIGURE 3.1
In the upper picture the two identical objects are approaching with equal but opposite velocities represented by the colored arrows. In the lower picture they have collided, stuck and stopped.

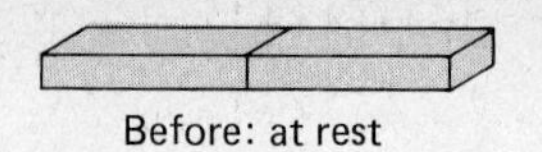

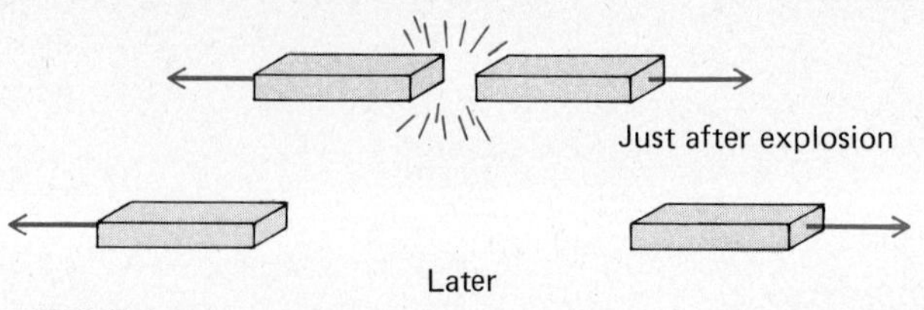

FIGURE 3.2
In the upper picture the two objects are at rest and coupled with an explosive cap. In the second picture the cap has exploded and they are moving away with equal and opposite velocities. In the third picture they have separated further. Velocity vectors are colored.

In our experiment, the two objects, being identical, must have the same mass. We take one of them as our "standard unit" of mass. In a moment we shall extend our definition of mass to other objects of different materials, but first let us describe a slight variation of this experiment.

Imagine that our two objects are coupled together and at rest as in Fig. 3.2. Between them is a small explosive cap that we can explode at will. This cap is designed to be left-right symmetric; that is, when it explodes, it acts equally on both objects. After the cap has been exploded, how will the two objects move? Using arguments identical to the preceding ones, we can conclude that, because the system is completely left-right symmetric, the two objects must have equal and opposite velocities after the explosion. You will observe that this case is nothing more than the first experiment run backwards. There is no fundamental difference between them except that the initial and final states have been reversed.

We shall now use these experiments to extend our definition of mass. First, consider another object with a different shape and perhaps made of a different material. We attach it to one of our standard objects with an ideal explosive cap between them and explode the cap. If they then move off with equal opposite velocities, we *define* them as having equal masses. We could also project them directly at each other with equal speeds as in Fig. 3.3 and see if they stop after colliding. (Here we arrange it so that they must stick together.) Again, if they are at rest after the collision, we define them as having equal masses. The two definitions are completely equivalent and will always give the same answer.

The standard unit of mass in the metric system is the kilogram (kg), which was originally defined by the French in 1799 as the mass of a cube

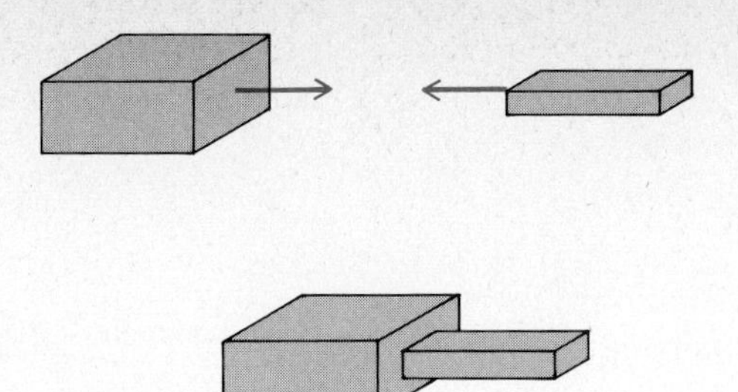

FIGURE 3.3

An experiment to test whether a large object has the same mass as our standard. In the upper picture the two masses approach with equal but opposite velocities indicated by the colored arrows. In the lower picture they have come to rest after the collision indicating they are of equal mass.

of water 10 cm to an edge. Thus, the volume of the cube is 1000 cubic centimeters (cm^3) or 1 liter. Fortunately, the French constructed a piece of platinum-iridium alloy that was supposed to agree with this definition; this model has been the fundamental definition of the kilogram ever since. Common subunits of the kilogram are the gram ($1 \text{ gm} = 10^{-3} \text{ kg}$) and the milligram ($1 \text{ mg} = 10^{-6} \text{ kg}$). In the English-speaking countries the pound has been used as a mass unit. *However, the pound is actually a unit of force, not mass,* and we shall define it in a later section.

The experiments that we have just described enable us to measure equal masses. In this sense, any mass equal to the standard kilogram in Sèvres, France, is defined as having a mass of 1 kg. If a given mass is equal to two 1-kg masses, it has a mass of 2 kg. If two identical objects together are equal to a 1-kg mass, they each have a mass of $\frac{1}{2}$ kg. In this manner, the definition can be extended to arbitrary masses. Another useful mass unit is the atomic mass unit (amu), which is defined as one-twelfth of the mass of the most abundant isotope of carbon (carbon 12). This unit is used extensively in chemistry and nuclear physics. One amu equals 1.66×10^{-27} kg.

Mass plays a fundamental role in dynamics. First of all, it is important because it is conserved. In any process, such as a collision or a chemical reaction, the total mass does not change. (This statement will be modified in Part IV, when we talk about relativistic physics, in which we shall find that mass can be changed into energy.) Because it is conserved, the mass of an object is an intrinsic property of the object. A man who weighs 180 lb on earth, 30 lb on the moon, and nothing in an orbiting space station will still have the same mass in each case.

Mass is also important because it enables us to define other dynamical quantities, such as momentum. Momentum **p** is defined as the product of mass times velocity or, in symbols,

$$\mathbf{p} = m\mathbf{v} \tag{3.1}$$

Like velocity, momentum is a vector quantity. Its magnitude is the mass of a particle times its speed, and it points in the direction of motion. Since we are considering only motion in one direction, the vector character of momentum does not concern us. However, once we have chosen a particular direction as positive, momentum will be positive for motion in that direction and negative for motion in the opposite direction.

EXAMPLE Suppose a 3-kg particle is moving with a speed of 5 m/sec in a certain direction as shown in Fig. 3.4. We would say that the particle has a momentum of (3 kg) × (5 m/sec) and write $p = 15$ kg-m/sec. Another particle that has a mass of 1.5 kg and moves in the opposite direction with a speed of 10 m/sec would have a momentum of (1.5 kg) × (−10 m/sec) = −15 kg-m/sec. The particle is moving in the opposite direction, so its momentum is negative. The total momentum of both particles taken together is zero, because the individual momentum vectors are of the same length and point in opposite directions.

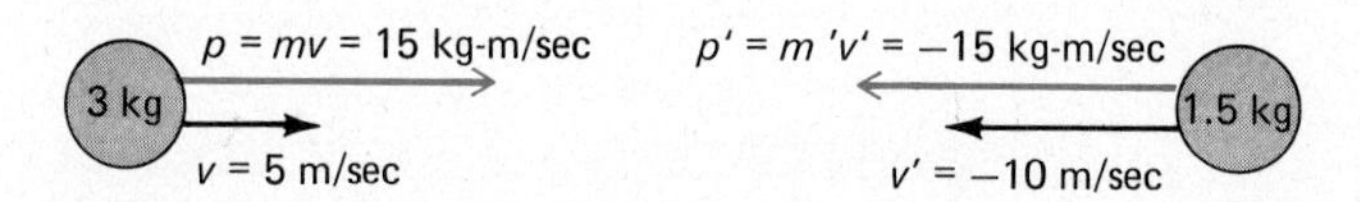

FIGURE 3.4
The velocity and momentum vectors for two objects moving in opposite directions. Note that although the velocity vectors (in black) are not of equal length, the momentum vectors (colored) are, and since they are in opposite directions their sum is zero.

As the preceding example indicates, the momentum of several particles is the vector sum of the individual momenta. For motion on a line, this description means that the momenta of the particles moving in the negative direction is subtracted from the momenta of those moving in the positive direction.

EXAMPLE Three particles, a, b, and c, are moving with velocities $v_a = 3$ m/sec, $v_b = 10$ m/sec, and $v_c = -6$ m/sec. All these velocities are along the same direction except that particle c is moving in the opposite direction from a and b. The masses are: $m_a = 20$ kg, $m_b = 5.5$ kg, $m_c = 4$ kg. We wish to find the individual momenta and the total momentum. Using $p = mv$, we find that $p_a = 60$ kg-m/sec, $p_b = 55$ kg-m/sec, and $p_c = -24$ kg-m/sec. Note that p_c is negative because v_c is negative. The total momentum P is $P = p_a + p_b + p_c = 60 + 55 - 24 = 91$ kg-m/sec.

Let us return to our collision experiment that we described at the beginning of this section. Two objects of equal mass are given equal and opposite velocities. When they collide, they stick together and stop. Because they have equal speeds and equal masses, their momenta are of the same magnitude. However, they are moving in opposite directions; therefore, their momentum vectors point in opposite directions and their initial total momentum is zero. After the collision, the final momentum is zero since they are both at rest. The initial momentum is equal to the final momentum; that is, the total momentum of the system is the same after the collision as it was before. This example illustrates a conservation law, the *conservation of momentum*. This law states that for a system free from external influences (gravity is an external influence but the effect of one of the objects on the other is not), momentum is conserved.

We have not yet proved this law. However, we can prove it for another simple case. Consider our same experiment: the collision of two equal mass objects moving with equal and opposite velocities. What will this experiment look like through a television camera that is moving along next to one of the objects? The camera does not participate in the collision; but, as shown in Fig. 3.5, it keeps on moving past when the two blocks stop. Before the collision the camera sees one block that is not moving, because

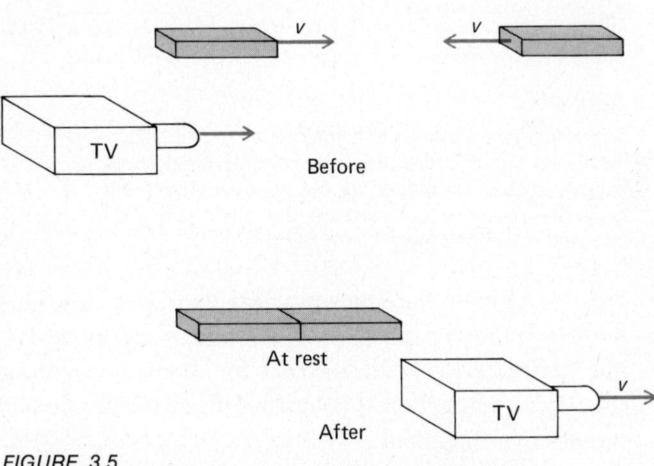

FIGURE 3.5
The same collision as shown in Fig. 3.1 with a TV camera observing. In the upper picture the two objects are approaching each other with equal and opposite velocities and the TV camera is moving at just the same speed as the left object. In the lower picture the two objects have collided and stopped, but the TV camera is moving with the same velocity it had in the upper picture.

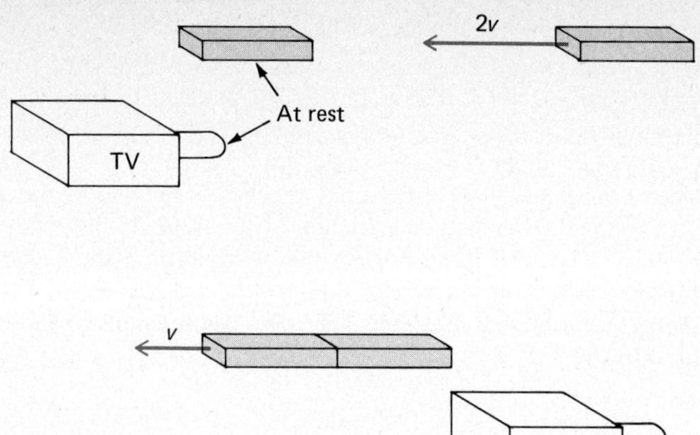

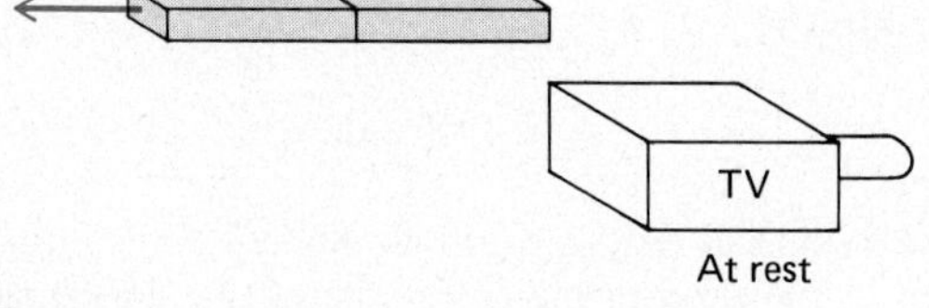

FIGURE 3.6

The same sequence of events as shown in Fig. 3.5 from the point of view of the TV camera. In the upper picture the camera and the left mass are at rest while the second mass is approaching with speed $2v$. In the lower picture the camera is still at rest but the two masses have collided and are moving away with speed v.

the camera is traveling along with it. Their relative velocity is zero as shown in Fig. 3.6. It sees the other block approaching with a speed of $2v$, which is just the relative velocity of the two objects before the collision. We have used the rules for addition of velocities in arriving at this result. After the collision the camera sees the two objects stuck together and moving away with a speed v. Remember that the camera originally was moving along with one of the objects and had velocity v relative to the laboratory. After the collision the two objects are at rest relative to the laboratory and the camera is still moving, so their relative velocity is still v. We conclude that if an object moving with a certain speed ($2v$ in this case) strikes an object of equal mass at rest and they stick together, the pair (with a mass of $2m$) will move off with half the original velocity. Since the moving camera's point of view is just as valid as that of a stationary camera, this description must always be true. Momentum is conserved here also. The initial momentum is P_i = (mass)(initial speed) = $(m)(2v) = 2mv$, whereas final momentum is P_f = (final mass)(final speed) = $(2m)(v) = 2mv$. Momentum is again conserved. Using similar arguments, we can show that momentum is conserved for all collisions of the type described regardless of the masses involved. The ingredients of the argument are symmetry and the rule for adding velocities. The conclusion is conservation of momentum. This is one of the

central conservation laws of physics. It remains valid even in areas where more restrictive laws, such as those of Newton, fail.

EXAMPLE Consider an 800-kg car moving down a highway at a speed of 30 m/sec (about 60 miles per hour). The car crashes into the rear of a 2200-kg truck parked by the side of the road. The car becomes attached to the rear of the truck, and they move off together. Obviously, the friction of the truck on the road will stop them quite rapidly. However, if we neglect this friction, or if we look for the speed of the combination just after the collision, we can use conservation of momentum to find the speed after the collision. The initial momentum of the car is its mass times its velocity:

$$P_0 = (800\text{ kg})(30\text{ m/sec}) = 24{,}000\text{ kg-m/sec}$$

The truck is at rest, so its initial momentum is zero. After the collision the momentum of the car and the truck will still be 24,000 kg-m/sec, because momentum is conserved. However, now the net mass is 3000 kg, the mass of the car plus the mass of the truck. Equation (3.1), $p = mv$, gives us

$$24{,}000 = 3000v$$

which can be solved to give $v = 6$ m/sec as the speed of the combination after the collision.

3.2 NEWTON'S LAWS OF MOTION

We all know intuitively that to start something moving or to change an object's velocity, it must be "pushed" or "pulled." To formalize this description, we note that if an object changes its velocity, then its momentum also changes. The amount by which the momentum changes depends on how large the "push" is and how long it acts. The change in the momentum also depends on the direction of the "push." In place of the intuitive word "push," let us use the word "*force*" and symbolize it by the letter **F**. We define the force acting on a body in terms of the change that it produces in the momentum. If a force acts for a time interval Δt and produces a change in momentum $\Delta\mathbf{p}$, then the force is defined by

$$\mathbf{F} = \frac{\Delta\mathbf{p}}{\Delta t} \qquad (3.2)$$

We are assuming, for the moment, that only one force is acting on the body in question. This relation is known as *Newton's Second Law of Motion*. The change in momentum $\Delta\mathbf{p}$ in Eq. (3.2) is given by $\Delta\mathbf{p} = \mathbf{p}_2 - \mathbf{p}_1$, where $\mathbf{p}_1$

is the initial momentum and $\mathbf{p}_2$ is the final momentum. Because $\mathbf{p} = m\mathbf{v}$ and m is constant, we can write $\Delta\mathbf{p} = m\mathbf{v}_2 - m\mathbf{v}_1$. Rearranging this equation gives us $\Delta\mathbf{p} = m(\mathbf{v}_2 - \mathbf{v}_1)$. By using the definition of acceleration, $\mathbf{a} = \Delta\mathbf{v}/\Delta t$, we find that Newton's Second Law can be written in the form

$$\mathbf{F} = m\mathbf{a} \tag{3.3}$$

which is the form usually seen in most text books.

We use Eq. (3.3) or Eq. (3.2) to define force. In words, unit force is the force that gives unit acceleration when applied to a body of unit mass. In the metric system, mass is measured in kilograms and acceleration in meters per seconds squared (m/sec^2). The unit of force in this system is called the newton (N), and it gives a body with a mass of 1 kg an acceleration of 1 m/sec^2. Only this unit of force is used in this book.

We have discussed a case in which just a single force acts on a body. This force is the net or resultant force on the body. In more complicated situations, several forces may act. Thus, the letter **F** in Eq. (3.3) stands for the net or resultant force acting *on* a body that has the mass m and acquires the acceleration **a**. Since both **F** and **a** are vectors, Eq. (3.3) tells us that they are parallel; that is, the acceleration is always in the same direction as the net force as shown in Fig. 3.7. This net force is just the vector sum or the resultant of the various forces acting on the object with mass m. The graphical methods explained in Sec. 2.4 are easy and are sufficiently accurate for any problem in this book. For the special case where the forces act along the same line, they are added or subtracted just like scalars in order to find the resultant. The following examples and the problems at the end of the chapter should help you become familiar with the application of Eq. (3.3).

At this point, let us consider the special case in which the net force on a body equals zero. From Eq. (3.3), we find that the acceleration is also zero. Acceleration is change in velocity per unit time; therefore, the velocity does not change. Thus, zero acceleration means constant velocity, and the motion of our object will be in a straight line at constant speed. We can now state *Newton's First Law of Motion* in the form:

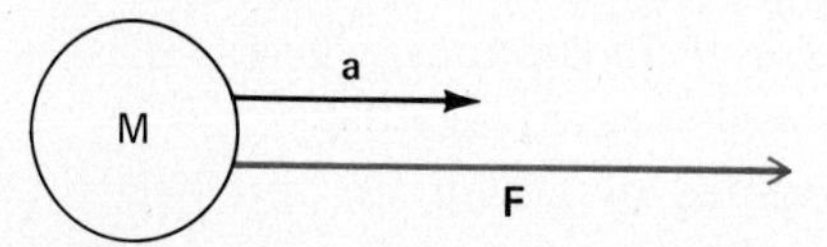

FIGURE 3.7
The force (colored arrow) on a body and the acceleration (black arrow) that it produces are always in the same direction.

> *If no net force acts on a body, it will remain at rest if it was originally at rest or it will continue moving in a straight line with constant speed.*

In this statement of Newton's First Law, the key word is the adjective "net" before the word "force." It is perfectly possible for two or more forces to act on a body in such a way that they cancel one another. In this case, no *net* force acts on the body, and it obeys the First Law. In place of the word "net" we could just as well have used "resultant" or "unbalanced"; they have the same meaning in this context. To sum up, the First Law is a special case of the Second Law of Motion, because it tells us what happens when zero net force acts on a body. In our definition of mass, we used a very special case of the First Law, which said that if *no* forces act on a body, it will move with constant velocity.

EXAMPLE We now take up several examples of problems based on the Second Law. Suppose that we wish to find the force required to give a body with a mass of 2 kg an acceleration of 5 m/sec^2. From Eq. (3.3) we find that $F = 2 \times 5 = 10$ N. On the other hand, we might wish to know the acceleration acquired by a body of 4-kg mass when it is subjected to a force of 15 N. We see from Eq. (3.3) that a equals $\frac{15}{4} = 3.75$ m/sec^2. In another case, we might have a force of 20 N acting to the right and a force of 8 N acting to the left on a body of 4-kg mass. The net force on the body is then $(20 - 8) = 12$ N. Consequently, the acceleration of the body is $\frac{12}{4} = 3$ m/sec^2.

In everyday life the words "weight" and "mass" are sometimes used interchangeably, but this is not the case in physics. *Mass* describes the inertial properties of bodies and is defined quantitatively in Sec. 3.1. Mass is a universal property of a body. It is the same anywhere in the universe and the same for any type of interaction in which it might take part. (Here we neglect the changes predicted by the theory of relativity, which have no practical effect on bodies moving at ordinary speeds.) *Weight,* on the other hand, describes the *force* exerted by one body (usually the earth or some other planet) on another due to gravity. We define the weight of a body as the force of gravity on the body. Examples are shown in Fig. 3.8. Thus, weight is measured in newtons, whereas mass is measured in kilograms.

The force exerted by gravity on an object is always proportional to the mass of the body. It does not depend on any other property, such as the type of material of which the body is made. If we let g stand for the constant of proportionality measured at some point on the earth and

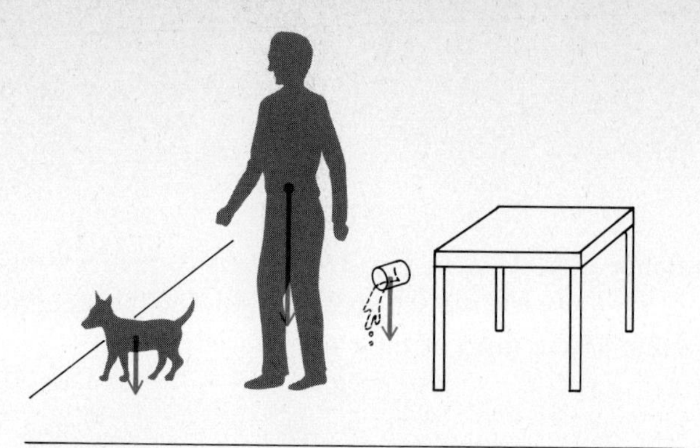

FIGURE 3.8
Weight is the force gravity exerts on an object. Its direction is always downward.

W stand for the weight of the body of mass m placed at that point, then the weight is given by the equation

$$W = mg \tag{3.4}$$

To interpret g, we consider an object in free fall. The sole force acting on it is the force of gravity, its weight. Combining Newton's Second Law, Eq. (3.3), with Eq. (3.4) for the force on the object, we find that $mg = ma$ and $g = a$; that is, the object falls with the acceleration of gravity, and g is the same g of Sec. 2.3. Thus, $g = 9.8\ \text{m/sec}^2$ or, for our purposes, approximately $10\ \text{m/sec}^2$. We have just used Newton's Second Law and Eq. (3.4) for the weight of an object to prove that all objects fall with the same acceleration in the earth's gravitational field.

It has been found experimentally that at a given location the acceleration of gravity g is the same for all bodies. Let us consider two bodies, labeled *1* and *2*, located at the same place. Equation (3.4) gives the weight of body *1* as

$$W_1 = m_1 g \tag{3.5}$$

Similarly, the weight of body *2* is

$$W_2 = m_2 g \tag{3.6}$$

Note that in Eqs. (3.5) and (3.6) we used the fact that the acceleration of

gravity g is the same for both bodies, because they are at the same point on the earth. If we now divide Eq. (3.5) by Eq. (3.6), we find:

$$\frac{W_1}{W_2} = \frac{m_1}{m_2} \tag{3.7}$$

A balance or scale as in Fig. 3.9 actually compares weights (forces). From Eq. (3.7), however, we see that the ratio of the weights of two objects is the same as the ratio of their masses.

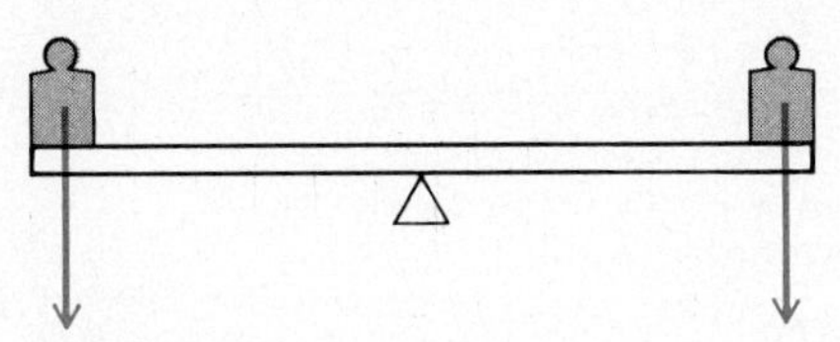

FIGURE 3.9
Equal arm balances measure weight not mass.

In practice, we use a set of calibrated masses on a balance in determining the mass of an unknown body. Although this method is not the same as the direct comparison of masses in an inertial experiment, as we discussed earlier in Sec. 3.1, a balance is commonly used because of its ease and accuracy. Note here that the English language has the two nouns, "weight" and "mass," but only a single verb, "to weigh." In laboratory work, when we weigh an object and state that it weighs 2 kg, we really mean that it has a mass of 2 kg and therefore a weight of 19.6 N. Assuming that we know the local acceleration of gravity, if we state the value of either the mass or weight for a body, we also imply a knowledge of the other quantity. Thus, you can see how two words are used differently in physics even though they are considered the same in everyday life.

Newton's Third Law of Motion is quite independent of the first two laws. It deals with forces acting between a pair of bodies, whereas the first two laws deal only with the forces acting *on* a given body. We can state the Third Law in the following form:

> *When body* A *exerts a force on body* B, *body* B *exerts an exactly equal but oppositely directed force on body* A.

In different words, forces between bodies always occur in equal and opposite pairs. If you push against the wall with a force of 10 N, as shown in Fig. 3.10, the wall pushes back on you with a force of 10 N. This law holds whether either of the bodies is at rest, moving, or undergoing an acceleration.

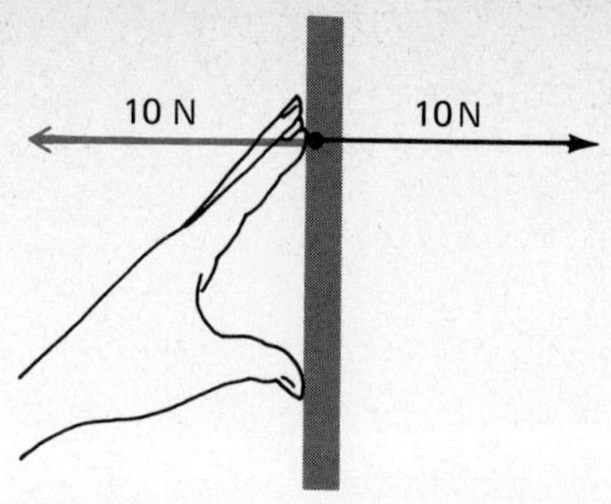

FIGURE 3.10

The black arrow represents the force the hand is exerting on the wall. The colored arrow represents the equal and opposite force the wall exerts back on the hand.

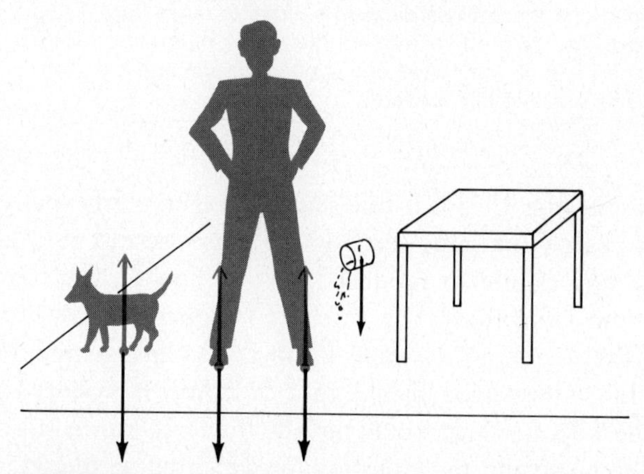

FIGURE 3.11

The man and the dog are exerting forces on the floor due to their weight. The floor pushes back on them (colored vectors) thus supporting them. What about the glass?

It may sound odd to say that you can walk across the floor only because the floor pushes on your shoes, but that is the case. Look at Fig. 3.12. If you push to the left on the floor, according to the Third Law, the floor will push to the right on you. But it is this force to the right that allows you to move, according to the Second Law and Eq. (3.3). Imagine that you are trying to walk across some ice so slick that there is very little friction. You could not exert a force to the left on the ice, so that the ice would not exert a force to the right on you in return; thus, you could not walk across the ice.

As another example of the Third Law, let us consider the following paradoxical question, which is based on the diagram of a horse and wagon

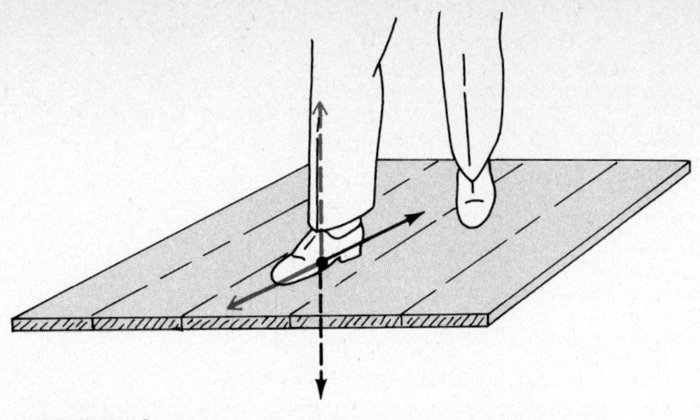

FIGURE 3.12

The black vectors represent the forces the man's foot exerts on the floor. The dotted vector is his weight and the solid one the backwards push needed to move him forward. The colored vectors are the forces of the floor back on his foot, the dotted one supporting him and the solid one pushing him forward.

shown in Fig. 3.13. Is it true that the wagon always pulls back on the horse with exactly the same force as the horse uses to pull the wagon forward? If so, how is motion produced? If not, how do you reconcile your answer with the Third Law? The Third Law is obeyed in all situations, so it is true that the wagon always pulls backward as hard as the horce pulls forward. Motion of the wagon, including acceleration, is produced by the forces acting *on* the wagon, which would include the force exerted forward on the wagon by the horse and the retarding force exerted on the wagon's wheels by the ground. Thus, the Second and Third Laws are applied independently in analyzing situations of this sort.

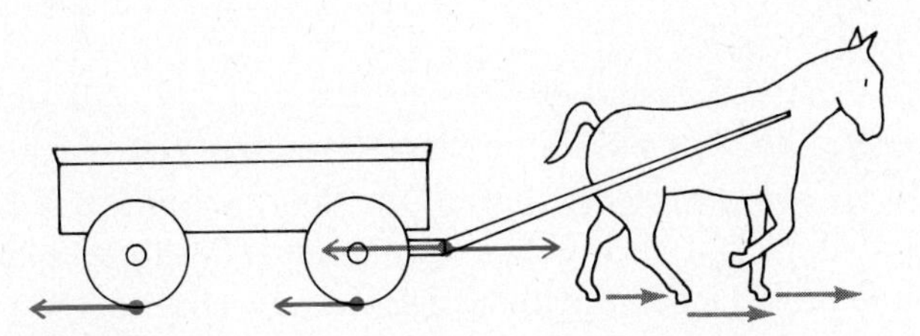

FIGURE 3.13

A horse pulling a wagon. The colored arrows are the forces on the wagon. The black arrows are the forces on the horse. Note that the force of the horse on the wagon is equal and opposite to the force of the wagon on the horse.

3.3 EQUILIBRIUM

When the net force on a particle is zero, we say that the particle is in *equilibrium*. In this case, because the acceleration of the particle is zero, it either remains at rest or continues moving in a straight line with constant speed. Most examples of equilibrium are static cases, for which the speed is zero, but there are important examples in which the speed is constant. For instance, when a car travels down a straight road at constant speed, the force exerted forward on the tires is just equal and opposite to the retarding forces exerted by air resistance and friction. Similarly, after a parachutist has fallen a reasonably short distance, he reaches a constant speed. In this case, the downward force of gravity and the upward retarding force of the parachute just balance one another.

In these two examples, the forces involved were all along the same straight line, so that they could be added and subtracted just like ordinary numbers. If the forces on an object are not along the same line, we must use the methods of vector addition discussed in Sec. 2.4. If a given set of forces is to produce equilibrium, their vector sum must be zero. Thus, if we use the graphical method of adding vectors, the head of the last arrow must end on the beginning point of the first vector. A typical vector diagram of this sort is shown in Fig. 3.14. Sometimes we know that an object is in equilibrium, but we do not know all of the forces acting on it. Then we might have to find the magnitude and direction of the unknown force that must act for equilibrium to hold. To find an unknown force, we add graphically all the forces that we do know. Then the force completing the chain of forces so as to give zero net force is the required unknown. This procedure is illustrated in Fig. 3.15.

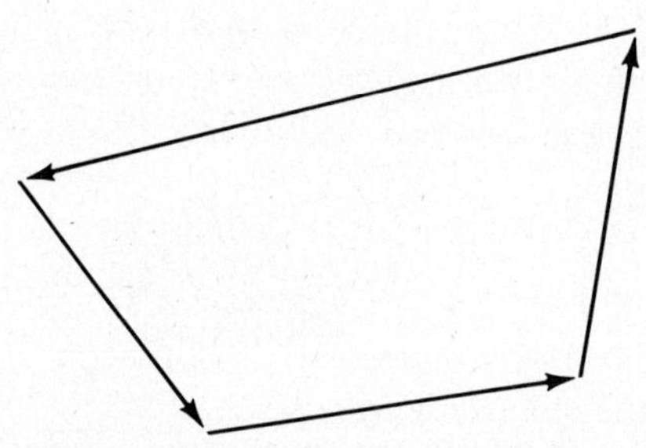

FIGURE 3.14
The four forces shown are in equilibrium because their vector sum has zero length, that is, it vanishes.

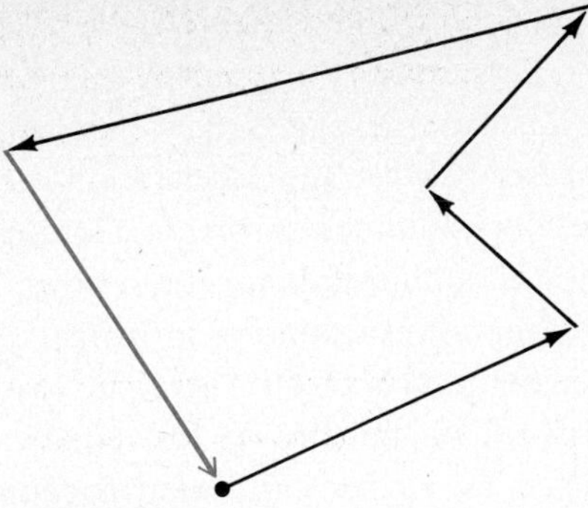

FIGURE 3.15
The black vectors represent the known forces acting on some object. The colored vector is the additional or unknown force needed for equilibrium to hold.

3.4 WORK, ENERGY, AND POWER

The related concepts of work and energy play a central role in physics. *Work* is done by a force acting over a distance. Suppose that a constant force F acts through a distance d parallel to the force, as shown in Fig. 3.16. We define the work W done by the force as the product of the force times the distance or, as an equation,

$$W = Fd \tag{3.8}$$

where F and d are the magnitudes of the parallel vectors **F** and **d**. Suppose, on the other hand, that a constant force F acts through a distance d that is perpendicular to the force, as shown in Fig. 3.17. In this case, the work done is defined as zero. Because any constant force acting for a distance can be represented as the vector sum of a force parallel to d and one perpendicular to d, these two special cases can be used to treat any constant force. (The general case in which the forces are not constant will not be

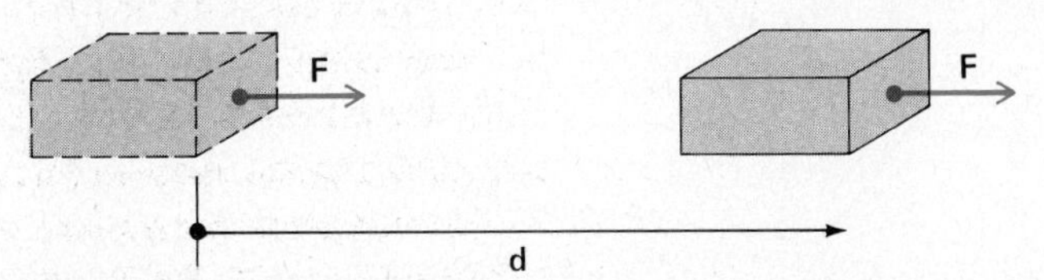

FIGURE 3.16
A constant force **F** (color) acts over a distance **d** (black). Since the force is parallel to the displacement vector the work done by the force is given by $W = Fd$.

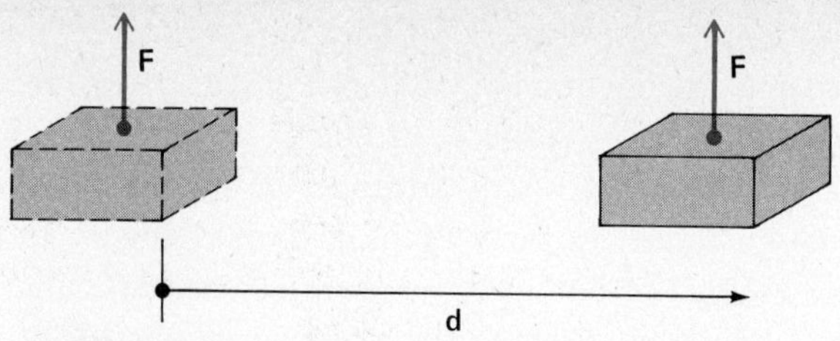

FIGURE 3.17

A force **F** (color) acts over a distance **d** (black). Since the force is perpendicular to the displacement the work done is zero.

discussed in this book.) You should note that this description is the technical meaning of the word "work," even though other meanings are used in everyday life. In the metric system, which we are using in this book, force is measured in newtons and distance in meters. Thus, the unit of work is 1 N-m. Generally, however, this unit is called the *joule,* after J. P. Joule, an English physicist who did research on thermodynamics and heat. We have then:

$$1 \text{ joule (J)} = 1 \text{ N-m}$$

The rate at which work is done by a force is also important. Again we restrict our discussion to a constant force. If the force does an amount of work ΔW during a time Δt, we define the *power,* P in this situation, by the equation

$$P = \frac{\Delta W}{\Delta t} \tag{3.9}$$

This definition is the specific technical meaning of the word "power," although there are other common meanings. In the metric system we measure work in joules and time in seconds, so the unit of power is the joule/sec. This unit occurs so frequently in physics that it is named the *watt,* after James Watt, who built one of the first steam engines. We can then write:

$$1 \text{ watt (W)} = 1 \text{ joule/sec}$$

When work is done by a force on a body, one or more properties of the body will change, such as its speed, its height, its temperature, and so forth. With each of these properties we associate a scalar quantity called

energy. This energy is defined in such a way that the work done in producing the change equals the increase in the energy. Thus, when work is done on a body, one or more forms of its energy will increase. In analyzing various forms of energy, we usually arrange the way in which force is applied so that only a single property of the body is changed. In this case, all of the work done on the body goes to increase one particular form of the body's energy. In many cases, the work done in increasing the energy can be recovered. We call these systems *conservative*. If the work cannot be recovered, the system is *nonconservative*. The energy associated with a temperature change is nonconservative, and only part of it can be recovered. The work "lost" in such nonconservative or dissipative systems is not truly lost; it is just not available to perform work. We shall discuss this concept in the chapter on thermodynamics.

Of the types of energy associated with conservative systems, the most important types are *kinetic* energy and *potential* energy. Kinetic energy is work stored in the motion of a system, and potential energy is energy stored in a system by virtue of its position. Let us now look at examples of these two types.

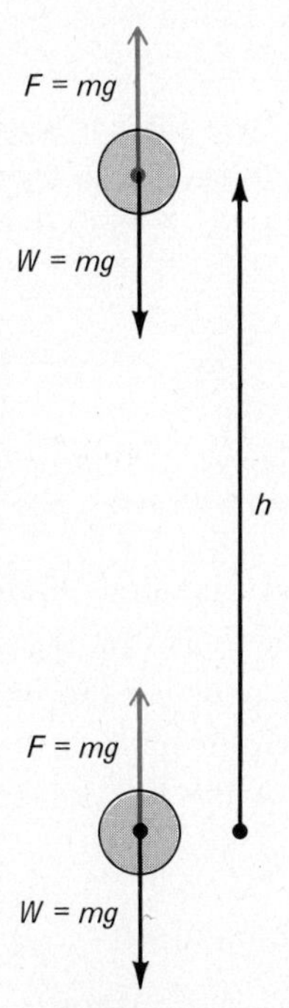

FIGURE 3.18

The work needed to lift an object against force of gravity. The weight is shown in black and the force needed to overcome gravity in color. The work done by the force against gravity is the increase in potential energy.

Suppose that you want to raise a body a vertical distance h, starting with the body at rest and ending with the body at rest, as shown in Fig. 3.18. The only property of the body that has changed is its vertical position. Because the work to accomplish this change has been done against the force of gravity, we say that you have increased the *gravitational potential energy* of the body. From Eq. (3.4) we know that the weight of the body whose mass is m is mg. Thus, the force required to raise the body a distance h against the force of gravity is numerically equal to mg, and the work done by this force is then $(mg)h$. From the general definition of energy in the preceding paragraph, we find for the increase in the gravitational potential energy of the body (abbreviated U_g) the value:

$$\text{Increase in } U_g = mgh \tag{3.10}$$

The work that has been done by the external force to raise the energy of the body is not lost but rather is stored and is available to do work. In this particular situation, we might think of the piledriver. Work is done to raise a heavy weight; the work is recovered when the weight drops and drives the piling into the ground.

Let us now consider a constant force F accelerating a body of mass m from rest to some final speed v during a time t, as shown in Fig. 3.19. In this case, the definition of acceleration given by Eq. (2.5) reduces to $a = v/t$, because $v_1 = 0$ and we have let $v_2 = v$. Thus, Eq. (3.3) takes the form:

$$F = ma = m\frac{v}{t}$$

The average speed in this case is $\bar{v} = v/2$, so that the distance covered is $d = (v/2)t$. The work done by the force is then given by

$$W = Fd = \frac{mv}{t} \times \frac{vt}{2}$$

$$W = \tfrac{1}{2}mv^2$$

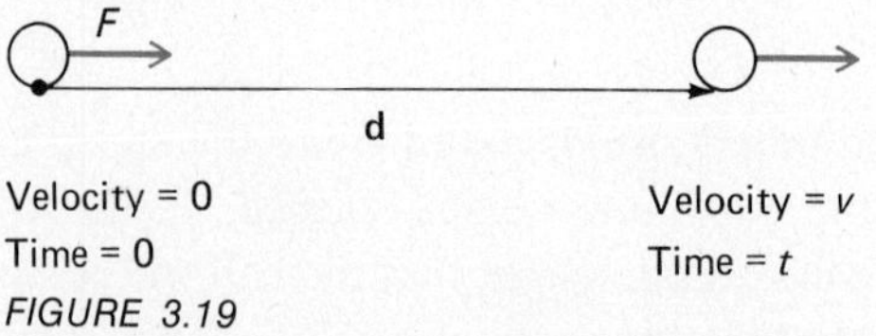

F

d

Velocity = 0
Time = 0

Velocity = v
Time = t

FIGURE 3.19

An object is accelerated from rest by a constant force shown in color. After a time t it has moved a distance **d** and has a speed v. The work the force has done is $W = Fd$ which is shown to be $mv^2/2$.

In this example, all the work was used to increase the speed of the body, without any other property of the body changing. We say, therefore, that the increase in the *kinetic energy* K of the body is

$$K = \tfrac{1}{2}mv^2 \tag{3.11}$$

Although this discussion of kinetic energy has dealt with only a constant force, the kinetic energy of a body is $\frac{1}{2}mv^2$, regardless of the way in which the speed v was obtained. The work done to accelerate a body to a speed v is not lost; it can be recovered. For instance, when you swing a hammer, the head of the hammer has kinetic energy, which is then recovered when the head hits the nail and drives it into the wood.

One of the most important laws of physics is the *Law of Conservation of Energy*. This law states that energy can be neither created nor destroyed. It can change forms and be transferred from one system to another, but if you can keep track of all of its different guises, you will never lose any. From this point of view, work is a way of transferring energy from one form into another.

As an example, consider a ball thrown vertically upward with an initial speed of v_0 as shown in Fig. 3.20. The energy of the ball just after it is thrown is all kinetic and is given by $\frac{1}{2}mv_0^2$. As the ball rises, gravity does work on it. We have already calculated the work that gravity does. If the ball has risen to a height h, this work is mgh. The kinetic energy of the ball will be decreased by just this amount. However, as the ball rises, its potential energy increases. When it has risen to a height h, its potential energy is just mgh, and the sum of its potential and kinetic energy is unchanged. When the ball has reached its maximum height H, it has no speed left, and all its energy is potential. If the total energy of the ball is E, at the bottom $E = \frac{1}{2}mv_0^2$. Part way up, when its speed has decreased to v, $E = \frac{1}{2}mv^2 + mgh$. At the top $E = mgH$. Conservation of energy then tells us that

$$\tfrac{1}{2}mv_0^2 = \tfrac{1}{2}mv^2 + mgh = mgH \tag{3.12}$$

where v_0 is the initial speed, v and h are the speed and height part way up, and H is the maximum height reached. Equation (3.12) can be used to calculate the maximum height that the ball will reach as well as its speed at any intermediate location. As the ball falls downward again, it gives up potential energy and gains kinetic energy. When the ball has returned to its starting point, it has lost all of its potential energy and has exactly the same kinetic energy as it had at the beginning (if we neglect losses due to

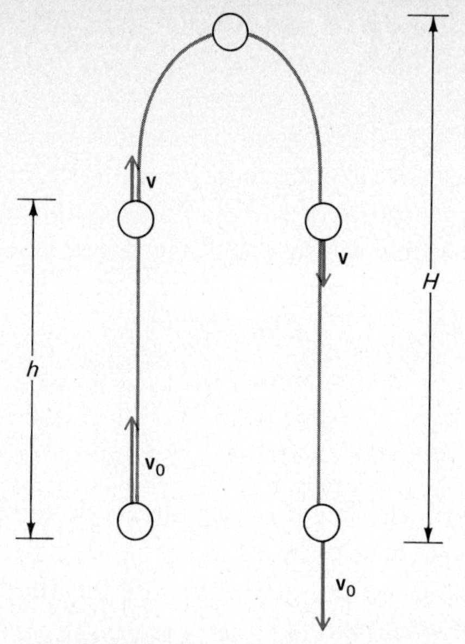

FIGURE 3.20

A ball thrown upwards with initial speed $\mathbf{v}_0$. Initially its energy is all kinetic. On the way up its energy is part kinetic and part potential and at the top it is all potential. The horizontal motion has been exaggerated.

air resistance). Its total energy, the sum of kinetic and potential, remained constant throughout, which is the content of Eq. (3.12).

EXAMPLE Here is another example of the Law of Conservation of Energy. Suppose that a particle with a mass of 2 kg is released from rest and falls a distance of 5 m. The change in the potential energy of the particle U_g is then given by:

$$\Delta U_g = mgh = (2 \times 10 \times 5) = 100 \text{ joules}$$

This decrease in potential energy is all converted into an increase in kinetic energy, so that we have

$$\Delta K = \Delta U_g = 100 \text{ joules} = \tfrac{1}{2}mv^2$$

The mass of the particle is 2 kg; therefore, we easily find that the speed gained by the particle is 10 m/sec. Again we have neglected any effects due to air resistance. Neglecting air resistance gives a very good approximation if the particle is made of a dense material such as a metal, but that would not be the case if the object were a rubber balloon.

There are many other types of energy besides those discussed here. It takes work to make an object rotate, and this work is stored in the kinetic energy of rotation. A spinning potter's wheel is an example of energy stored in this form. To compress a spring also requires work. This work, as we

shall see in the next section, is stored in the spring's potential energy. Most watches are driven by the energy stored in their main springs. As another example, the nucleus of an atom is held together by a tremendous amount of energy, which is stored as a form of potential energy. It is this energy that we tap in a nuclear reactor or, more dramatically, in a hydrogen bomb. The sun has the same source of energy. Probably the energy stored in the nuclei of hydrogen atoms will be one of the principle sources of energy in the 21st century. We shall return to the subject of nuclear energy in the last part of this book.

3.5 ELASTICITY

In many cases, solid bodies made of, say, steel seem to be completely rigid. We all know, however, that it is easy to bend a steel paper clip or knitting needle. In this section, we are concerned with the effects of forces on the shapes and sizes of solid objects.

When forces applied at the opposite ends of a solid tend to make the body longer, we say that a *tension stress* has been applied to the body. Such stress would be the stretching of a guitar string when the instrument is tuned. If forces applied at the ends of a solid tend to decrease its length, a *compression stress* has been applied to the body. An example might be the compression of a truss supporting a bridge when traffic on the bridge is heavy. In both of these cases, the forces act along the same line. If the forces do not act along the same line, the body experiences a *shear stress,* which has the effect of changing the shape of the body without changing its volume. If you push horizontally on the top of a block of Jello, you will be exerting a shear stress. Examples of the three types of stresses are shown in Fig. 3.21. Note that the actual deformations produced on real solids are usually much smaller than the ones shown in the diagram.

Let us next consider the relation between the magnitude of a stress applied to a body and the deformation that results. This problem is complicated because it involves not only the material of the body but also its

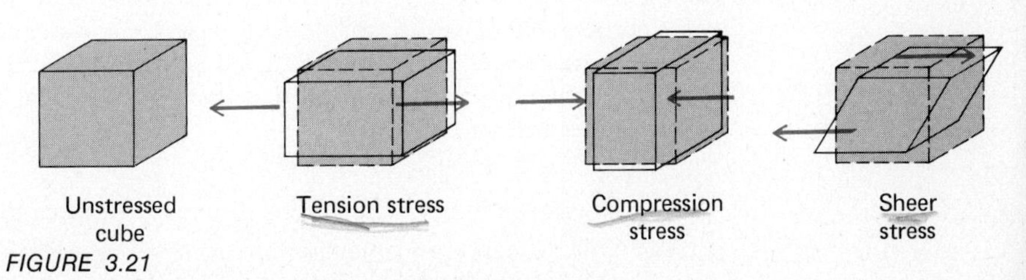

FIGURE 3.21

Various types of stress. The colored vectors are the applied forces.

temperature, shape, and other factors. In the 17th century, Robert Hooke found that if the stress applied to the body were small, then the resulting deformation was proportional to the stress. This discovery is known as *Hooke's Law,* although it is actually not a law but an empirical approximation. It is commonly used, however, because it is simple and is a good approximation for many materials.

Figure 3.22 shows how we can apply a number of forces in turn to a piece of wire. The increases in the length of the wire compared to its unstretched length are shown in the diagram, although they are exaggerated for clarity. We see that the increases in length (deformations) are proportional to the applied forces (stresses). If we introduce a constant of proportionality, k, we can write the following equation between the applied force F and the resulting elongation x:

$$F = kx \tag{3.13}$$

The constant k is known as the Hooke's Law constant of this particular wire. But the value of k varies. It is much larger for steel than it is for rubber, for instance; and it is larger for a thick wire than for a thin wire if both are made of the same material.

We know that all wires break when a sufficiently large stress is applied to them, so you might wonder how a real wire behaves as the stress applied to it is gradually increased. The graph of tension stress versus resulting elongation for a typical material is shown in Fig. 3.23. As long

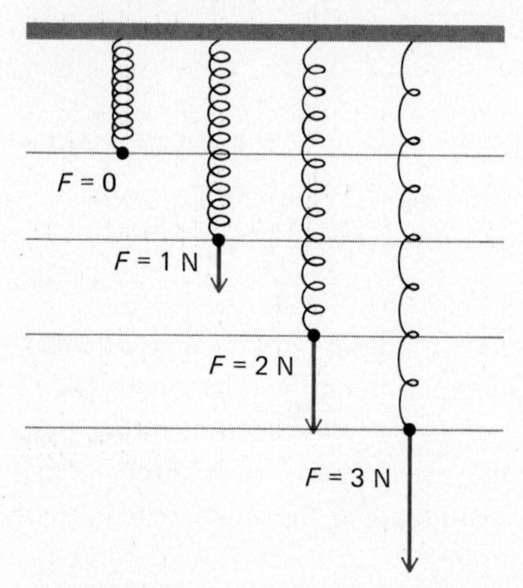

FIGURE 3.22

The length increase of a spring for various applied forces. Note that the increase is directly proportional to the applied force.

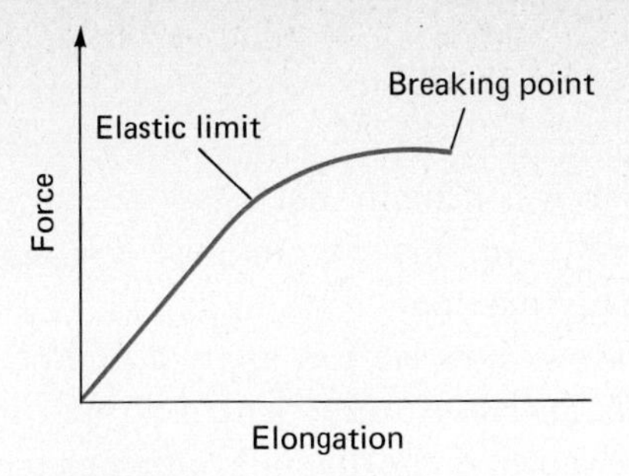

FIGURE 3.23
The tension force applied to a wire versus the elongation it produces.

as the tension is small, the elongation is proportional to the tension, in accord with Eq. (3.13). However, at a point known as the *elastic limit,* the material begins to yield and Hooke's Law no longer applies. In addition, if the tension is now removed, the length of the wire does not return to its original value, as would have been the case at any point on the straight portion of the graph. At an even higher value of the tension, the breaking point is reached, and suddenly we have two pieces of wire.

If a material obeys Hooke's Law, we can easily calculate the work required to lengthen the wire a given amount. As we gradually increase the force applied to the ends of the wire, the force takes on all values between $F_1 = 0$ (initially) and $F_2 = kx$ (finally). The average force applied to the wire during this process is

$$\bar{F} = \frac{F_1 + F_2}{2} = \frac{0 + kx}{2} = \frac{kx}{2}$$

The work done in producing the final elongation is then

$$W = \bar{F}x = \frac{kx}{2}x = \frac{1}{2}kx^2 \qquad (3.14)$$

To compress a material an amount x with a given force, the same constant k is found to be valid experimentally. Thus, both Eq. (3.13) and Eq. (3.14) are correct when x stands for a contraction rather than an elongation. In most actual cases, the elastic body is a spring, and the contraction or elongation is small compared to the unstretched length of the spring, so that Hooke's Law applies.

EXAMPLE In order to demonstrate these ideas, suppose that a force of 100 N is required to lengthen or compress a certain spring a distance of 5 cm. Then, the spring constant

is $k = F/x = 100/0.05 = 2000$ N/m. In addition, the work required to produce this change in length of the spring is given by $W = \frac{1}{2}kx^2 = \frac{1}{2}(2000)(5 \times 10^{-2})^2 =$ 2.5 joules. Because we know the value of k for this particular spring, we can easily compute the force and work required to produce different changes in its length.

As we discussed in Sec. 3.4, whenever work is done on an object, an equal amount of energy is stored that can later be recovered. In the case of an elastic body, we call this energy *elastic potential energy* and write it as

$$U_s = \tfrac{1}{2}kx^2$$

If we refer to the example in the preceding paragraph, the elastic potential energy stored is 2.5 joules. This type of energy is displayed in a watch spring or in a slingshot when it is pulled back. There are many other examples of the use of elastic potential energy.

EXAMPLE Suppose that a car manufacturer who is filming an advertisement for a new bumper stands a car whose mass is 1000 kg (approximately 2200 pounds) on its front end and observes that the bumper is depressed 0.1 m (about 4 inches). He plans to run the car into a brick wall at a speed of 2 km/sec (about 5 miles per hour) and film the amount by which the bumper is compressed. We wish to figure out this amount in advance. From the car's mass of 10^3 kg, we calculate that its weight mg is 10^4 N. Because that force compresses the bumper 10^{-1} m, we can use Eq. (3.13) to find the spring constant:

$$k = \frac{F}{x} = \frac{10^4}{10^{-1}} = 10^5 \text{ N/m}$$

When the car strikes the wall, its kinetic energy is converted into potential energy in the bumper's spring. Thus, by conservation of energy we have:

$$\tfrac{1}{2}mv^2 = \tfrac{1}{2}kx^2$$

where x is the distance that the bumper is compressed. After cancelling $\frac{1}{2}$ from both sides of the equation and solving for x, we find

$$x = v\sqrt{\frac{m}{k}} = 2\sqrt{\frac{10^3}{10^5}} = \frac{2}{\sqrt{10^2}} = 0.2 \text{ m}$$

This figure corresponds to a deformation of about 8 inches. You should note, however, that our solution to this problem was idealized. Some of the kinetic energy would be transformed into heat, sound, and other forms of energy, whereas we assumed that all of the energy was transformed into elastic potential energy.

DISCUSSION QUESTIONS

1. Why are larger hammerheads used to drive larger nails and spikes?

2. The engineer of a railroad train puts on the brakes and the train comes to a stop. Why do you fall forward?

3. When you push hard against the wall of a room, you produce no motion, yet you do get tired. Is there something inconsistent between the definition of work used in physics and its physiological meaning? (Note: Your body does use up energy in this situation.)

4. The engine of your car continues to burn gasoline even when your car is moving with constant speed along a straight, horizontal highway. Why is this process necessary if Newton's First Law is true? Discuss any energy interchanges involved in this process.

5. Is it possible for a body to have energy without having any momentum? Can a body have momentum without having any energy?

6. Explain why a rocket can produce thrust outside the earth's atmosphere where there is no air for the exhaust to push against.

7. Is it possible for two forces of magnitudes 5 and 8 units, respectively, to maintain a particle in equilibrium?

8. A closed laboratory is located in a spaceship. The ship is in interstellar space and accelerating at 10 m/sec^2. Discuss the laws of motion obeyed by objects in this laboratory, and compare them with the laws in a similar laboratory at rest on the surface of the earth.

9. Substances that flow, such as plastics, are known as rheological substances. Would you expect them to obey Hooke's Law?

10. If you squeeze a ball of putty, are you doing work? Could you recover this work?

11. When you blow up a balloon, you do work. Is it possible to recover this work?

12. Assume that a spring obeys Hooke's Law for small compressions. Is there an obvious limit to the compression for which this law fails? What is it?

PROBLEMS

1. Compute the force required to give a mass of 3 kg an acceleration of 5 m/sec^2.

2. When a force of 25 N is applied to a certain particle, it acquires an acceleration of 2 m/sec^2. What is the mass of the particle? (*Ans.:* 12.5 kg)

3. A force of 19.6 N is applied to a particle with a mass of 2 kg. What is the acceleration of the particle?

4. A rocket with a mass of 100 kg accelerates vertically upward with an acceleration of 5 m/sec^2. Compute the upward thrust produced by the rocket's engine. (*Ans.:* 1480 N)

5. Compute the weight in newtons of 1 lb.

6. If a force of 20 N is applied to a particle with a mass of 4 kg, what is the acceleration of the particle? (*Ans.:* 5 m/sec^2)

If, at the start, the particle had a velocity of 3 m/sec, what is its momentum 2 sec after the force is applied?

7. A force of 20 N acts through a distance of 3 m. Compute the work done by this force. (*Ans.:* 60 joules)

If the force acts during a time of 4 sec, compute the power exerted by this force.

8. Compute the kinetic energy of a boat with a mass of 10,000 kg moving at a speed of 20 km/hr. (*Ans.:* 1.6×10^5 joules)

What force is required to bring this boat to rest in a time of 50 sec?

Assuming that the applied force is constant, compute the distance traveled by the boat in coming to rest. (*Ans.:* 142 m)

9. A particle is acted upon by an upward force of 20 N and a horizontal force of 30 N. Compute the magnitude and direction of a third force that will hold the particle in equilibrium.
(*Ans.:* 36 N downward and making an angle of 56.3 deg with the vertical)

10. A force of 15 N acts through a distance of 2 m. Compute the work done by this force.

If the force acts during a time of 3 sec, compute the power exerted by the force. (*Ans.:* 10 W)

11. Compute the kinetic energy of a car with a mass of 1500 kg traveling at a speed of 20 m/sec.

Calculate the height from which the car would have to fall from rest in order to gain the same amount of energy. (*Ans.:* 20 m)

12. A ball is projected vertically upward with an initial speed of 10 m/sec. Compute the maximum height to which the ball rises.

At the top of its path, what is the acceleration of the ball?
(*Ans.:* 10 m/sec^2 downward)

13. Compute the speed that a car with a mass of 1000 kg must have in order for it to have the same momentum as a truck with a mass of 25,000 kg traveling at a speed of 15 m/sec.

Compute the kinetic energies of the car and the truck.
(*Ans.:* car: 7.05×10^7 joules; truck: 2.82×10^6 joules)

14. Compute the initial upward speed of a particle that reaches a maximum height of 10 m above its point of release.

15. Compute the kinetic energy of a car with a mass of 1500 kg traveling at a speed of 80 km/hr. (*Ans.:* 3.72×10^5 joules)

If all of the kinetic energy of the car is used up in 20 sec, what is the average rate at which the car gives up energy, measured in watts?

16. When an object weighing 5 kg is suspended by a spring, the length of the spring changes by 2 cm. Compute the force constant of this spring.

17. Compute the force required to stretch a spring with a force constant of 10 N/m a distance of 3 cm. (*Ans.:* 0.3 N)

Compute the work necessary to produce the specified elongation, if the initial elongation is zero.

18. A force of 100 N is used to compress a spring with a force constant of 5 N/cm. Compute the amount of the compression. (*Ans.:* 0.2 m)

Compute the work required to produce the specified compression.

19. You wish to store an energy of 2 joules in a spring that is to be compressed 4 cm. What must the force constant of the spring be?
(*Ans.:* 2.5×10^3 N/m)

20. Compute the force necessary to give a mass of 5 kg an acceleration of 10 m/sec^2.

21. When a force of 30 N is applied to a certain particle, it acquires an acceleration of 4 m/sec^2. What is the mass of the particle?
(*Ans.:* 7.5 kg)

Compute the momentum of the particle after the particle has experienced the force for 2 sec, if the particle started at rest.

22. A 5-kg mass moving with a speed of 15 m/sec strikes a 20-kg mass at rest, and they stick together. Find the speed of the two objects after the collision.

23. An 8-kg mass moving to the right at 6 m/sec strikes an unknown mass that is moving to the left at 4 m/sec. After the collision, both objects are at rest. Find the unknown mass. (*Ans.:* 12 kg)

24. A 100-kg astronaut on a space walk pushed himself with a speed of 2 m/sec away from his spacecraft. If the mass of the spacecraft is 4000 kg, find its recoil velocity.

4 Gravity

By the middle of the 17th century, scientists had gathered a lot of experimental information about the motions of the planets. This evidence was summarized by Johannes Kepler in his three laws of planetary motion. These laws state that:

> 1. *Each planet moves around the sun in an elliptical orbit with the sun as the focus.*
> 2. *As a planet moves around the sun, its radius vector (the vector from the sun to the planet) sweeps out equal areas in equal times (see figure on next page).*
> 3. *The square of a planet's period (the time needed to complete one complete orbit) is proportional to the cube of the planet's average distance from the sun.**

With this heliocentric theory of the solar system, Kepler could explain the observations better than with any previous theory. However, his system was

*This distance is equal to the *semimajor* axis of the orbital ellipse.

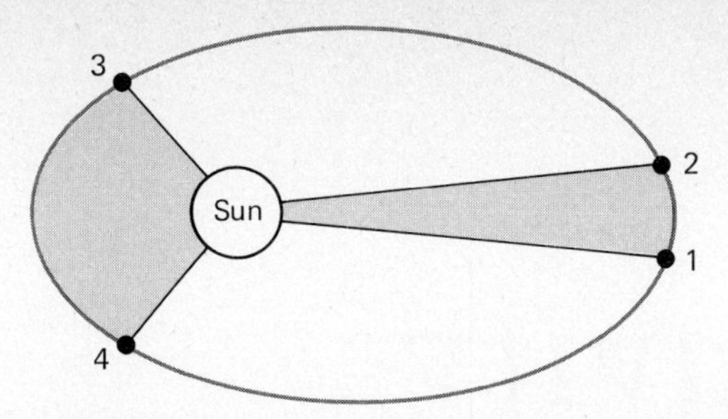

Kepler's Second Law illustrated. The time needed for the planet to move from 1 to 2 is just the same as that needed to move from 3 to 4. The distance moved is fixed by requiring that the shaded areas are equal. Thus the planet moves more slowly when far from the sun.

not based on any fundamental laws. He merely set forth rules that accounted for the data.

It was Sir Isaac Newton who solved the problem of predicting these empirical laws from the basic laws of motion. For this task, Newton developed three tools. The first tool was his set of laws of motion, which we studied in the last chapter. These laws provide an explanation for how objects move. The second tool, which is the main subject of this chapter, was his theory of universal gravitation. This theory explains the forces that pull the planets into their curved paths around the sun. The last tool was calculus, developed by Newton and by Baron von Leibnitz. Calculus provides the rigorous mathematics for treating the curved paths and variable forces of the planetary problem. The *Principia,* in which Newton explained these theories, is one of man's greatest single creative efforts.

4.1 UNIVERSAL GRAVITATION

In formulating his Law of Universal Gravitation, Newton reasoned that both the magnitude and the direction of a planet's velocity are continually changing, because the planet's orbit is elliptical. Thus, the planet is accelerated. But, as he stated in his Second Law of Motion, a force is always required to accelerate a body. It was not possible to measure directly the force between a planet and the sun; therefore, Newton had to guess the nature of the force and then see if this force would actually predict the observed motions of the planets. He concluded that every object in the universe exerts an attractive force on every other object. This force is proportional to each of the masses and is inversely proportional to the square of the distance between them. Thus, for a pair of masses m and M separated

by a distance r, as shown in Fig. 4.1(a), the force was proportional to mM/r^2. If we introduce a constant of proportionality G, we can write the law of gravitation in the form:

$$F = G\frac{mM}{r^2} \qquad (4.1)$$

This equation gives the magnitude of the force. The direction of the force is always along the line joining the centers of the masses, and it is always attractive. According to Newton's Third Law, Eq. (4.1) actually gives the magnitude of two forces, one on each mass. Both of these forces are shown in Fig. 4.1(a).

This law represented a tremendous intuitive leap on Newton's part because it predicted the existence of a force that did not *seem* to exist. Our everyday experiences give us no indication that objects of ordinary size attract each other. The reason is that the force of attraction is exceedingly small. Two 1-kg masses separated by a meter exert a force of less than 10^{-10} N on each other. Because gravitational forces between ordinary-

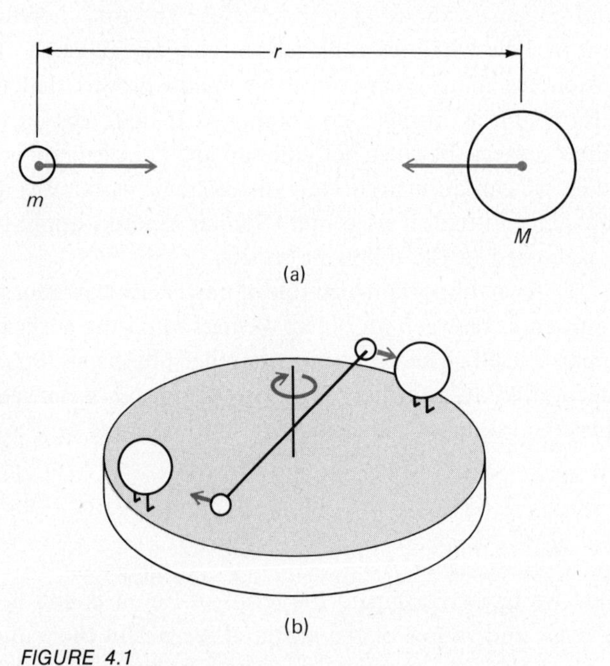

FIGURE 4.1

(a) The equal but opposite force vectors are the attractive gravitational force between the two objects. (b) The Cavendish balance.

sized objects are so small, Newton had no way to verify Eq. (4.1) directly. However, this law of force did predict correctly the motions for planets and satellites and so was indirectly confirmed.

In 1798, an English scientist, Henry Cavendish, devised a delicate device to study Eq. (4.1), and his results agreed with this equation. The apparatus used by Cavendish is shown in Fig. 4.1(b). Two small masses attached to the ends of a light crossbar were attracted towards the larger masses. The forces between the masses twisted the fiber supporting the crossbar through a measurable angle. By a fairly simple calculation, the force between each small mass and its neighboring large mass could be computed. In this way, Cavendish verified Eq. (4.1) and determined the value of the constant G to within an accuracy of about 1 per cent of the best modern value, which was a remarkable achievement in view of the crudity of his apparatus. If we use the metric system, with mass measured in kilograms, distance in meters, and force in newtons, the constant G is found to have the value of 6.67×10^{-11} N–m²/kg².

So far, our consideration of gravitational forces has been restricted to objects that are very small compared to their distance of separation. Consequently, the objects could be treated as mathematical points. The diameter of the sun is about 1 million miles, whereas the earth's distance from the sun is about 93 million miles. The sun, therefore, is not really a point particle when we consider its force on the earth. This fact bothered Newton for many years, until he finally proved that two spherical and uniform objects attract one another as if their masses were concentrated at their centers. Because both the sun and the earth are very nearly uniform bodies, we can quite accurately discuss their motions as if the mass of each were concentrated at its center. Similar remarks apply to satellites, which are discussed in the following section.

As an important example of universal gravitation, consider the force of attraction between an object of mass m on the surface of the earth and the earth itself. The earth attracts other objects as if its entire mass were concentrated at its center; therefore, the distance between the two masses is just the radius of the earth. We find the force of attraction is

$$F = \left(\frac{GM_e}{R_e^2}\right) m \tag{4.2}$$

where we have rearranged the terms of Eq. (4.1) and used M_e and R_e for the mass and radius of the earth. If we put in the values for G, M_e, and R_e, we find that the quantity in parentheses is just 9.8 m/sec², and the equation reads $F = (9.8)\,m$. After combining this equation with Newton's Second Law, $F = ma$, we conclude that if gravity is the only force acting

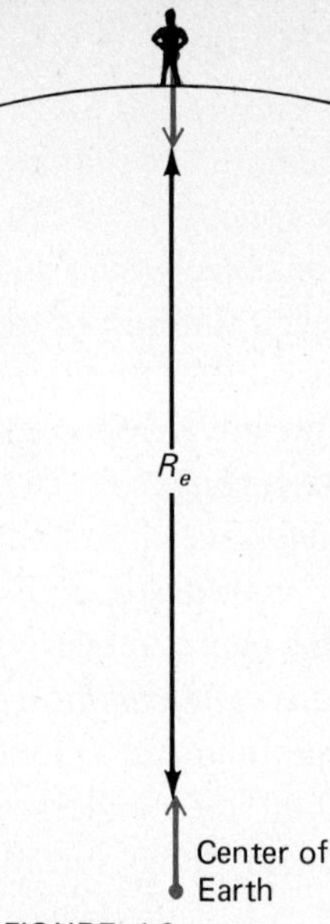

FIGURE 4.2

The gravitational force of the earth on a man at the surface.

on the body, it will fall towards the earth with an acceleration of 9.8 m/sec^2. Because the radius of the earth is very large compared to ordinary heights, this acceleration will be essentially constant.

We have just derived the fact that all objects fall with the same acceleration in the earth's gravitational field, a fact that was presented as an empirical law in the second chapter. However, using similar reasoning, we can also find the acceleration of gravity on any other planet or moon. Since Eq. (4.2) gives the force of gravity, which we have called weight, acting on an object of mass m, we conclude that

$$W = mg \tag{4.3}$$

where

$$g = \frac{GM}{R^2} \tag{4.4}$$

is the weight of the object on the surface of any spherical mass M with radius R. Furthermore, using $F = ma$, Eq. (4.4) gives the acceleration of gravity for any planet. If we use the mass and radius of the moon, we would find the acceleration of gravity on the moon to be about one-sixth that on the earth, a fact that the astronauts could have tested had they not had more important things to do.

4.2 UNIFORM CIRCULAR MOTION

Before we can use Newton's Law of Gravitation to explain the motion of the planets, we must describe the kinematics of that motion. Kepler's First Law (stated at the beginning of this chapter) tells us that any object orbiting the sun follows an elliptical orbit. For objects such as Halley's comet, this ellipse is very long and narrow. However, for most of the planets, the ellipse is almost a circle. Describing the general case of elliptical orbits is beyond the scope of this book; it is treated in advanced texts on mechanics. We consider here only the case of circular orbits.

Fortunately, most of the planets have approximately circular orbits, so we introduce little error by this simplification. For a circular orbit, Kepler's Second Law tells us that the planet moves with constant speed. Motion in a circle with constant speed is called *uniform circular motion,* and it appears in many situations other than planetary motion. Any fixed point on a rigid body rotating at a constant rate undergoes uniform circular motion. Because our world is filled with things turning, uniform circular motion is quite common.

Suppose that a particle moves in a circular path of radius r at constant speed v. Figure 4.3 shows the position and velocity vector of the particle at various times as it moves around the circle. The velocity vector always points in the direction of motion, but it remains the same length, although its direction is constantly changing. Thus, even though the speed is constant, the particle is continually accelerated. To find the acceleration we must calculate how much the velocity changes in a given time. Consider

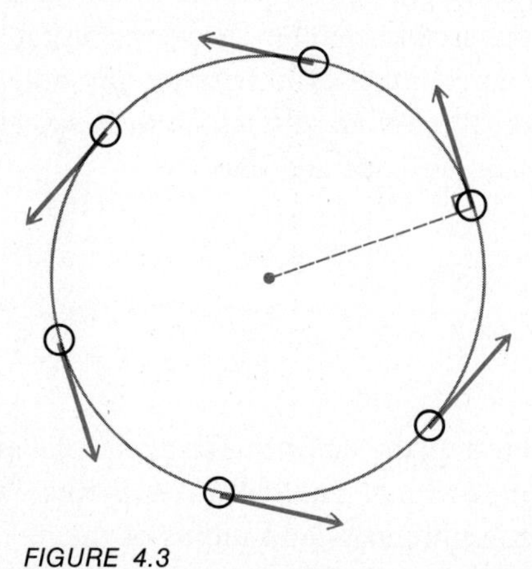

FIGURE 4.3

A particle moving in a circle at constant speed. The colored velocity vectors all point in different directions but have the same length.

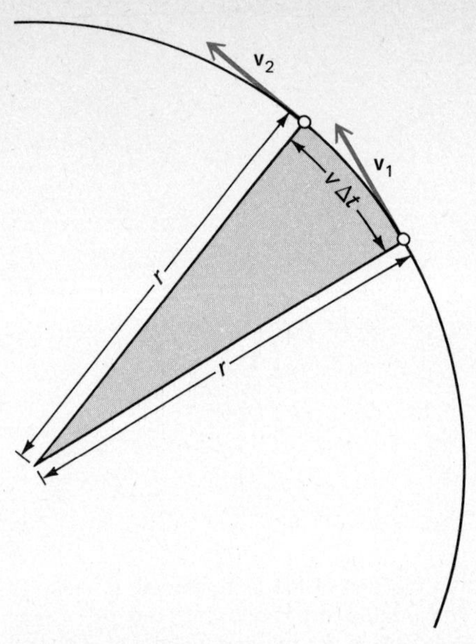

FIGURE 4.4

The position and velocity of a particle at the beginning and end of a time interval during which it has moved a distance $v\,\Delta t$.

a time interval Δt that is short compared to the time it takes the particle to make a complete revolution. During this time the particle will move a distance $v\,\Delta t$ around the circle. Figure 4.4 shows the motion of the particle during the time Δt. At the start of the interval the particle's velocity is $\mathbf{v}_1$ and at the end, $\mathbf{v}_2$. These vectors are shown in the figure. Note that they have the same length v, because the speed is constant. Acceleration is defined as the change in velocity divided by the change in time. The change in velocity $\Delta\mathbf{v}$ is given by $\Delta\mathbf{v} = \mathbf{v}_2 - \mathbf{v}_1$ or, adding $\mathbf{v}_1$ to both sides,

$$\mathbf{v}_2 = \mathbf{v}_1 + \Delta\mathbf{v} \tag{4.5}$$

Figure 4.5 shows the vector addition indicated in Eq. (4.5). The vector $\Delta\mathbf{v}$ must be added to the velocity at the beginning of any short time interval to give the velocity at the end of the interval. Notice that the vector $\Delta\mathbf{v}$ must always point toward the center of the circle to keep the particle on the circle and to keep the speed constant. Comparing Fig. 4.4 to Fig. 4.5, we see that the colored triangles are similar, and we can write

$$\frac{\Delta v}{v} = \frac{v\,\Delta t}{r} \tag{4.6}$$

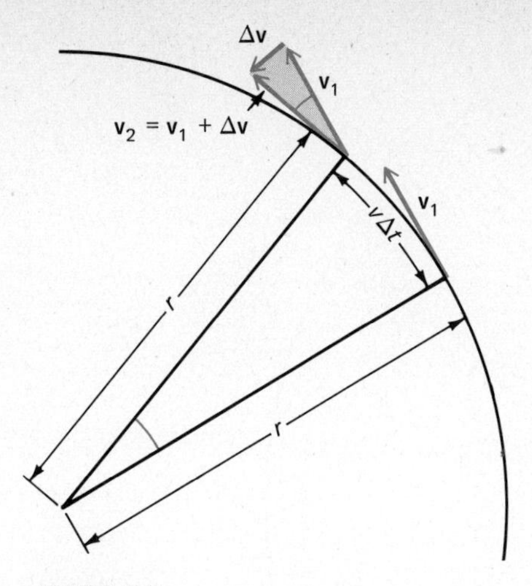

FIGURE 4.5

The velocity vector $\mathbf{v}_2$ at the end of the time interval is given by $\mathbf{v}_1 + \Delta\mathbf{v}$. This vector sum is shown in red. The shaded triangle is similar to the shaded triangle of Fig. 4.4.

where we have used v for the length of both $\mathbf{v}_1$ and $\mathbf{v}_2$. Of course, $v\,\Delta t$ is really the length of the arc in Fig. 4.4, not the chord. However, if Δt is very small, the chord and the arc will be almost the same and we can neglect the difference. Now if we multiply Eq. (4.6) by $v/\Delta t$ on both sides, we find

$$\frac{\Delta v}{\Delta t} = \frac{v^2}{r}$$

by Eq. (2.5), however, acceleration is defined by $\Delta v/\Delta t$, so we have

$$a = \frac{v^2}{r} \tag{4.7}$$

The acceleration experienced by a particle in uniform circular motion is called *centripetal acceleration.* Its magnitude is given by Eq. (4.7). Its direction is parallel to the vector $\Delta\mathbf{v}$ and thus is directed towards the center of the circle.

Newton's Second Law tells us that where there is acceleration, there must be force. We use the equation $F = ma$. Then the force necessary to

hold a mass m in uniform circular motion is

$$F = m\frac{v^2}{r} \quad (4.8)$$

This force is called the *centripetal force.* Whenever an object is observed moving in a circle at constant speed, the net force on the object is given by Eq. (4.8) and the direction of the force is towards the center of the circle. An example would be provided by the string holding a ball that you swing around your head. This centripetal force is perpendicular to the motion of the object. It does no work. Furthermore, the kinetic energy of the object should be constant. Indeed, this is the case, because kinetic energy is $\frac{1}{2}mv^2$ and the speed v is constant.

4.3 ORBITAL MOTION

We are now in a position to explain orbital motion. The net force acting on a planet moving in a circle at uniform speed is given by Eq. (4.8). This net force is constant in magnitude and is always directed towards the center of the circle. Newton's Law of Gravitation provides just such a force. For a circular orbit, r (and of course each mass) does not change; therefore, the magnitude of the force of gravity is constant. Furthermore, because the force on the smaller mass m is towards the larger mass M, located at the center of the circle, the direction of the gravitational force is correct. If we set the force needed for uniform circular motion [Eq. (4.8)] equal to the force of gravity [Eq. (4.1)], we find

$$G\frac{mM}{r^2} = m\frac{v^2}{r} \quad (4.9)$$

After multiplying this equation by r^2/m, we have

$$v^2r = MG \quad (4.10)$$

This equation is the basic one for a circular orbit. It tells us what speed a satellite must have to maintain a circular orbit. Equation (4.10) actually contains Kepler's Third Law, although we will have to rearrange it. In making one complete trip around its orbit, a satellite travels a distance $C = 2\pi r$, the circumference of the orbit. The time necessary to complete

one orbit is called the *period* T of the satellite. Because distance equals speed times time, we can write

$$C = vT$$

If we solve for v and let $C = 2\pi r$, we have

$$v = \frac{2\pi r}{T}$$

Substituting this equation into Eq. (4.10) gives

$$\left(\frac{2\pi r}{T}\right)^2 r = MG$$

or, if the terms are rearranged,

$$T^2 = \left(\frac{MG}{4\pi^2}\right) r^3 \tag{4.11}$$

which is Kepler's Third Law. This equation is nothing but Eq. (4.10) written in terms of the period instead of the speed.

One of the interesting things about Eq. (4.10) is that the speed of a satellite in a given circular orbit does not depend on the mass of the satellite. This is one way to understand the phenomenon of weightlessness. Because an astronaut, the capsule he is in, and his toothbrush are all essentially in the same orbit, they all move with the same speed and follow the same path at the same time. Consequently, there is no relative motion between them, and the astronaut and his toothbrush appear to float in the capsule.

Another way of understanding orbital motion and weightlessness is to imagine the following experiment. You are standing on the top of a high mountain on the moon. We choose the moon because there will be no air resistance. You throw an object horizontally and observe where it lands. This motion is shown in Fig. 4.6. The faster you throw the object, the further it will travel before it hits the surface of the moon. Eventually, if you throw the object fast enough, the surface of the moon will start to recede below the object because of the moon's curvature. As you further increase the speed of the object, this effect will become more pronounced until you reach a speed at which the surface of the moon recedes just as fast as the object falls, and the object maintains the same altitude. The object then has a circular orbit. An object in orbit is actually falling all the time.

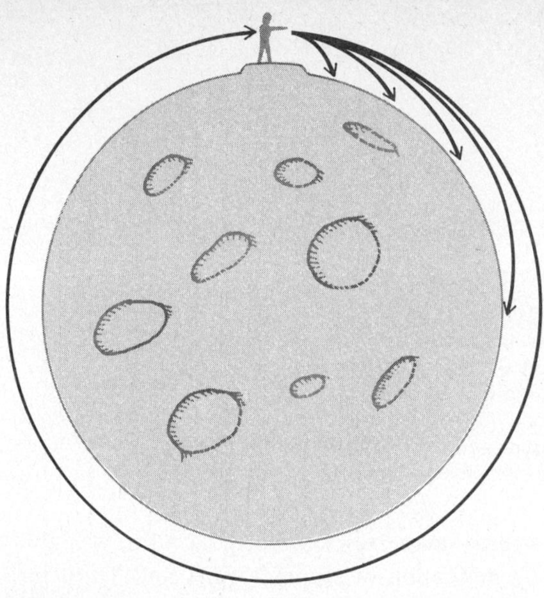

FIGURE 4.6
A man, standing on a mountain on the moon, throws an object with increasing speed.

It has just enough horizontal velocity so that it never gets any closer to the mass around which it is orbiting.

All of these ideas and equations apply equally well to a planet in orbit around the sun, to a moon or satellite in orbit around a planet, or to an astronaut in orbit around the moon. In each case, however, M must be the mass of the object at the center of the orbit. We have been assuming that m, the mass of the orbiting object, is much smaller than M, the mass around which m is moving. This assumption holds for an artificial satellite orbiting the earth. However, it is not very good for the earth-moon system and cannot be used at all for a binary star system with two stars orbiting each other. In our approximation, we have neglected the effect of the force that the satellite exerts on the object it circles. If M is much larger than m, the force on M (which is, of course, equal to the force on m) produces a negligible effect. If, however, M is comparable in size to m, this force will cause M to move. This phenomenon is treated in more advanced books. If you are interested, you might try to find the form of Kepler's Third Law for two equal mass objects in a circular orbit around each other, as shown in Fig. 4.7. Each of the masses moves around the same circle at opposite ends of a common diameter.

In this section we have seen how Newton's laws of motion and gravitation can be used to show that a satellite in a circular orbit will obey

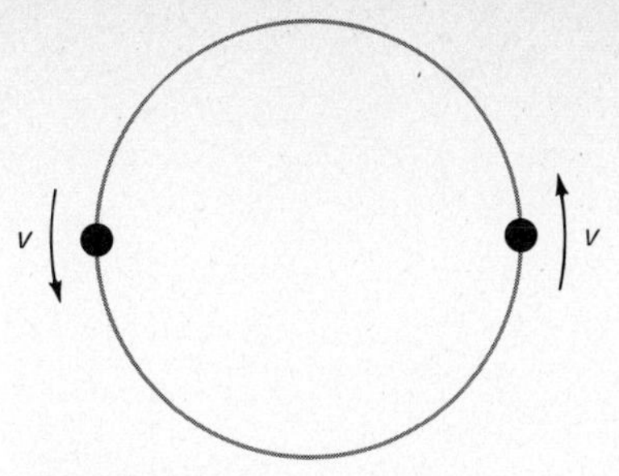

FIGURE 4.7

Two equal masses in a circular orbit around each other. Their centripetal acceleration is determined by their speed *v* and by the radius of their common orbit. The gravitational force holding them in orbit is determined by their masses and the distance between them, in this case the *diameter* of the circle.

Kepler's three laws. Newton, using calculus, was able to treat the general case. The derivation of Kepler's First and Third laws is complex and is presented in advanced texts on mechanics. His Second Law, however, can be understood in qualitative terms. It states that as a satellite moves around a planet in its elliptical orbit, its radius vector sweeps out equal areas in equal times. Thus, as it moves closer to the planet, it gains speed.

There are two effects at work here. First, as the radius of an orbit decreases, the satellite moves in towards the object it is circling. This motion is in the direction of the gravitational force, which then does work on the satellite, increasing its kinetic energy and hence its speed. Second, the satellite speeds up as its radius decreases in obedience to a conservation law. This law, which we have not mentioned, is *conservation of angular momentum*.

Along with the laws concerning conservation of energy and conservation of linear momentum, this law forms the foundation of all physics. Every physical theory, from the simple laws that we study in this book to the complexities of quantum electrodynamics (which is perhaps the most complete, precise, and powerful theory yet produced in physics), contains these three conservation laws as fundamental postulates. Angular momentum is too complex a dynamical quantity for us to discuss quantitatively. However, we can see the effect of its conservation in rather simple terms.

If a particle moves under the influence of a force that is always directed towards the same point (remember that the gravitational force always points from the satellite to the object around which it is orbiting), then the motion of the particle will obey Kepler's Second Law. The radius vector of the particle will sweep out equal areas in equal times. This behavior shows conservation of angular momentum. Clearly, as the radius vector of

a satellite decreases, the satellite must move faster to sweep out the same area in a fixed time interval. As a simple test of this law, try swinging a stone on a string around your head. The centripetal force exerted by the string on the stone will always point at your hand. If you now shorten the string without moving your hand, the stone will move faster and faster, illustrating this conservation law. The angular momentum of a mass in uniform circular momentum is $pr = mvr$, the linear momentum times the radius of the circle. When the motion is not circular, the component of the momentum perpendicular to the radius vector appears in place of the momentum. It is a simple exercise in geometry to show that conservation of this quantity is equivalent to Kepler's Second Law.

DISCUSSION QUESTIONS

1. When a man is put into orbit in a satellite, does either his weight or his mass change? Which is more fundamental, mass or weight?

2. Explain why a satellite put into orbit thousands of miles from the surface of the earth can stay in the orbit almost indefinitely, even though the satellite's fuel is exhausted.

3. Discuss the procedures necessary to move a space station from a small circular orbit to a larger one.

4. When a car is driven in a circle, passengers seem to be thrown to the outside. However, when a space station circles a planet, its occupants move with the station. Discuss these two situations and explain the difference.

5. The gravitational pull of the moon on the earth is responsible for ocean tides. Discuss this effect and explain why the interval between high tides is about $12\frac{1}{2}$ hours.

PROBLEMS

1. Use Eq. (4.4) to calculate g on the surface of the earth. The mass of the earth is 6×10^{24} kg, and its radius is 6400 km.

2. Find the mass of the sun. The average radius of the earth's orbit is 1.5×10^8 km, and one year equals roughly $\pi \times 10^7$ sec.

3. Compute the force of attraction between a man of mass 80 kg and a girl of mass 50 kg. Assume that they are separated by 20 cm.

4. Calculate the radius of the circular orbit that a satellite must have to travel once around the earth every 24 hours and thus remain over the same spot. (This *synchronous* orbit is useful for communications satellites.)
(*Ans.*: 26,400 mi (from center of earth))

5. A good racing car can sustain a lateral force roughly equal to its weight. Find the maximum speed that such a car can travel around a flat track of 40-m radius.

6. Find the acceleration of an object moving in a circle of 6-m radius at a speed of 9 m/sec. If its mass is 5 kg, find the net force on it.

7. A moon circles a planet with a period of 2 days. Find the period of a satellite in a circular orbit four times as large.

8. A 1500-kg car is being driven at 20 m/sec around a curve whose radius of curvature is 80 m. Find its centripetal acceleration. Find the force needed to keep it from skidding off the road.

9. Find the acceleration of a cockroach sitting on the edge of a 12-in. long-playing record that is rotating at $33\frac{1}{3}$ revolutions/sec. If the mass of the cockroach is 500 mg, find the frictional force.

10. A 90-kg astronaut is in a circular orbit around a 5000-kg space station. Find the gravitational force on the astronaut. Find the period of his orbit.

11. A 2-kg ball is constrained by a string as it moves in a circle. If the tension in the string is 24 N and its length is 3 m, find the speed of the ball.
(*Ans.:* 6 m/sec.)

12. A typical neutron star would have a mass of 2×10^{30} kg and a radius of 12 km. Find g on the surface of such a star.

13. A satellite in a circular orbit around the earth has a period of 3 hr. Find the radius of its orbit.

14. Find the force of attraction between two 6-gm masses separated by 72 cm. If they were released from rest, find their initial acceleration and the speed they would have after 2 sec.

Part Two
Thermodynamics

5

Heat and the First Law of Thermodynamics

Thermodynamics is the branch of physics that deals with matter in bulk, particularly with its thermal properties. Historically, this branch developed somewhat independently of mechanics. Not until the end of the 19th century was its connection with mechanics gradually understood. Today we view the laws of thermodynamics as statements about large numbers of elementary systems (perhaps as many as 10^{23}), each of which obeys the laws of mechanics. These elementary systems might be the molecules in a gas or the atoms in a solid. From the laws binding each constituent, physicists can infer what the bulk properties of the system will be by using statistical arguments. The branch of physics that makes these arguments is called *statistical mechanics*.

The thermodynamics of bulk matter are based on statistical mechanics. For example, the precise form of the internal energy (which we shall define shortly) can be derived for certain sufficiently simple systems. In this book we shall only describe the thermodynamics of large systems and, for the most part, shall not attempt to derive our laws from first principles. However, you should keep in mind that all the laws presented in the next three chapters can be deduced from the microscopic behavior of the individual constituents.

5.1 TEMPERATURE

We all have some sort of a feeling for the meaning of temperature in terms of hot and cold. However, the human senses are not very reliable in estimating temperatures. When our hands are cold, even lukewarm water feels quite hot. If a metal pan and wooden spoon are heated to the same temperature in an oven, the pan will feel hotter than the spoon because it conducts heat to the skin more rapidly. Obviously to assign temperatures we must use some property of nature that is not as variable as our human senses.

It is a fact of ordinary experience that various physical properties of materials vary with temperature. When you drive a car on a hot day, the tires heat up and the air pressure in the tires increases. On a cold day it is harder to start the car's engine, not only because the oil is more viscous ("thicker") but also because the battery delivers energy less readily. Builders of bridges must allow for the increase in length of metal rods with an increase in temperature. Similarly, the volume of a liquid usually increases with temperature, so that oil tanks are not filled full to allow for this expansion. These are just a few examples of physical properties which depend on temperature.

In principle, at least, the variation of any physical property could be used to define a temperature scale, although the various scales would not agree very well. First, we arbitrarily choose two temperatures that can easily be reproduced in the laboratory and assign numbers to these temperatures. In the Celsius (or centigrade) scale, named after Anders Celsius, a Swedish astronomer, the temperature at which pure water freezes is defined as 0°C and the temperature at which pure water boils is defined as 100°C. In each case, the measurement is to be made when the pressure of the surrounding air is 14.7 lb/in.^{2}, which is standard atmospheric pressure. The interval between the freezing and boiling points of water is subdivided into 100 degrees. Scientists and the majority of the world's inhabitants use this scale.

The Fahrenheit scale is used in the English-speaking countries. Here the freezing point of water is defined as 32°F and the boiling point of water as 212°F. On the Fahrenheit scale the interval between the freezing and boiling points of water is subdivided into 180 degrees. The relation between the Fahrenheit and Celsius temperature scales is shown in Fig. 5.1. From the diagram we note that each Celsius degree is $\frac{180}{100} = \frac{9}{5}$ as great as a Fahrenheit degree. Because the freezing point is defined as 32°F, we must subtract 32 degrees from a Fahrenheit temperature before converting it to a Celsius temperature. We then have, as the relation between the temperature F on the Fahrenheit scale and the corresponding temperature C on

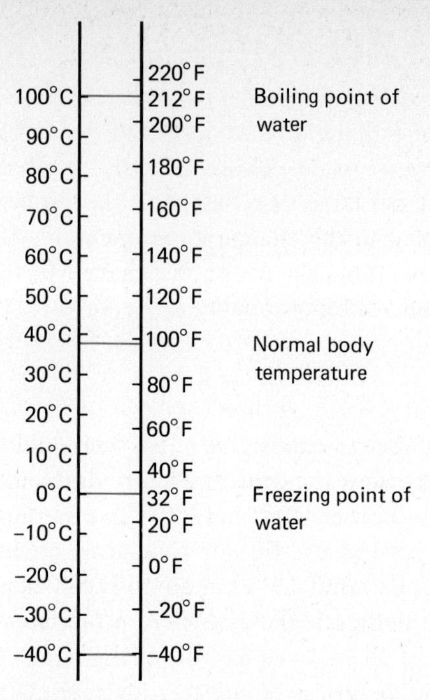

FIGURE 5.1
The Fahrenheit and Celsius temperature scales.

the Celsius scale, the equation

$$C = \tfrac{5}{9}(F - 32) \tag{5.1}$$

If we wish to convert a Celsius temperature to a Fahrenheit temperature, we can write Eq. (5.1) in the form

$$F = \tfrac{9}{5}(C) + 32 \tag{5.2}$$

EXAMPLE Let us convert 68°F into the equivalent Celsius temperature. From Eq. (5.1) we have

$$C = \tfrac{5}{9}(68 - 32) = \tfrac{5}{9}(36) = 20°C$$

If we wish to convert 25°C into the equivalent Fahrenheit temperature, we calculate from Eq. (5.2) that

$$F = \tfrac{9}{5}(25) + 32 = 45 + 32 = 77°F$$

As we shall see later, negative temperatures are possible in both scales, so

that Eqs. (5.1) and (5.2) are valid for both positive and negative temperatures.

Once the freezing and boiling points of water are fixed, the variation of some property of a material with temperature can be used to establish a practical temperature scale. Experimentally, as long as the range of temperatures is not too large, the change in the property of the material is always proportional to the change in temperature. If we let ΔT be the change in the temperature, let Δy be the change in some property, and introduce a constant of proportionality k, we have

$$\Delta y = k\,\Delta T \tag{5.3}$$

To decide what number to choose for k, we must find out how much the property y changed over a known temperature difference, say 100 degrees. If Y is the change in y between 0°C and 100°C, we use Eq. (5.3) and evaluate k as $Y/100$. Then we can use Eq. (5.3) either to predict the change in y for some other ΔT or to find ΔT from an observed change in y.

This latter method is the essence of a thermometer. By using the change in pressure of an enclosed gas, the change in length of a metal rod, or the change in volume of a liquid, we can devise a temperature scale. In fact, all of these properties and many others are used in actual practice. However, the various temperature scales derived from these and other properties do not agree perfectly except at 0°C and 100°C, because the various properties do not change in the same way with temperature. However, as we shall see in Sec. 6.1, the changes in pressure or in the volume of all gases will yield the same temperature scale if conditions are properly chosen. This ideal gas temperature scale is then used to calibrate thermometers such as the common mercury-in-glass thermometer, that utilize the changes in other physical properties.

How are very low and very high temperatures measured? Helium may be used to measure temperatures down to about −269°C. Below that temperature helium is a liquid. However, the magnetic properties of certain substances vary systematically with temperature. Thus, the magnetic properties of one of these materials can be calibrated against the helium gas thermometer. At temperatures at which helium is a liquid, the measured value of the magnetism of the calibrated material can be used to provide a temperature scale by extrapolation. Similarly, at very high temperatures the glass used in a gas thermometer would melt. The electrical resistance (see Sec. 8.3) of many materials varies regularly with temperature. If the variation in electrical resistance of a material is measured at moderate temperatures, the value of the electrical resistance at high temperatures can be used to calculate the temperature. Alternatively, the wavelength at which

the emission from a very hot body is a maximum can be used to determine its temperature, as is discussed in Sec. 13.4.

5.2 HEAT

When energy is added to a body, its temperature generally rises. This energy is not lost but, as we shall see in the next section, is stored within the body. There are many ways of adding energy to or removing energy from a body. Some of these ways involve the performance of mechanical work, such as expanding or compressing a gas with a piston. In this case, either work is done on the gas (in compression) and its temperature rises or the gas does work on its surroundings (in expansion) and the temperature of the gas drops. For this reason, a bicycle pump grows warm when used. A rise in temperature is always an indication that energy has been added to a body.

Energy can also be added in ways that do not involve the performance of work. For instance, when you place a pot on a stove and light the gas, the temperature will rise, indicating that energy is being added to the water. However, no mechanical work is performed on the pot. In general, when energy is added to a body without the performance of work, we call this energy *heat*.

The flow of energy between bodies in the form of heat was studied before it was known that heat is a form of energy. Therefore, units of heat are defined in terms of the heat required to raise the temperature of a unit mass of water 1 deg on some temperature scale. The calorie is defined as the heat required to raise the temperature of 1 gm of water by 1 deg C. The kilocalorie (kcal) is defined as the heat required to raise the temperature of 1 kg of water by 1 deg C, as in Fig. 5.2. (The kilocalorie will be used

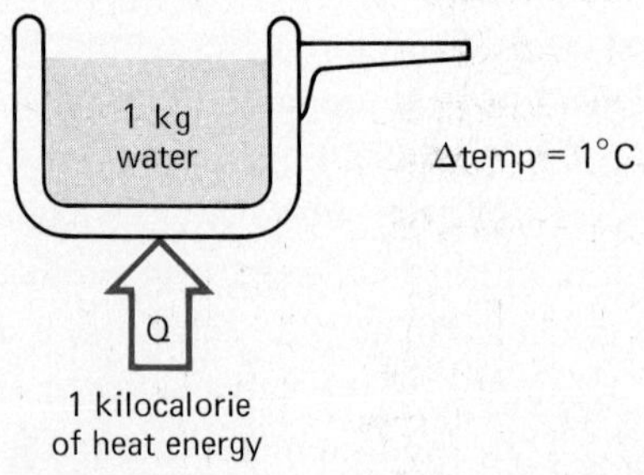

FIGURE 5.2

One kilocalorie of heat will raise the temperature of one kilogram of water one degree Celsius. One kilogram of water has a volume of essentially one liter or a little more than a quart.

throughout this book. It is also the unit used by nutritionists, who omit the prefix "kilo.") In the English-speaking countries, the British thermal unit (Btu) is defined as the heat required to raise the temperature of 1 lb of water by 1 deg F. Note that 1 Btu equals 0.252 kcal, although we shall not refer to Btu again.

Historically, the connection we have assumed between energy and heat was discovered rather late in the development of thermodynamics. Most of the pioneering work was done under the assumption that heat is a fluid, called *phlogiston,* that is conserved in much the same way as water. Speculations concerning the relation between heat and other forms of energy were made during the first part of the 19th century by Count Rumford, H. L. F. von Helmholtz, J. R. von Mayer, and others. However, the quantitative experiments that James Prescott Joule, an English physicist, made in the period 1840–1860 provided the most convincing proof of these ideas.

Joule found that whenever energy was added to a substance, the same amount of energy always effected the same change, regardless of the form of energy. He raised the temperature of objects, using mechanical, acoustical, and electrical energy, and always had identical results; that is, the same amount of energy produced the same change in temperature. After two decades of experimentation he concluded that 4180 joules of any sort of energy would have the same effect as 1 kcal or heat. (Of course, he did not call his unit of energy a *joule;* it was named for him later.) Thus, he showed that heat can be treated as a form of energy on a par with other forms of energy. Naturally, in a given situation, the same unit of measure must be used for all of the energies appearing in a single experiment. The energies may be measured entirely in joules or all in kilocalories. To do this measuring, we use the experimental relation that 1 kcal equals 4180 joules, which is known as the *mechanical equivalent of heat.*

From the definition of the kilocalorie as the amount of heat needed to raise the temperature of 1 kg of water 1 deg C, it is evident that if 2 kcal of heat are transferred to 1 kg of water, its temperature will change by 2 deg C. Similarly, if 2 kcal of heat are added to 4 kg of water, its temperature will only change by $\frac{2}{4}$ or 0.5 deg C.

Suppose, however, that the substance to which we add heat is not water. In general, the amount of heat needed to produce the same change in temperature in the same mass of another substance is different. We define the *specific heat* of a substance as the ratio between the amount of heat needed to produce a given change in temperature in the substance and the amount of heat needed to produce the same temperature change in an equal mass of water. Because specific heat is the ratio of two amounts of heat, it is a pure number (i.e., without physical units) and has the same value for a given substance regardless of the heat unit and the temperature scale

that might be used. The heat required to raise the temperature of a mass m of water by ΔT is, by our definition of the unit of heat, simply $m\,\Delta T$. If the heat Q is required to raise the temperature of the same mass of the substance through the same temperature, then from the definition of specific heat, we can write

$$s = \frac{Q}{m}\Delta T \qquad (5.4)$$

The amount of heat needed to raise the temperature of 1 kg of water 1°C is 1 kcal; therefore, Eq. (5.4) says that the specific heat s of a substance is *numerically* equal to the amount of heat needed to raise 1 kg of the substance 1°C as in Fig. 5.3. Once we know the specific heat of a substance, we can find the heat needed to raise the temperature of a mass m through a temperature interval ΔT by rewriting Eq. (5.4) as

$$Q = ms\,\Delta T \qquad (5.5)$$

When several bodies that initially have different temperatures are placed in contact with each other, we know from common experience that energy, in the form of heat, will flow from the hotter to the cooler objects and that this process will continue until all the bodies are at the same

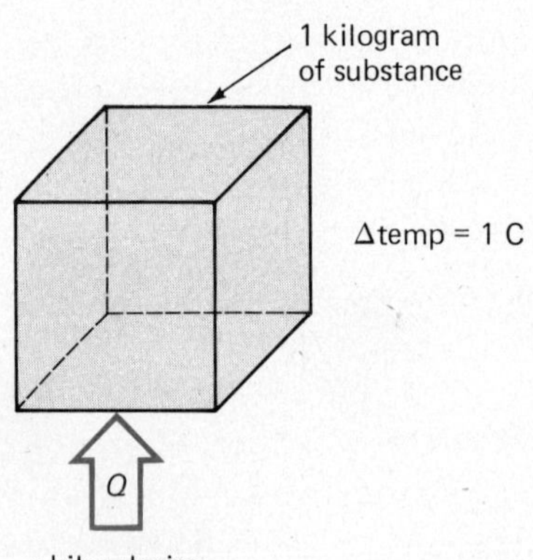

FIGURE 5.3

The specific heat s of a substance is numerically equal to the amount of heat energy needed to raise the temperature of one kilogram of the substance one degree Celsius.

temperature. If this conversion occurs in an insulated container so that no energy can escape, we know from the conservation of energy law that the amount of heat energy given up by the hotter objects must be exactly equal to the amount of heat energy absorbed by the cooler ones. This rule is the basic principle of *calorimetry*. Most laboratory work in calorimetry is done in an insulated container, such as the Thermos bottle shown in Fig. 5.4, so that heat is transferred only among a small number of substances, usually two or three. The heat gained or lost by each substance is calculated from Eq. (5.5). The amounts of heat gained and lost are then equated, and the resulting equation is solved for the unknown quantity, which is usually either the final temperature of the mixture or the specific heat of one of the substances. The following examples will help you understand how to apply this principle.

EXAMPLE Suppose that 2 kg of water at 80°C are mixed with 4 kg of water at 40°C. Let us call the final temperature of the mixture T. Then the change in temperature of the hot water is $(80 - T)$°C and the change in temperature of the cool water is $(T - 40)$°C. If we use Eq. (5.5) to calculate the heats gained and lost and then equate these heats, we find that

$$2 \times 1 \times (80 - T) = 4 \times 1 \times (T - 40)$$

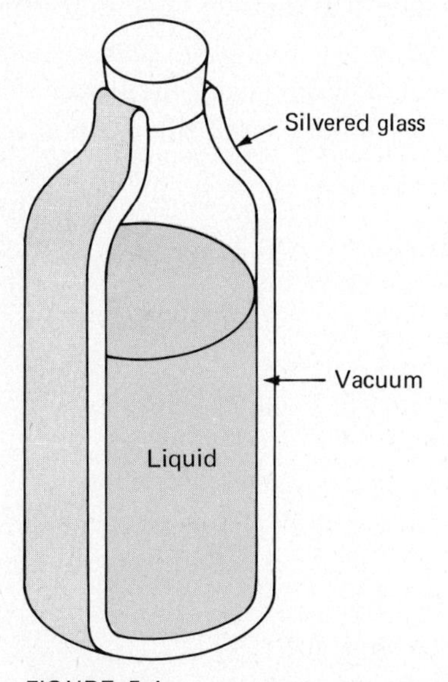

FIGURE 5.4
A cutaway view of a thermos bottle. It is constructed from two silvered glass bottles joined at the opening. The air has been evacuated from between the two bottles. Dewar flasks, used in low temperature physics (cryogenics), have essentially the same construction.

When we multiply out the parentheses in this equation, we obtain

$$160 - 2T = 4T - 160$$

which yields $T = 53°C$.

EXAMPLE As a second example of calorimetry, let us consider mixing together 2 kg of copper of specific heat 0.1 and temperature 200°C with 3 kg of water at a temperature of 20°C. As before, we shall call the final temperature of the mixture T. In this case, the temperature of the copper decreases by $(200 - T)°C$ and the temperature of the water increases by $(T - 20)°C$. If we use Eq. (5.5) to calculate the heats gained and lost, we find the following equation:

$$2 \times 0.1 \times (200 - T) = 3 \times 1 \times (T - 20)$$

After we multiply out the factors in this equation we obtain

$$40 - 0.2T = 3T - 60$$

As the solution, $T = 100/3.2 = 31°C$.

5.3 THE FIRST LAW OF THERMODYNAMICS

Thus far in this chapter, we have used the Law of Conservation of Energy as one of our motivating principles. When energy is added to a body, it is not lost. It is stored in the body. Energy stored in this way is called *internal energy*. If the internal energy of a body is increased by adding energy in some way, the properties of the body change. Perhaps its temperature increases or it expands; perhaps both. However, the change produced depends only on the amount of energy added. In other words, under fixed circumstances (such as constant pressure or temperature) if a certain amount of energy is added to a body, whether by doing mechanical work or by allowing heat to flow, the change in the internal energy and in the other properties of the body will be the same. This proposition is, of course, subject to experimental verification. Suffice it to say, this verification has occurred many, many times. If it had failed, it would have called into question the entire concept of conservation of energy.

Joule's experiments and the subsequent development of physics support our generalization that the Law of Conservation of Energy applies not only to the simple forms of mechanical energy discussed in Chapter 3 but also to the multitude of forms by which energy is now known. *This law states:*

In a system isolated from the rest of the universe the total of the various forms of energy in the system is conserved.

Naturally, one form may be converted into another form, but the sum of the various forms of energy in the system remains the same. This law, as we have said before, is fundamental to the science of physics.

In our earlier discussion of Joule's experiments on the equivalence of various forms of energy with heat, we tacitly assumed that all the energy added to a substance was used to increase the temperature. This situation may not be true. For instance, the substance may expand and thus do mechanical work, as when a gas expands and moves a piston. If we pass electrical energy through a battery, some of this energy may act to raise the temperature of the battery and some of it will be stored in the form of chemical energy, another form of internal energy. Thus, if we add energy to a system, some of it is stored as internal energy, perhaps reflected by a change in temperature, perhaps stored in some internal form of potential energy. Or part of it may reappear as work done on the surroundings. The internal energy is a quantity that cannot be measured directly. Its value must be inferred from other data. For instance, if we change a substance from one state to another, we can find the change in internal energy by keeping track of how much heat and work was required to produce the change. This change will be independent of the method producing the change. It will depend only on the initial and final states of the system.

Let us put these ideas more exactly. If ΔQ is the *net* amount of heat added to a system, then ΔQ is the heat that enters the system minus the heat that leaves. In addition, work may be done on the system or the body may do work on its surroundings. Let ΔW be the *net* work done by the system on the rest of the universe; then ΔW is the work done *by* the system minus the work done *on* the system. Conservation of energy now tells us that the heat added must show up as either work or increased internal energy. If we let ΔU be the *increase* in internal energy, then we have

$$\Delta Q = \Delta W + \Delta U \qquad (5.6)$$

In words, *the net heat added to a system is equal to the work done by the system plus the increase in internal energy; energy is conserved.* This equation is known as the First Law of Thermodynamics.

In using this law, you must remember that heat is positive when it is added to the system, whereas work is positive when it is done by the system on its surroundings.

The concept of energy conservation has been very useful in the

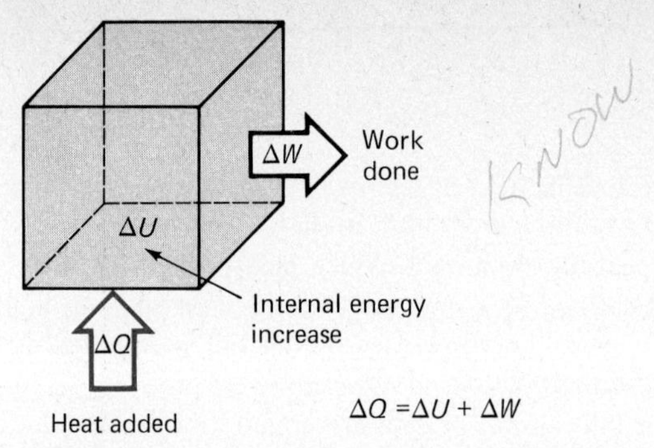

FIGURE 5.5

The first law of thermodynamics states that the heat added to a system becomes either internal energy or work done by the system.

development of science. When apparent violations of the principle occur or new phenomena are discovered, new forms of energy are introduced in order to make the principle continue to hold. For instance, in 1905, Einstein predicted theoretically that mass could be converted into energy. This conversion was soon verified experimentally by nuclear physicists. The most striking example of "mass-energy" is the atomic bomb. Only if mass is considered as a form of energy do nuclear reactions preserve the principle of conservation of energy.

DISCUSSION QUESTIONS

1. How would you suggest measuring the temperature of inaccessible objects, such as the sun and the upper atmosphere?

2. When a mercury-in-glass thermometer is dipped into hot water, the mercury column first falls and then rises. Why?

3. Explain the difference between the scientific usages of the words "temperature" and "heat."

4. Give one or more examples of the conversion of some form of energy into heat energy.

5. Show that for an isotropic substance the coefficient of volume expansion should be three times the coefficient of linear expansion for the same substance.

6. Most substances expand when they are heated. Hang a weight with a rubber band and gently heat the rubber band with a match. Discuss the result.

PROBLEMS

1. Compute the value of 98.6°F on the Celsius scale. (*Ans.:* 37.0°C)

2. Compute the value of 0°C on the Fahrenheit scale.

3. Determine the temperature at which the Fahrenheit and Celsius scales have the same value. (*Ans.:* -40)

4. Compute the heat required to raise the temperature of 5 kg of water by 15°C. (*Ans.:* 75 kcal)

5. The specific heat of lead is 0.03. Calculate the amount of heat necessary to raise the temperature of 20 kg of lead by 30°C.

   Repeat for aluminum, which has specific heat 0.20. (*Ans.:* 120 kcal)

6. A kilogram of water falls a distance of 20 m. If you assume that all the mechanical energy caused by this fall is converted into heat, compute the rise in temperature of the water.

7. Five kilocalories of heat are added to a gas that does 10^4 joules of work. Compute the change in internal energy of the gas. Is it an increase or decrease? (*Ans.:* 1.09×10^4 joules; increase)

8. How much heat must be added to a body if it does 20,000 joules of work and its internal energy increases by 30,000 joules?

9. Two kilocalories of heat are added to a body and its internal energy decreases by 10^4 joules. How much work is done by the body or on the body in this process? (*Ans.:* 18,360 joules of work done by the body)

10. Compute the heat required to raise the temperature of 3 kg of water from 20°C to 25°C. (*Ans.:* 15 kcal)

11. When 20 kcal of heat are added to a certain amount of water, the temperature of the water is raised by 4 deg Celsius. Compute the mass of the water.

12. The specific heat of copper is 0.093. Calculate the amount of heat required to raise the temperature of 8 kg of copper from 20°C to 25°C. (*Ans.:* 3.72 kcal)

13. Compute the distance that water must fall vertically so that the temperature of the water will be raised by 0.1°C on striking the ground. Assume that half of the mechanical energy dissipated is converted into heat.

14. Eight kilocalories of heat are added to a gas that does 20,000 joules of external work. Compute the change in internal energy of thc gas and state whether this change is an increase or decrease. (*Ans.:* 13,400 joules increase)

15. Compute the value of 80°F on the Celsius scale. (*Ans.:* 26.7°C)

16. Compute the value of 27°C on the Fahrenheit scale.

17. Compute the amount of heat in kilocalories that must be added to a gas that does 30,000 joules of external work and has its internal energy decrease by 10,000 joules.

18. Compute the temperature at which the Fahrenheit value is twice the Celsius value. (*Ans.:* $C = 160°$; $F = 320°$)

19. Twenty grams of lead at 100°C is dropped into 150 gm of water at 25°C. Assuming that all the heat lost by the lead is gained by the water, find the equilibrium temperature. The specific heat of lead is 0.03.

20. A 15-gm metal block initially at 80°C is dropped into 100 gm of water at 0°C. If the final temperature of the system is 10°C, find the specific heat of the metal.

6 Properties of Matter

In this chapter we shall discuss the thermodynamic behavior of certain simple types of matter. The physical laws described here not only illustrate the ideas of the preceding chapter but also explain a wide range of everyday phenomena.

6.1 THE IDEAL GAS

Consider a confined gas. Suppose that the gas exerts a force F on an area A of the wall of the container. If the force had any portion parallel to the surface of the wall, an equal and opposite force would be exerted by the wall of the container on the nearby gas, according to Newton's Third Law. Presumably, the gas would move parallel to the wall of the container. However, if enough time has elapsed for the gas to be in equilibrium, no such motion is observed. We conclude, therefore, that in equilibrium the force exerted by the gas on the wall of its container is at right angles to the surface of the wall. We define the pressure p exerted on the gas on the surface as the ratio of the magnitude of F to A as shown in Fig. 6.1. According to the Third Law of Motion, this ratio is also the pressure exerted on the gas. In the form of an equation, we can write

$$p = \frac{F}{A} \tag{6.1}$$

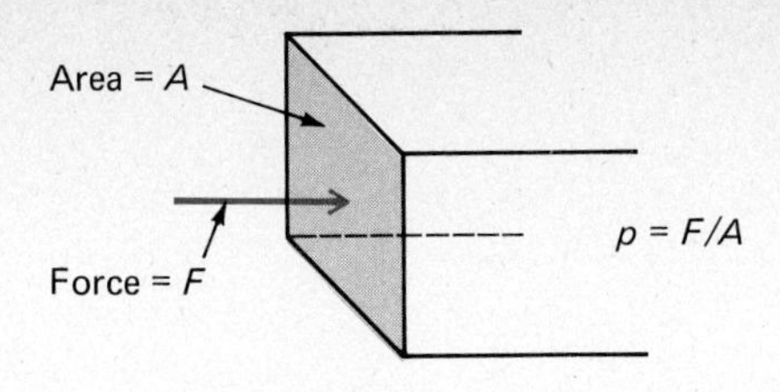

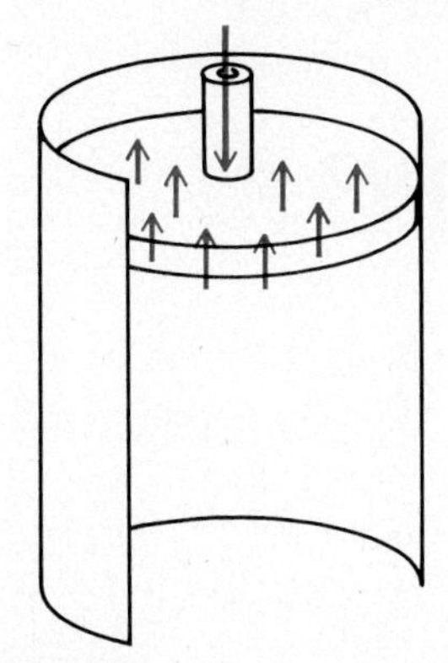

FIGURE 6.1
In the upper picture a single force is acting on a surface. The pressure is $p = F/A$. In the lower picture a gas confined by a piston is exerting many little forces upward on the piston. They are exactly balanced by the single force exerted downward by the weight.

In the metric system, pressure is measured in the unit of newtons per square meter (N/m^2), although other units, such as pounds per square inch ($lb/in.^2$), are commonly used.

We live under a blanket of air, so we experience an average pressure due to the weight of the atmosphere of about 15 $lb/in.^2$ or 10^5 N/m^2. For many purposes, it is more useful to know the difference between the actual pressure of a gas and the surrounding atmospheric pressure. This difference in pressure is known as the *gauge pressure.* For instance, the pressure in the tires of your car may be 28 $lb/in.^2$ in terms of gauge pressure, whereas the total or absolute pressure in the tires would be about 43 $lb/in.^2$ In the remainder of this section "pressure" will always mean total pressure unless otherwise specified.

In Sec. 5.1 we said that the pressure of a confined gas changes with temperature. To make our explanation simpler, we will keep the gas in a rigid container, as shown in Fig. 6.2, so that its volume cannot change. For any gas at sufficiently low pressure we find the relation between the pressure of the gas and its temperature in degrees Celsius to be as shown in Fig.

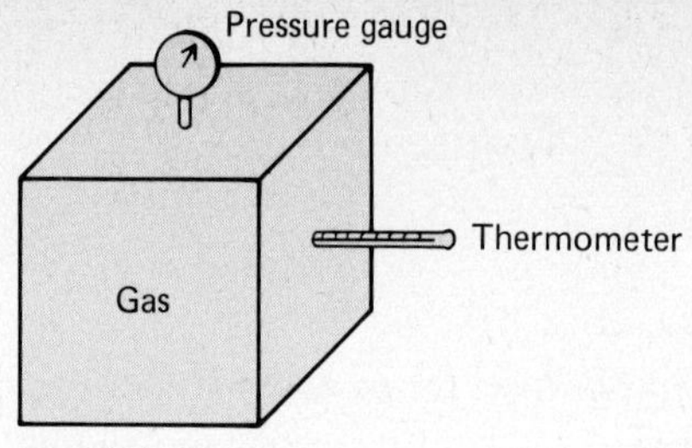

FIGURE 6.2

A quantity of gas is contained in a fixed volume. The gas can be heated, raising the temperature and pressure, or cooled, lowering the temperature and pressure. The amount of gas in the container can also be changed.

6.3. Furthermore, we notice that for all gases at low pressure the straight line when extrapolated to the left would predict zero pressure for the gas at $-273\,^\circ$C. (The gas would actually become a liquid at temperatures well above $-273\,^\circ$C, but the following remarks apply as long as the substance is still a gas.) This behavior suggests that we define a new temperature scale with its zero at $-273\,^\circ$C, so that the relation between the new scale T and the Celsius scale t is as follows:

$$T = t + 273^\circ \tag{6.2}$$

This equation defines the Kelvin scale, named for the British physicist. The size of the degree is the same in both scales, but their zeros are taken at

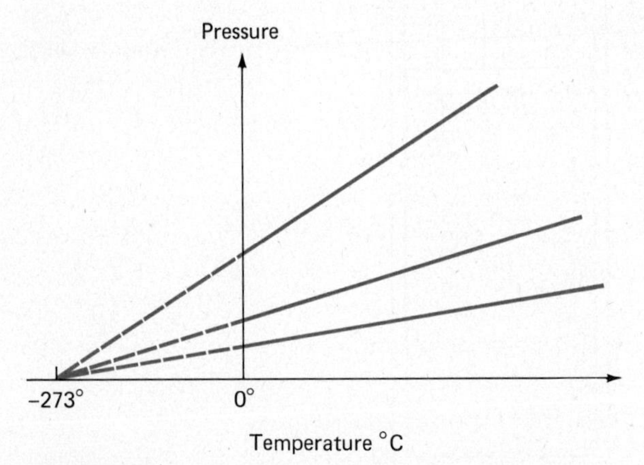

FIGURE 6.3

Pressure versus temperature for the experiment shown in Fig. 6.2. The different lines come from using different amounts and different types of gas in the container. The lower lines come from using less gas.

different points. In terms of this new scale, the pressure of the gas p and its temperature T are seen to be proportional. If we introduce a constant of proportionality c, we can write

$$p = cT \quad \text{(constant volume)} \tag{6.3}$$

As another simple case, let us discuss the behavior of a gas at low pressure when the pressure is kept constant and the volume changes with temperature, as shown in Fig. 6.4. For any gas we find that the relation between volume and temperature at constant pressure is as shown in Fig. 6.5. Amazingly, if we extrapolate the straight line to the left, we find that the volume would become zero at $-273\,^{\circ}\mathrm{C}$ if the substance were still a gas at such a low temperature. In this case, we find that we should again define a new temperature T as given in Eq. (6.2). If we use this new temperature scale, we can say that the volume V of a gas is proportional to the temperature T. After introducing a constant of proportionality c', we have

$$V = c'T \quad \text{(constant pressure)} \tag{6.4}$$

When we compare Eq. (6.3) with Eq. (6.4), we see that the same temperature scale is found experimentally for all gases at low pressure. A

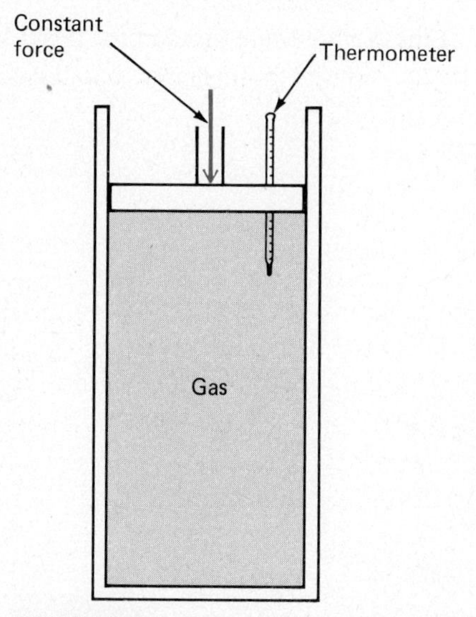

FIGURE 6.4

A quantity of gas is contained in a cylinder with a movable piston. The net force downwards on the piston is constant and thus the pressure is constant. As the gas is cooled and heated the piston will move and the volume will change.

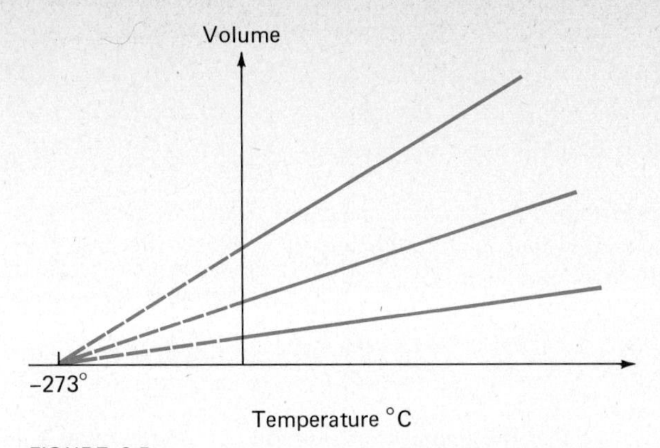

FIGURE 6.5

Volume versus temperature for the experiment shown in Fig. 6.4. The different lines come from using different amounts of gas, and again the lower lines from less gas.

gas that obeys these two laws is called an *ideal* gas. Equation (6.2) defines the temperature scale for such a gas; this scale is known as the *ideal gas temperature scale*. In the laboratory it is easily and accurately approximated with gases such as hydrogen or helium at low pressures. Thermometers based on this temperature scale are then used to calibrate such common thermometers as the mercury-in-glass thermometers.

A second look at Eqs. (6.3) and (6.4) shows us that the two equations can be combined into a single equation. We introduce a new constant of proportionality R and write

$$pV = RT \tag{6.5}$$

In order to obtain Eq. (6.3) from Eq. (6.5), we put $c = R/V$. To get Eq. (6.4) from Eq. (6.5), we put $c' = R/p$. Thus, Eq. (6.5) describes the behavior of ideal gases and is known as the *ideal gas law*. In solving problems involving ideal gases, we usually have an initial state given by p_1, V_1, and T_1, and a final state given by p_2, V_2, and T_2. If we solve Eq. (6.5) for R, we can write the following equation:

$$R = \frac{p_1 V_1}{T_1} = \frac{p_2 V_2}{T_2} \tag{6.6}$$

Thus, we do not have to know the value of R in solving most problems. Examination of Figs. 6.3 and 6.5 indicates that the constant R depends on the amount of gas present. In fact, as we shall see in the next section, it

is proportional to the number of gas molecules present. You should keep in mind when using Eqs. (6.5) or (6.6) that the temperature T must be the ideal gas temperature defined by Eq. (6.2) and that the pressure p must be the total pressure on the gas.

EXAMPLE Suppose that a fixed volume of gas is heated from $-73°C$ to $127°C$. Using Eq. (6.2), we find that $T_1 = 200°K$ and $T_2 = 400°K$. Because $V_1 = V_2$, we can rewrite Eq. (6.6) as

$$p_2 = \left(\frac{T_2}{T_1}\right)p_1 = \frac{400}{200}p_1 = 2p_1$$

and we see that the pressure will double. If, on the other hand, we had kept the pressure constant, then the volume of the gas would have doubled.

6.2 KINETIC THEORY OF GASES

The ideal gas law expressed in Eq. (6.5) can be derived by using the *kinetic theory,* which makes a few simple assumptions about gas molecules and derives results using statistical arguments. We assume that an ideal gas consists of point molecules, each of which occupies no volume and which do not exert forces on each other except at the moment of impact during a collision. All of the molecules move in a random motion that is continually changing as molecules collide as in Fig. 6.6. We also assume that these collisions conserve energy and momentum.

From the preceding paragraph, we might conclude that the irrational movements of the molecules would not let us make any calculations about them. This conclusion would not be true, because the number of

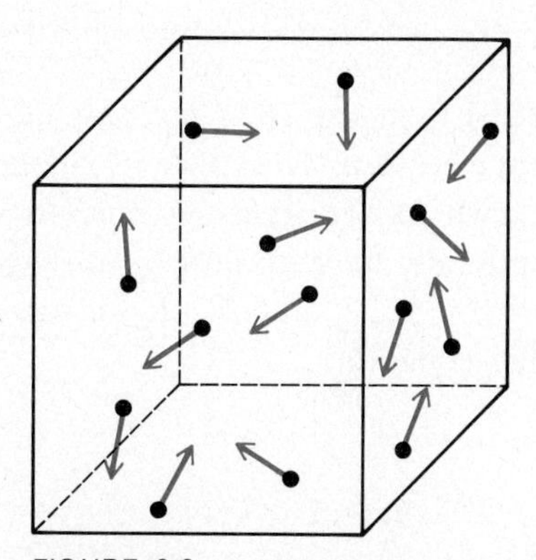

FIGURE 6.6
The random thermal motion of the gas molecules in a container. The colored arrows are the velocity vectors.

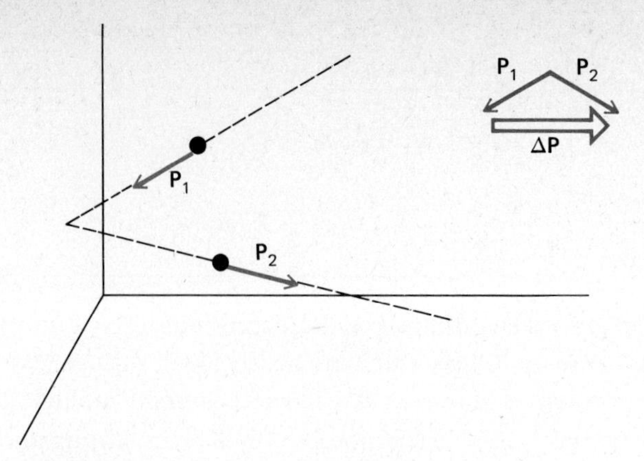

FIGURE 6.7

A gas molecule collides elastically with the wall of a container. Its speed is unchanged but the part of its momentum (and velocity) perpendicular to the wall is reversed by the collision. Thus there is a change in momentum as shown and a force results.

molecules involved is large and averages can be used. For instance, if a million coins are dumped on a pavement, we could not expect to predict whether a given coin would turn up heads or tails, but we could predict fairly confidently that about half a million coins would have heads showing. In a similar way, the kinetic theory does not let us predict what an individual molecule will do, but it does allow us to calculate the effect of the entire collection.

Let us consider a gas molecule striking the wall of the container and rebounding, as shown in Fig. 6.7. According to Eq. (3.2), the change in momentum of the molecule requires that the wall exert a force on the molecule. By Newton's Third Law, which is discussed in Sec. 3.2, the molecule must exert an equal and opposite force. Thus, the impact of the molecule exerts a small force on the wall of the container. If this idea is carried through mathematically for the case of N molecules, it is found that the pressure p, volume V, and average kinetic energy $\bar{K} = \overline{\frac{1}{2}mv^2}$ are related by the equation:

$$pV = \tfrac{2}{3}N\bar{K} \qquad (6.7)$$

In this equation, we see that averaging all of the N molecules has led to a definite result, even though individual molecules have different speeds and energies.

When we compare Eq. (6.7) with Eq. (6.5), we see that they can

only be consistent if the following relation holds:

$$\begin{aligned}\bar{K} &= \frac{3}{2}\left(\frac{R}{N}\right)T \\ &= \frac{3}{2}kT = \frac{1}{2}\overline{mv^2}\end{aligned} \tag{6.8}$$

In Eq. (6.8) the constant $k = R/N$ turns out to be a universal constant that is the same for all gases that are nearly ideal, such as hydrogen and helium. This constant is known as *Boltzmann's constant,* and its value is $k = 1.38 \times 10^{-23}$ joules/deg K. [The absolute or Kelvin temperature scale is defined in Sec. 7.2 and corresponds to the ideal gas temperature defined by Eq. (6.2).]

As an example of the use of Eq. (6.8), let us compute the average energy of a molecule of a gas at 27°C. This temperature corresponds to $T = 27 + 273 = 300$°K. We have then

$$\bar{K} = \tfrac{3}{2} \times 1.38 \times 10^{-23} \times 300 = 6.20 \times 10^{-21} \text{ joules}$$

Although this figure may seem like a small amount of energy, the kinetic energy of all the molecules in 1 gm of air at this temperature turns out to be about 130 joules. If the atmospheric pressure has its standard value of about 10^5 N/m^2, the average speed of an air molecule under these conditions will be about 510 m/sec = 1140 mph. For comparison, it should be noted that the energy of an air molecule under these conditions is about one-thirtieth as much as the energy released when oxygen and hydrogen combine to form a water molecule and about 3×10^{-11} as much as the energy released when a single uranium nucleus undergoes fission. (See Sec. 14.3, for a discussion of nuclear fission.)

The kinetic theory also provides us with a model that we can use to find the form of the internal energy U of an ideal gas. If we assume, as we have done earlier, that a gas is composed of point masses that do not interact except through brief collisions, then the only way that they can store energy is in their individual kinetic energies. If $\bar{K}$ is the average kinetic energy per molecule, then the total kinetic energy stored by N molecules is $N\bar{K}$. By Eq. (6.8), $\bar{K} = \tfrac{3}{2}kT$, and so we can write

$$U = \tfrac{3}{2}NkT \tag{6.9}$$

for the internal energy of the gas. For more complex molecules that have real internal structure, such as oxygen (which is composed of two oxygen

atoms chemically bound together), this relation for the internal energy must be modified somewhat. However, as long as the molecules are considered not to interact (except by inelastic collisions) the internal energy, U, depends only on the temperature and the number of molecules and is directly proportional to their product. However, the constant of proportionality changes from $\frac{3}{2}k$ to $\frac{5}{2}k$ or some larger value, depending on the type of gas.

In the case of real gases, the simple assumptions of the kinetic theory of ideal gases evidently are inadequate. Molecules do exert forces on one another as they approach each other. They are not points and do occupy volume. When these factors are taken into account, as was first done by Johannes van der Waals, a Dutch physicist, Eq. (6.5) takes the modified form given below:

know

$$\left(p + \frac{a}{V^2}\right)(V - b) = RT \tag{6.10}$$

In this equation, a and b are constants characteristic of the gas being studied. The constant a corrects the observed pressure for the attractions between molecules, and the constant b corrects the observed volume for the volume occupied by the molecules. Although still an approximation, this equation describes real gases quite well. Other even more refined equations, however, offer better descriptions.

As we discussed earlier, individual molecules of a gas have different speeds. The distribution of speeds for an ideal gas was first calculated by a Scottish physicist, James Clerk Maxwell, about a century ago. A typical distribution of molecular speeds for a gas at a given temperature is shown in Fig. 6.8. Note that many molecules have speeds greatly differing from

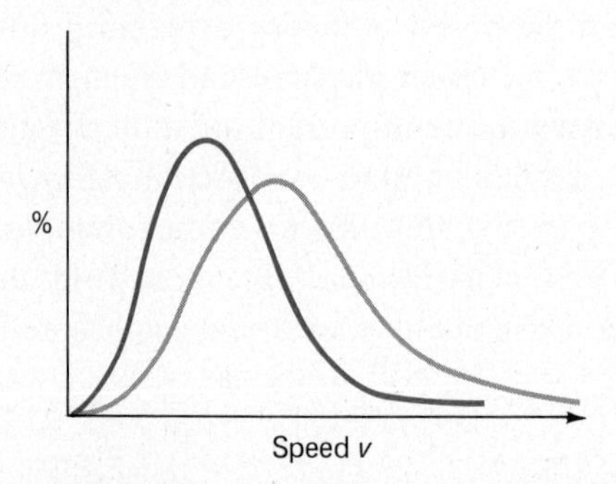

FIGURE 6.8

The percentage of molecules in a gas having a particular speed. The two curves are for different temperatures, the light curve being the higher one.

the average or most probable speed. Because a light molecule, such as hydrogen, has a greater speed at a given temperature than a heavy molecule, like oxygen, the earth's gravitational attraction is sufficient to hold oxygen in our atmosphere, whereas hydrogen escapes the earth's gravitational pull. This fact helps to explain why there is little hydrogen in our atmosphere, even though it is one of the most common elements. If the temperature of the gas is raised, the whole distribution curve on the diagram moves to the right and so does the most probable speed. When the theory is carried through quantitatively, we find that a relatively small increase in temperature (proporitional to the average energy per molecule) greatly increases the number of molecules with speeds exceeding a given value. If the temperature of our earth rose only slightly, the oxygen and nitrogen in the atmosphere would soon leave our planet, because the molecular velocities would enable molecules to escape the earth's gravitational influence. Fortunately, our source of heat, the sun, seems to have been quite constant for a very long time.

The kinetic theory can be directly verified in a simple way discovered by a Scottish botanist, Robert Brown, in the early 19th century. He found that small particles suspended in a liquid and viewed with a microscope moved about erratically. We can easily observe this *Brownian movement* in air by studying cigarette smoke illuminated transversely under a microscope. If we watch a given particle, we find that it moves about quite unpredictably. The particles that we are talking about are very small (perhaps 10^{-4} cm), but they are still large compared to the size of the liquid molecules. By chance, one side or another of the suspended particle may receive more impacts by the liquid molecules and thus the particle will receive an impulse in a particular direction. A moment later, another side of the particle may receive the majority of the impacts, so that the particle will move in another direction. This behavior causes the erratic motion. In 1905, Albert Einstein proposed a theory explaining this motion. It was verified by J. B. Perrin, a French physicist and chemist, several years later. Their combined work was quite important in establishing the atomic theory and in determining the mass of atoms. (In fact, Einstein received the Nobel prize in physics in part for this theory, rather than for his theories of relativity, which are now more famous.) By about 1915, the scientific world was convinced of the existence of atoms and molecules.

6.3 GASES, LIQUIDS, AND SOLIDS

In the preceding section we discussed the properties of the ideal gas, which can be closely approximated experimentally by using a gas at low pressure. If we study a real gas, we find that as we subject the gas to higher pressures

and lower temperatures, the gas will condense into a liquid. Similarly, as we lower the temperature of a liquid, it eventually becomes a solid. This process can be explained qualitatively in terms of the attractive forces between the molecules. As the temperature is decreased, the random thermal motion of the molecules in the gas decreases. The molecules begin to collect together in groups or droplets, which then collect to form the liquid. As the temperature is further decreased, the thermal motion becomes sufficiently low that the individual molecules can form chemical bonds and the liquid solidifies. In this section we shall consider the relations between these three states of matter and how they depend on such influences as pressure and temperature.

Consider some liquid that partially fills a closed container, as shown in Fig. 6.9. At a given temperature the liquid will evaporate until we find a constant vapor pressure above the liquid. This pressure is known as the *equilibrium vapor pressure* of this particular liquid at this temperature. If we raise the pressure on the vapor or lower the temperature of the combination, some of the vapor will condense into liquid. If we measure the equilibrium vapor pressure p of the vapor of a given liquid at various temperatures T, we find a curve similar to that shown in Fig. 6.10. Consider the point marked A in Fig. 6.10. For this particular pair of values of pressure and temperature, the liquid and its vapor can exist in equilibrium in any proportions. However, if we raise the pressure, all of the substance will become a liquid. On the other hand, if we raise the temperature, all of the substance becomes a vapor. Thus, the equilibrium vapor pressure curve shown in Fig. 6.10 separates the region of liquid from the region of vapor. This curve is known as a *phase diagram*.

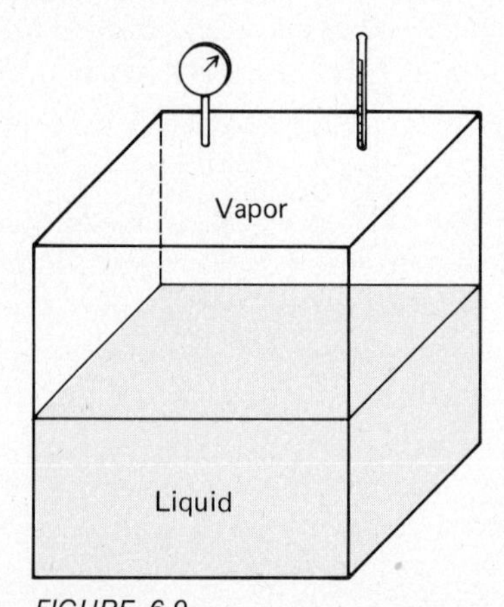

FIGURE 6.9

A liquid and its vapor in equilibrium in a closed container. There is no air above the liquid, only its vapor.

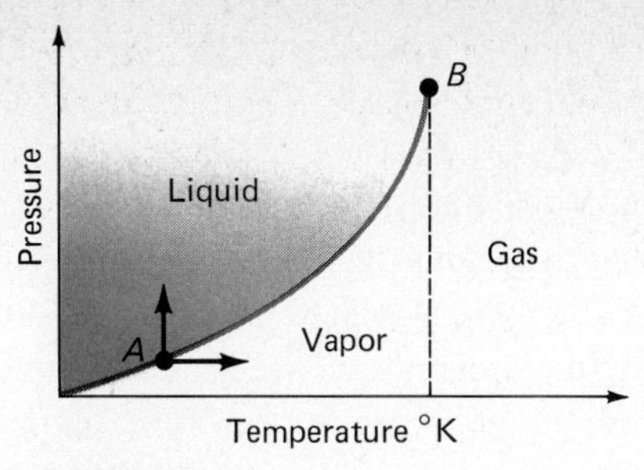

FIGURE 6.10

Phase diagram for a substance. The line shows the pressures and temperatures for which the liquid can exist in equilibrium with its vapor.

From this discussion you might presume that we could always change a vapor into a liquid at any temperature by applying a large enough pressure. But we cannot. At temperatures above a certain temperature, known as the *critical temperature,* no amount of pressure will change the substance from the gaseous state into the liquid state. In Fig. 6.10 the critical temperature is shown as the point marked *B*. Note here that it is customary to call a substance a *gas* when it is at a temperature above its critical temperature and a *vapor* when it is at a temperature below its critical temperature. Clearly, for a substance to approximate an ideal gas, it must be at a temperature well above its critical temperature.

Using a similar diagram we can investigate the relation between pressures and temperatures at which a solid and its liquid could exist together in equilibrium in any proportions. A typical equilibrium curve for melting or freezing is shown in Fig. 6.11. At the point marked *C* in Fig.

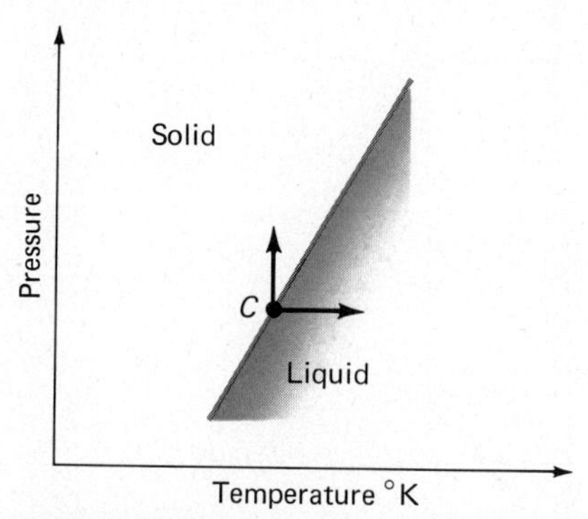

FIGURE 6.11

A phase diagram showing the fusion or melting curve. For points on the line the liquid and solid can exist together in equilibrium.

6.11, the solid and liquid exist together in equilibrium. If we now raise the pressure or lower the temperature, all of the substance will become a solid. (Note here that the melting curve for water is unusual, because it slopes to the left rather than to the right, showing that ice melts under high pressure.) Thus, the melting or fusion curve on a phase diagram separates the region in which all of the substance is a solid from the region in which all of the substance is a liquid.

In some cases, a solid changes directly into a vapor. Common examples are the evaporation of mothballs and dry ice (frozen carbon dioxide). This process is called *sublimation.* As we have seen earlier in this section, there are combinations of pressure and temperature at which the solid and its vapor can exist together in equilibrium in any proportions whatsoever. If we make a graph of these combinations, we obtain the sublimation curve of the substance.

You might now think it worthwhile to plot the vapor pressure, fusion, and sublimation curves all on a single graph of pressure against temperature. Unfortunately, the changes in pressure and temperature that are required to show all three of these curves are very large, so that it is not possible to show all of them drawn to correct scale on one diagram. However, Fig. 6.12 does illustrate the three curves for changes of state in a qualitative way. The most striking feature of Fig. 6.12 is the point at which there is a particular combination of pressure and temperature (marked *T* on the diagram). Here the vapor, liquid, and solid can exist together in equilibrium in any proportions. This convergence is known as the *triple point.*

Everyone realizes that a given mass of ice has a greater cooling effect on a drink than the same mass of ice water. The ice absorbs heat as it melts;

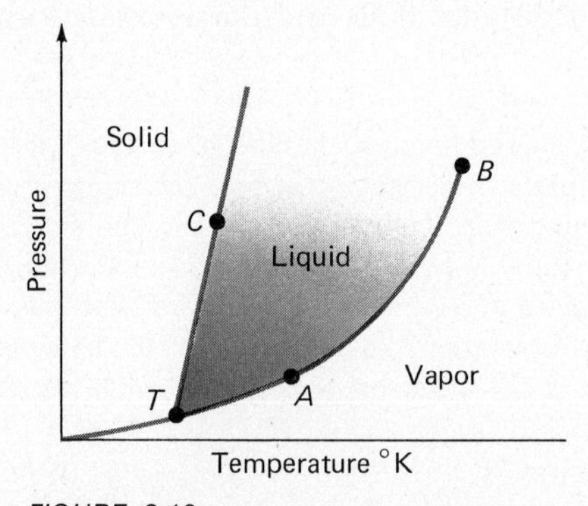

FIGURE 6.12

A complete phase diagram, not drawn to scale. At the point *T*—called the triple point—solid, liquid, and vapor can all coexist in equilibrium.

thus, the resulting ice water cools the drink more. The heat required to melt a unit mass of a substance without changing its temperature is known as the *heat of fusion* of the substance. For ice, the heat of fusion is about 80 kcal/kg.

Similarly, if we let rubbing alcohol evaporate from our skin, the skin is cooled. Evidently, heat is required to change the liquid alcohol into vapor, and this heat is supplied by our skin. The heat required to change a unit mass of a liquid into vapor is called the *heat of vaporization* for the substance. For water, the heat of vaporization depends on temperature and is about 540 kcal/kg at 100°C. The *heat of sublimation* of a substance is similarly defined as the heat required to change a unit mass of the substance into vapor.

You might well ask at this point where the energy supplied in melting ice goes if it does not show up as increased temperature. The answer is that it goes into breaking the bonds between the water molecules. The individual molecules of both ice and water at 0°C have the same average kinetic energy and thus the same temperature. However, the molecules in ice are bound together by chemical bonds, and these bonds must be broken to melt the ice. The 80 kcal of heat energy necessary to melt 1 k of ice goes into breaking these bonds. This is an example of increasing the internal energy of a substance without producing a change in temperature. If we ignore the small volume change that occurs when ice melts, the First Law of Thermodynamics tells us that the internal energy of 1 kg of water is 80 kcal more than that of 1 kg of ice.

The heat of vaporization can be discussed similarly in terms of the energy needed to separate the molecules of the liquid without increasing their average thermal energy. But, in this case, the change in the internal energy cannot be calculated so simply because of the large volume change that occurs when a substance boils and the associated work it must do on the environment.

EXAMPLE As an example of the energy involved in change of phase, let us consider adding 100 kcal of heat to a piece of ice having a mass of 2 kg at a temperature of 0°C. We find that the amount of ice melted is $\frac{100}{80} = 1.25$ kg. The result is that we now have 1.25 kg of water and 0.75 kg of ice, both at a temperature of 0°C. If we begin again with the same piece of ice but add 200 kcal of heat to it, we use up 160 kcal in melting the ice and obtaining 2 kg of water at 0°C. The remaining 40 kcal will raise the temperature of the water by $\frac{40}{2} = 20$ deg. Thus, the final temperature of the water will be 20°C.

EXAMPLE Let us consider another example of the relation between temperature and heat added to water when changes of phase are involved. What happens when we add heat at a constant rate to 1 kg of ice that is initially at 0°C? The temperature of the material is shown as a function of the heat added in Fig. 6.13. We see that

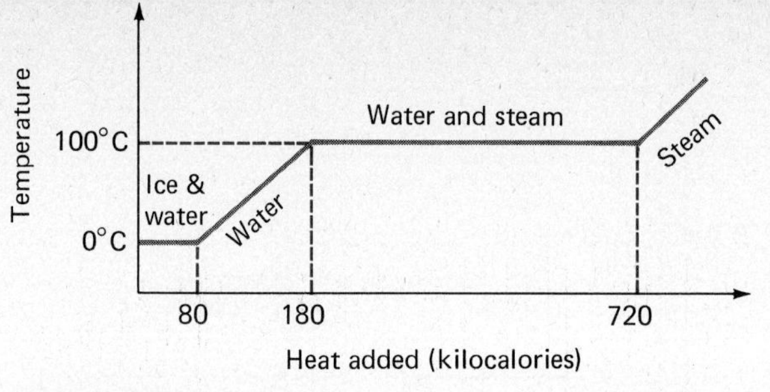

FIGURE 6.13

One kilogram of ice at 0°C is heated until it becomes steam. The line shows the temperature of the substance as a function of the heat added.

a total of 720 kcal of heat must be added to change 1 kg of ice at 0°C into 1 kg of steam at 100°C.

The conservation of energy law tells us that when a liquid freezes, it must give up heat per unit mass equal to its heat of fusion. Thus, when 1 kg of water freezes, 80 kcal of heat are given up. This fact is often useful. Such heat may keep the temperature of the surroundings from dropping below 0°C. Heat is also given up when a vapor condenses into a liquid. Here the amount of heat is equal to the substance's heat of vaporization. This factor is used in steam radiators, where each kilogram of steam gives up 540 kcal of heat. The heat of sublimation is illustrated when snow is formed from the change of water vapor into small ice crystals, thus warming the air somewhat.

The energy changes that we have just described are utilized in the common household refrigerator to transfer heat from the food inside the refrigerator to the air of the room. We use a liquid, usually Freon, which vaporizes readily at room temperature. When some of this liquid evaporates in the cooling coils of the refrigerator, as shown in Fig. 6.14, the heat of vaporization of the liquid is absorbed from the food, thus cooling the food. The warm vapor is now compressed by the pump into a liquid, and the heated liquid is cooled by the air of the room passing over the cooling fins. The liquid is then ready to be vaporized again and to cool the food further. As the net result of this cyclic process, heat is transferred from the food to the room, leaving the liquid basically unchanged. The whole cost of the process is based on operating the pump that transforms the vapor into a liquid. In the following chapter, we shall study more carefully the energy balance in the cyclic process undergone by a heating engine.

There are various ways in which heat can be transferred from one

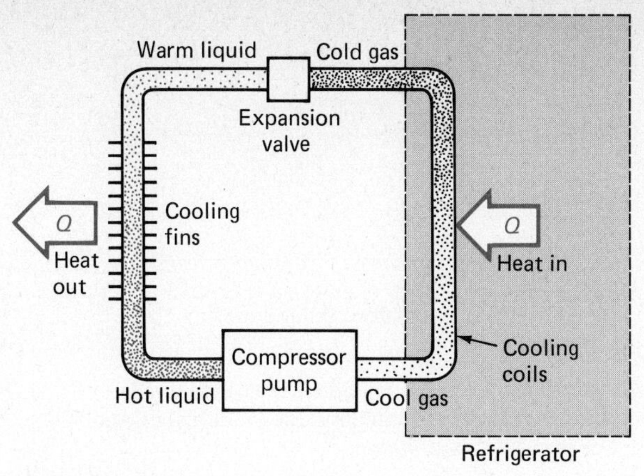

FIGURE 6.14

A schematic diagram of a refrigerator cycle.

body to another. In the process called *convection,* heat is gained by a material; the heated material physically moves and then warms a colder object. For example, lava flowing from a volcanic eruption carries heat from the interior of the earth by convection. In the case of hot-air furnaces, the heating unit in the furnace warms the air. Because this warm air is less dense than the cool air in the house, it rises, heating the house.

Heat may also be transferred by *conduction,* in which the material does not actually move. In this case, the heat energy is passed along from one atom or molecule to its neighbor in somewhat the same way that one bowling pin knocks down the next one. A familiar example of heating by conduction is the electric stove, in which the heating element transfers heat to the pan and then the pan transfers heat to the food.

The third way in which heat energy can be transferred is by *radiation.* As we shall see in Chaps. 10 and 11, waves transmit energy through a medium without affecting the medium. Heat is carried by waves similar to light waves. The most common example of heat transfer by waves is the transmission of energy from the sun to the earth. Note that the transmission of energy by radiation does not require the presence of a material medium. Refer to Chap. 11 for a more detailed discussion of the idea.

6.4 THERMAL EXPANSION

When the temperature of a substance is changed, usually its physical dimensions also change. For nearly all substances, an increase in temperature produces an increase in size. This behavior is known as *thermal expansion.*

In an ordinary thermometer, the expansion of the liquid causes it to rise inside the glass tube. Bridges have expansion joints to allow for the contraction and expansion of the bridge during the changing seasons.

If we measure the change in length (ΔL) of a solid bar when the temperature changes by an amount ΔT, we find to a good approximation that the change in length is proportional to both the change in temperature and the initial length of the bar L_0. If we introduce a constant of proportionality a, we can then write

$$\Delta L = aL_0 \, \Delta T \tag{6.11}$$

The coefficient of linear expansion, a, is therefore the fractional change in length, $\Delta L/L_0$, per unit change in temperature. Typical values for metals are in the range of 10^{-5}/deg C.

EXAMPLE As an example of the use of Eq. (6.11), let us consider the change in length of a steel rail initially 100 ft long and at a temperature of 20°C. For steel, $a = 1.2 \times 10^{-5}$/deg C approximately. If the temperature increases to 30°C, the increase in the length of the rail is given by

$$\Delta L = 1.2 \times 10^{-5} \times 100 \times 10 = 1.2 \times 10^{-2} \text{ ft} = 0.144 \text{ in.}$$

Thus, the change in length is not insignificant. If the rail is not free to expand this much, according to Hooke's Law (which we discussed in Sec. 3.5), it will exert a very large force on the rails at its two ends. Usually, therefore, railroad tracks are laid with small gaps between successive rails.

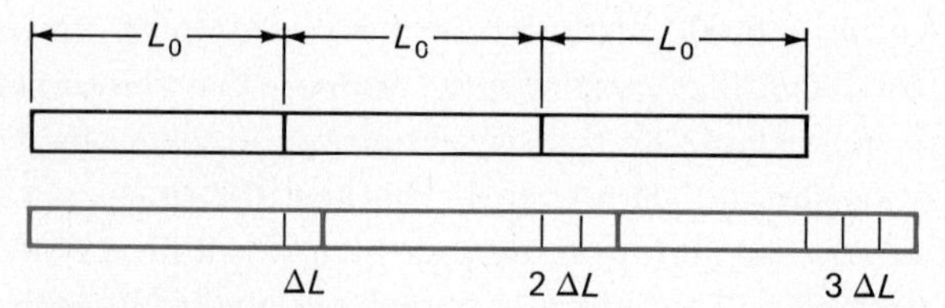

FIGURE 6.15
Thermal expansion. Notice that three rods end to end expand three times as much as one rod. The amount of expansion is proportional to the length of the rod as well as to the temperature change.

An example of a material that contracts when it is heated is rubber. If you hang a small weight with a rubber band and then gently heat the rubber with a match flame, you will see that the weight rise as the rubber contracts. Water also contracts as its temperature is raised from 0°C to 4°C. Thus, water near its freezing temperature is less dense than slightly warmer water. As a pond begins to freeze, the denser and warmer water sinks to

the bottom, whereas the coldest water is at the top. Therefore, ice forms at the top of the pond rather than at the bottom. Because water expands on freezing, ice is less dense than water, so the pond freezes from the top down. Ice is not a very good conductor of heat; usually, therefore, a pond will not freeze all the way to the bottom. Otherwise, fish could not survive a cold winter.

Changes in the volume of a substance are also associated with changes in temperature. If the change in volume is ΔV and the change in temperature is ΔT, it is found that ΔV is approximately proportional to ΔT and the original volume V_0. If we introduce as a constant of proportionality the coefficient of volume expansion b, we can make the equation

$$\Delta V = bV_0 \Delta T \qquad (6.12)$$

In words, b is the fractional change in volume, $\Delta V/V_0$, per degree change in temperature ΔT. If the material is isotropic (that is, it has the same properties in all directions), it is found that b equals approximately $3a$.

EXAMPLE Suppose that a liquid has a coefficient of volume expansion of 10^{-3}/deg C, which is typical. If the original volume of the liquid in a tank was 100,000 gal at a temperature of 10°C, when the temperature rises to 30°C, the increase in volume would be

$$\Delta V = 10^{-3} \times 10^5 \times 20 = 2000 \text{ gal}$$

In this example, the volume increases by 2 per cent.

As a final illustration of the expansion of materials with temperature, let us consider shrink-fitting. Suppose that you want to secure a circular collar on a cylindrical rod. If you make the collar slightly too small, it will not fit on the rod. However, if you heat the collar, it will expand enough so that you can slip it on the rod; then it will fit very tightly when it cools and contracts. This process is used quite a bit in industrial applications. We shall leave it up to you to figure out how to remove the collar once you have put it on this way.

DISCUSSION QUESTIONS

1. Explain how a pressure cooker works. Use Fig. 6.10.

2. The fusion curve of water slopes to the left instead of sloping to the right, as shown in Fig. 6.11 for most substances. Explain what effect this might have on ice skating.

3. Although the average energy of molecules in a gas at a certain temperature is the same, all the molecules do not have the same speed. Explain this apparent paradox.

4. Suppose that the temperature of a given amount of gas is raised a certain number of degrees; first, while the pressure of the gas is kept constant, and second, while the volume of the gas is kept constant. In which process would more heat be required?

5. Can heat be added to a substance without raising its temperature?

6. Discuss the most important requirements for a material out of which you hope to make an excellent frying pan.

7. When a thin piece of metal is flexed repeatedly, its temperature rises. Why?

8. As a gas molecule moves around a room, its gravitational potential energy changes. Do you think that this energy change is significant? Justify your answer quantitatively.

9. Why should a substance like rubber contract when it is heated, whereas most solids expand?

10. Water has the uncommon property of contracting as its temperature is raised from 0°C to about 4°C and then expanding as its temperature is raised further. Use this information to explain why ice skating is possible on the surface of a deep lake.

11. The energy in a room full of air is quite large. Explain why you cannot use this energy to heat your house.

12. Basically, temperature is defined only for substances in thermal equilibrium. Explain how kinetic theory could be used to define a temperature for particles in the interior of the sun.

13. Does the kinetic theory apply to an object such as a car?

PROBLEMS

1. Suppose that each of your shoes has an area of 10 in.^{2} and your weight is 180 lb. Compute the pressure exerted on the soles of your shoes by the ground.

2. A gas is originally at an absolute pressure of 10^5 N/m^2 when its volume is 2 m^3 and its temperature is 27°C. The temperature of the gas is raised to 87°C and its volume increases to 2.5 m^3. Compute the pressure of the gas under these conditions. (*Ans.:* 9.6×10^4 N/m^2)

3. A car weighs 4000 lb, and each of its tires is inflated to a gauge pressure of 25 lb/in.^{2} Compute the area of each tire that is in contact with a horizontal roadway when the car is at rest. (*Ans.:* 40 in.^{2})

4. A gas is at a pressure of 1 atmosphere when its volume is 3 m^3 and its temperature is 27°C. If the pressure on the gas is raised to 3 atmospheres and its temperature rises to 97°C, compute its volume under these conditions.

5. A gas has an initial pressure of 10^4 N/m^2, initial volume of 3 m^3, and initial temperature of 20°C. If the gas is compressed so that its pressure is 10^5 N/m^2 and volume is 1 m^3, compute its temperature. (*Ans.:* 704°C)

6. At a temperature of 27°C and a total pressure of 30 lb/in.^{2}, a certain gas has a density of 10^{-3} gm/cm^3. Compute the density of the gas when its temperature is raised to 77°C and its pressure raised to 75 lb/in.^{2}

7. How much ice at 0°C must be added to 0.5 kg of water at a temperature of 40°C so that the temperature of the resulting mixture will be 10°C? (*Ans.:* 0.167 kg)

8. If 0.1 kg of ice is added to 0.3 kg of water at a temperature of 20°C, what is the temperature of the resulting mixture?

9. Compute the heat required to convert 2 kg of ice at 0°C into steam at 100°C. (*Ans.*: 1440 kcal)

10. If 0.1 kg of steam condenses on 2 kg of aluminum at a temperature of 20°C and specific heat 0.20, compute the final temperature of the aluminum.

11. Take the average mass of a molecule of air to be 4.76×10^{-23} gm. Compute the average speed of such a molecule at a temperature of 27°C.

12. Take the coefficient of thermal expansion of steel to be 1.2×10^{-5} per degree Celsius. Compute the change in length of a bridge that is 6000 ft long at 20°F when the temperature increases to 100°F. (*Ans.*: 3.2 ft)

13. The volume of a gas kept at constant pressure increases from 4 m^3 at 0°C to 5 m^3 at 100°C. Find the temperature at which the volume of this gas would be 4.65 m^3.

14. The volume of a gas kept at constant pressure is 6 m^3 at 100°C. Compute the temperature at which this gas would have a volume of 4.20 m^3.

15. The tires on your car have a gauge pressure of 30 lb/in.^{2} on a day when the atmospheric pressure is 15 lb/in.^{2} and the temperature is 20°C. After a long drive the temperature of the tires has risen to 60°C, whereas the volume of the tires has not changed appreciably. Compute the gauge pressure in the tires now.

16. One-tenth kilogram of ice at 0°C is added to 2 kg of water at 80°C. Compute the final temperature of the mixture. (*Ans.*: 72.5°C)

17. Compute the heat given off when 3 kg of steam at 100°C condenses into water and then cools down to 60°C.

7 The Second Law of Thermodynamics

In Chapter 5 we discussed the transfer of energy in the form of heat and reformulated the Law of Conservation of Energy to account for this form of energy flow. The First Law of Thermodynamics is simply the Law of Conservation of Energy stated in a form applicable to bulk matter. In this chapter we shall consider the problem of converting heat energy into work. You will find that it is never possible to convert a given amount of heat energy into work without producing other changes in the environment (a process that will be stated more precisely later). The impossibility of complete conversion of heat into work is the gist of the Second Law of Thermodynamics. We begin this topic by considering the properties of heat engines.

7.1 HEAT ENGINES

In its simplest form a heat engine consists of a device that takes in heat ΔQ_1 at a high temperature T_1, does some external work ΔW, and exhausts some heat ΔQ_2 at a low temperature T_2. The operation of a heat engine is shown schematically in Fig. 7.1. The working substance in the engine is taken through a series of pressure values, temperature values, and other variables. When the conditions of the working substance have been returned to the initial values of pressure, temperature, and so on, we say that the

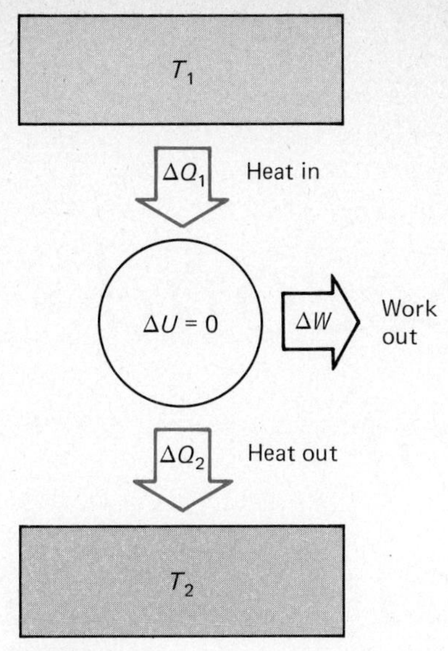

FIGURE 7.1

A heat engine. The engine takes heat ΔQ_1 from a reservoir at T_1, does work ΔW, and exhausts heat ΔQ_2 at temperature T_2.

substance has gone through a *cycle*. All theoretical discussions of heat engines involve taking the working substance through a complete cycle, so that the condition of the substance is unchanged at the end of the cycle.

As an example of a practical heat engine, let us consider a simple steam engine. It is shown in Fig. 7.2. Water is converted into steam in the boiler by the addition of heat, which corresponds to the heat ΔQ_1 in Fig.

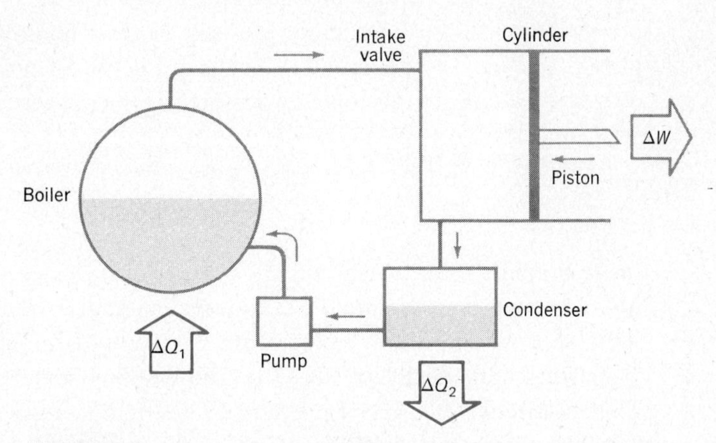

FIGURE 7.2

Schematic diagram of a real steam engine.

7.1. The steam expands in the cylinder and pushes the piston, which does mechanical work corresponding to ΔW in Fig. 7.1. The steam then condenses into water and gives up an amount of heat corresponding to ΔQ_2 in Fig. 7.1. The water has now been taken through a complete cycle and is ready to enter the boiler where a new cycle will begin.

The state of the working substance of a heat engine is the same at the beginning and at the end of a cycle; therefore, the internal energy of the working substance is also unchanged. Thus, in Eq. (5.6), $U_1 = U_2$. In Sec. 5.3 we treated the heat added to a body as positive and the heat given up by a body as negative. For the heat engine shown in Fig. 7.1, the First Law of Thermodynamics, Eq. (5.6), takes the form

$$\Delta Q_1 - \Delta Q_2 = \Delta W \tag{7.1}$$

You should note here that not all of the work ΔW is useful because of losses within the engine. The work ΔW is simply the amount of energy that has been converted from heat energy into some other form, whether useful or not.

An important term is the *efficiency* of a heat engine, which we define as the ratio of the work output of the engine to the total energy input to the engine. We can write the efficiency e of an engine in the form of the following equation:

$$e = \frac{\Delta W}{\Delta Q_1} \tag{7.2}$$

When we substitute the value of W from Eq. (7.1) into Eq. (7.2), we find

$$e = \frac{\Delta Q_1 - \Delta Q_2}{\Delta Q_1} = 1 - \frac{\Delta Q_2}{\Delta Q_1} \tag{7.3}$$

Thus, we can only make the efficiency of conversion of heat 100 per cent if the amount of heat, ΔQ_2, given out by the engine, is zero.

7.2 SECOND LAW OF THERMODYNAMICS

The First Law of Thermodynamics tells us that it is impossible to have a heat engine with an efficiency greater than 1. In terms of the 19th century's preoccupation with perpetual motion machines, this law says that it is impossible to build a *perpetuum mobilum* of the first kind, i.e., one that actually generates energy out of nothing. However, we still might hope to

build one with an efficiency equal to 1, which would not violate the Law of Conservation of Energy. Nevertheless, all such attempts have failed. Evidently, any heat engine must give out some heat, so that its efficiency can never be 100 per cent.

During the 19th century, Lord Kelvin and Max Planck summarized the impossibility of converting all of a given amount of heat into other forms by stating that no engine acting in a cycle can take in a given amount of heat and convert it entirely into other forms of energy. This is one statement of the Second Law of Thermodynamics.

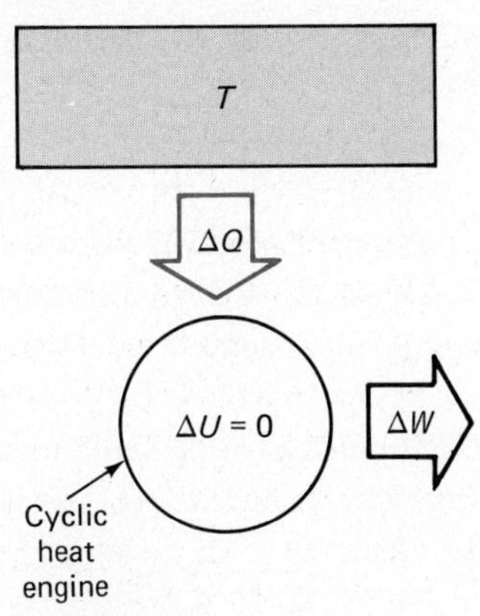

FIGURE 7.3
An impossible heat engine. This engine takes heat and converts it entirely to work, which is precluded by the Second Law of Thermodynamics.

To restate the law in a form that does not involve the somewhat artificial construct of a heat engine, we must define a *thermodynamic process.* This process is simply some change involving the flow of heat, the performance of work, and other changes in the states of the various systems involved. We must also define a *heat reservoir,* which is a source of heat that is at a definite temperature and is large enough to supply heat without changing that temperature. Using these two definitions, we can state the Second Law of Thermodynamics as follows:

> *It is impossible to have a process whose sole effects are the removal of heat from a single heat reservoir and the appearance of work in the surroundings.*

The words “sole effect” in this statement mean that no other changes occur in any part of the universe except those mentioned. A heat engine achieves

this process by operating in a cycle so that the state of the working substance in the engine is returned to its original state after each cycle. There are many other ways of stating the Second Law of Thermodynamics, but all of them can be proved equivalent to this one. This law has many important consequences, some of which we shall mention at the end of this chapter.

A reversible engine is an engine in which all the processes can be run backward, reversing the flows of work and heat and thus changing the signs of ΔW, ΔQ_1, and ΔQ_2. In practice, there is no such thing as a reversible engine because of energy losses due to friction and other factors. Thus, a reversible engine is an ideal heat engine, which at best can be only approximated. Early in the 19th century, Nicolas Carnot, a French physicist, showed that all reversible heat engines operating between the same temperatures, T_1 and T_2, must have the same efficiency. He also showed that the efficiency of any other engine had to be less than the efficiency of a reversible engine operating between the same pair of temperatures. Of course, this fact is not surprising. Generally speaking, all ideal engines should have the same properties, including their efficiencies, and ideal engines should be more efficient than nonideal engines. Even though it is impossible in practice to build an ideal engine, the theoretical properties of such engines set an upper limit on the performance of real engines, so the concept is a useful one.

From Eq. (7.3) we see that the efficiency of a heat engine depends on the amounts of heat taken in and given out during a cycle. Because Carnot had shown that the efficiencies of all reversible (ideal) engines operating between the same intake and exhaust temperatures are the same, it is evident that the efficiency of a reversible engine depends on the temperatures at which it takes in and gives out heat, T_1 and T_2, and on no other factors. Thus, the efficiency of a reversible engine is a property of nature that depends on temperature and can be used to construct a temperature scale.

Kelvin established the Kelvin, or absolute, temperature scale by stating that the temperature ratio $T_2/T_1 = \Delta Q_2/\Delta Q_1$. From Eq. (7.3) we see that the efficiency of a reversible engine in terms of this temperature scale is given by

$$e = 1 - \frac{T_2}{T_1} \tag{7.4}$$

Kelvin also showed that if an ideal gas were used as the working substance in the heat engine, the ideal gas temperature defined in Sec. 6.1 would be identical to the absolute temperature defined by Eq. (7.4). Although no ideal engine can be constructed, the absolute temperature scale can be approximated as closely as we please by using a gas at very low pressure. (Other

substances can also be used if they are similarly restricted.) The ideal gas temperature scale and the absolute temperature scale are identical, so temperatures on these scales are expressed in degrees Kelvin (deg K). We have then:

$$T(\text{deg } K) = 273 + t(\text{deg C}) \tag{7.5}$$

In all theoretical work in thermodynamics, the Kelvin scale is used, although many measurements are made with the Celsius scale.

EXAMPLE Let us compute the ideal efficiency of a heat engine operating between the temperature of 127°C and 27°C. The corresponding Kelvin temperatures are found from Eq. (7.5) to be 400°K and 300°K. The efficiency of this real engine is

$$e = 1 - \frac{300}{400} = 0.25 = 25 \text{ per cent}$$

A real engine operating between this pair of temperatures would have a lower efficiency because of losses, but at least we have an upper limit on the engine's possible efficiency.

Because the lower temperature T_2 is usually approximately room temperature (20°C), the efficiency of a real engine is best increased by increasing the temperature T_1 at which heat is taken into the engine. For instance, steam engines are operated at high pressures so that the temperature of the steam will be much higher than 100°C. The other way in which we can increase the efficiency of a real engine is to reduce the losses.

7.3 ENTROPY AND THE SECOND LAW

In the preceding section, we stated the Second Law of Thermodynamics in terms of the negative principle that no heat engine could convert all of the heat energy that it receives into some other form of energy. If we introduce the concept of *entropy,* we can use it to state the Second Law of Thermodynamics in a different way. The formulation of the law in terms of entropy is more mathematical and is thus better suited to theoretical studies in thermodynamics.

Suppose that a small amount of heat ΔQ is added slowly to a system at temperature T. We define the increase in entropy S of the system by the equation:

$$\Delta S = \frac{\Delta Q}{T} \tag{7.6}$$

The heat must be added slowly so that the system has a well defined temperature throughout, that is, it remains in equilibrium. The amount of heat must be small so that the temperature does not change appreciably. If a large amount of heat is to be added, it must be done in many small amounts; the *total* entropy change is the sum of the small increases each calculated using Eq. (7.6). If the heat ΔQ flows out of the system then the entropy change ΔS is negative.

It is also possible to change the state of a system in ways that do not involve the flow of heat. If the state of a system is changed by performing work, and if this process is reversible (the work can be recovered by reversing the process), then the entropy of the system does not change. Using these two types of processes—slow heat flow with the entropy change given by Eq. (7.6), and reversible work with no heat flow and no entropy change—it is possible to determine the difference in entropy between any two states of a system; hence we find the change in entropy when a system goes from one state to another. This is done by selecting a process (involving only the two types of change discussed) that connects one of the two states with the other. The entropy difference between the two states is then calculated. A fundamental theorem in thermodynamics guarantees that this entropy difference does not depend on the particular choice of the process connecting the two states. It can also be shown, using statistical mechanics, that the entropy of any system will vanish at absolute zero and thus the methods we have just outlined would in principle enable us to find the entropy of any system by warming it up from absolute zero. At the end of this section we will give a physical interpretation of entropy but first we will restate the Second Law in terms of entropy change.

Let us now apply the concept of entropy to the ideal heat engine discussed in Sec. 7.1 and shown diagrammatically in Fig. 7.1. The heat ΔQ_1 flows out of the hot reservoir at the temperature T_1; therefore, the entropy of this reservoir decreases by an amount $-\Delta Q_1/T_1$. In addition, the cold reservoir gains an amount of heat ΔQ_2 at a temperature T_2, so that its entropy increases by an amount $\Delta Q_2/T_2$. No other heat is flowing in the universe in this particular situation. Thus, the change in entropy of the universe is given by

$$\Delta S = \frac{\Delta Q_2}{T_2} - \frac{\Delta Q_1}{T_1} \tag{7.7}$$

The engine is an ideal engine, so the definition of absolute temperature tells us that $T_2/T_1 = \Delta Q_2/\Delta Q_1$. If we use this relation in Eq. (7.7), we find that the entropy change of the universe during the operation of an ideal heat engine acting through a cycle is zero.

Now consider a nonideal (actual) heat engine operating between

the temperatures T_1 and T_2 and taking in the same amount of heat ΔQ_1 at the temperature T_1 as the ideal engine discussed in the preceding paragraph. According to Carnot's theorem stated in Sec. 7.2, the nonideal engine has a lower efficiency than the ideal engine. Therefore, for a given heat intake ΔQ_1, it does less work and gives out more heat ΔQ_2 at the lower temperature T_2. In Eq. (7.7) every quantity is the same on the right side for the ideal and the nonideal engines, except ΔQ_2, which is larger for the nonideal engine. Thus, ΔS is positive for the nonideal engine, and the entropy of the universe increases when heat flows through such a nonideal engine.

As another example of entropy change, consider the flow of heat from a hot object to a cold one. This process is clearly irreversible. Heat never flows spontaneously from a cold to a hot object. Suppose an amount of heat ΔQ flows from a hot object at temperature T_1 to a cold object at temperature T_0. The entropy of the hot object decreases by an amount $\Delta S_1 = -\Delta Q/T_1$, whereas the entropy of the cold object increases by an amount $\Delta S_0 = \Delta Q/T_0$. The net change in the entropy of the universe is the sum of these changes, and we find that $\Delta S = \Delta S_1 + \Delta S_0 = \Delta Q/T_0 - \Delta Q/T_1$. Because T_1 is larger than T_0, this change is always positive, and so entropy has increased. If the heat flowed in the opposite direction, from cold to hot, then we would have had a net decrease in entropy. This development never occurs however.

The preceding discussion shows that heat flowing in various processes causes the entropy of the universe to be either unchanged (if the process is reversible) or increased. We can, therefore, state the Second Law of Thermodynamics in another way:

> *For any process occurring in an isolated system, the entropy never decreases.*

An isolated system is one that has absolutely no interaction with its surroundings. The universe is a good example of an isolated system. This statement can be shown to be entirely equivalent to the Kelvin-Planck statement of the same law, but it has the advantage of being stated in terms of a mathematically defined quantity, entropy. For actual, nonideal substances there is always an increase in entropy in any process.

By this time, you may have noticed that we have not actually defined "entropy." What we have done is defined the change in entropy for a certain class of processes involving the flow of heat. The formal definition of entropy is very complex and can only be given in the context of statistical mechanics, the branch of physics that derives thermodynamics from the basic laws of

the individual constituents of the bulk systems. This formal definition can, however, be described qualitatively.

First, it states that entropy, like internal energy, is an intrinsic property of the substance. Any given quantity of a substance under the same conditions (such as temperature, pressure, and volume) always has a given value for its entropy. When you change the state of a substance, you change its entropy. We know how to calculate this change for certain types of processes involving heat flow. However, for the same change of condition, regardless of the way the change is achieved, the change in entropy is the same.

The second part of the formal definition of entropy states that entropy is a measure of the state of disorganization of the system. The more disorganization, the more entropy. Ice clearly has an organized structure because its molecules are arranged in a regular crystal lattice. Water is less

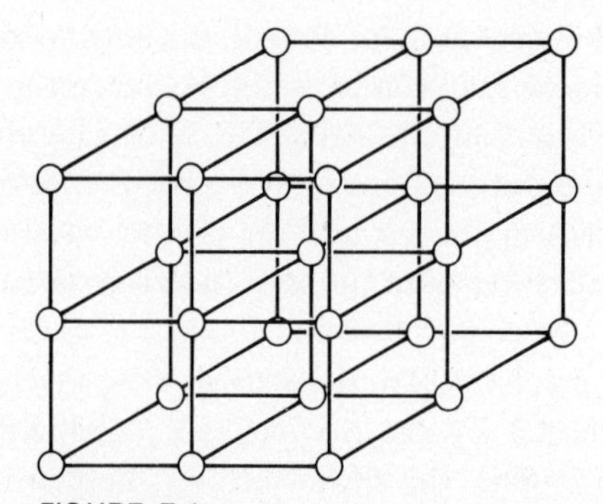

FIGURE 7.4
Part of a crystal lattice having a high degree of organization and a low entropy.

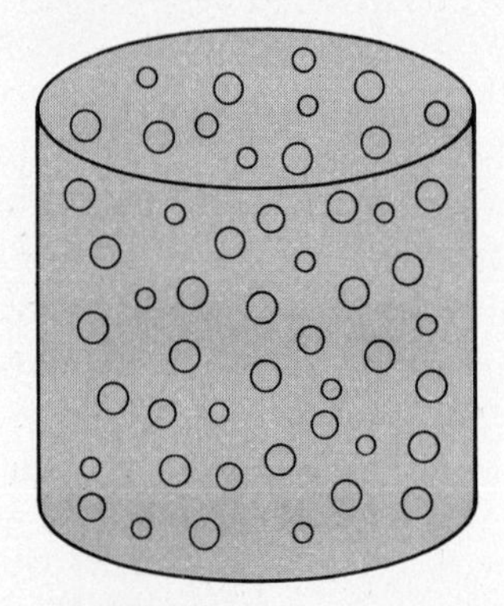

FIGURE 7.5
A closely packed but disorganized liquid having about the same density and thermal agitation as the crystal in Fig. 7.4, but more entropy due to its randomness.

organized because the molecules do not have any regular ordering, although they are still quite near each other. Steam has the least organization. The molecules of steam, according to the kinetic theory, are moving at high speed in a completely random fashion. Thus, the entropy of 1 gm of ice is less than that of 1 gm of water and much less than that of 1 gm of steam.

Of course, as discussed in Chapter 6, to change ice to water and then to steam, we must add heat, thus, by our definition of entropy change, increasing the entropy. Popularizations of scientific ideas often state that life results in highly organized systems and so violates the Second Law by producing a net decrease in entropy. This deduction is incorrect. These writers have read the statement of the Second Law incompletely. There is nothing in the Second Law that says the entropy of a given system must increase if it is in contact with other systems. Only the entropy of an isolated system must increase. Life is not an isolated system. Life draws heat and energy from many sources (including the sun). If you consider the net entropy change of the environment as well, the entropy decrease produced by living systems is more than compensated by the entropy increase in their environment, thus preserving the Second Law of Thermodynamics.

We must point out here one global consequence of the Second Law of Thermodynamics. A major result of the technological changes that have shaped the 20th century has been the tremendous increase in the consumption of energy. To produce power, men have had to develop many complex systems, principally mechanical and electrical. The machines that produce this power are governed by the Second Law of Thermodynamics. The internal combustion engine, the steam turbine, the thermonuclear reactor, and the rocket engine all draw their power from heat produced at a high temperature. This heat comes from the burning of fossil fuel, chemical fuel, or nuclear fuel. The Second Law tells us that this heat cannot be converted completely into usable energy. Some of it must be exhausted as waste heat at a lower temperature. Furthermore, the energy produced cannot be used completely efficiently; ultimately some is lost to friction or heat. This necessary loss is the source of thermal pollution. Until we learn to convert the potential energy that nature has stored for us directly into usable energy without the intervening step of producing heat, our civilization will be faced with the problem of disposing of the waste heat that the Second Law tells us is inevitable.

DISCUSSION QUESTIONS

1. In a *Carnot cycle* the working substance takes in an amount of heat ΔQ_1 at a constant temperature T_1 and then does work without any further heat transfer. The cycle continues with the substance giving off heat ΔQ_2 at a constant temperature

T_2, and the cycle is completed by having work done without heat transfer in such a way that the substance is brought back to its initial state of pressure, volume, temperature, and so on. If the substance is a gas, sketch this cycle on a diagram on which pressure is plotted against volume.

Sketch the same cycle on a diagram on which temperature is plotted against entropy.

2. It is said of a heat pump used to heat houses that more heat is delivered to the house than is taken from the ground. Does this statement violate any of the laws of thermodynamics?

3. Are there heat transfers that violate the Second Law of Thermodynamics? If so, explain the apparent paradox.

4. Could you cool your kitchen on a hot day by leaving your refrigerator door open?

5. Is there a change in entropy when a ball is thrown up into the air and then caught?

PROBLEMS

1. An ideal engine operates between the temperatures of 300°C and 20°C. If it is to do 10^4 joules of external work, how much heat does it take in? How much heat does this engine exhaust? (*Ans.:* 4.92 kcal)

2. An ideal engine exhausts heat at a temperature of 20°C. Compute its intake temperature if its efficiency is to be 90 per cent. (*Ans.:* 2657°C)

3. A heat engine takes in 5 kcal of heat at a temperature of 337°C and exhausts 3 kcal of heat at a temperature of 27°C. Compute the amount of external work done by this engine and its efficiency.

Compute the change in entropy in this process.
(*Ans.:* 0.00181 kcal/°K)

4. An inventor claims that his heat engine takes in 5 kcal of heat, gives off 2 kcal of heat, and does 15,000 joules of work. What is the efficiency of the engine? Would you invest money in this engine?
(*Ans.:* 71.6 per cent; no.)

5. An ideal engine is to take in 3 kcal of heat and do 10^4 joules of work. How much heat is exhausted by the engine and what is its efficiency?

If the intake temperature is 127°C, what must its exhaust temperature be?
(*Ans.:* −192°C)

6. Two kilograms of ice melt and give off 160 kcal of heat at a constant temperature of 0°C. Compute the change in entropy in this situation. Does this change violate the Second Law of Thermodynamics?

7. A heat engine takes in 4 kcal of heat at a temperature of 227°C, does 5000 joules of work, and exhausts heat at a temperature of 27°C. Compute the efficiency of this engine and the amount of heat exhausted.
(*Ans.:* 29.9 per cent; 2.81 kcal)

Compute the change in entropy in this process.

8. If the exhaust temperature of an ideal engine is 27°C, what must be its intake temperature if the efficiency of the engine is to be 20 per cent?
(*Ans.:* 102°C)

9. A steam engine takes in heat at 500°F and exhausts it at 70°F. If the efficiency of this engine is 20 per cent of the efficiency of an ideal engine, compute the efficiency of the steam engine.

10. An ideal engine takes 5 kcal of heat and does 10^4 joules of external work. How much heat is exhausted by the engine and what is its efficiency?
(*Ans.:* 2.60 kcal; 48 per cent)

If the exhaust temperature of the engine is 27°C, what is its intake temperature?

Part Three
Electrodynamics and Waves

8 Electricity

The age in which we live has been called the *electronic* age. It is characterized by such devices as the computer, telephone, television, and radio. Without such systems Western civilization as we know it would not be possible. These devices and countless others have been developed because of man's mastery of electricity and electromagnetic radiation. In this part of our book we shall study some of the simple laws of *electrodynamics,* the branch of physics dealing with electricity and magnetism. We consider first the source of all electromagnetic effects: *charge.*

8.1 CHARGE

The study of electric charges at rest is called *electrostatics.* Electrostatic effects are quite common in everyday life. After you have walked across a rug in dry weather, you often get a shock and notice a small spark when you touch something metal like a doorknob. When you comb your hair, you find that the comb often picks up small pieces of lint and paper. Under certain conditions an enormous spark, which we call a lightning bolt, passes between a cloud and the earth. Although all these examples differ, each one is a case of electrification by friction.

Suppose that you rub a hard rubber rod with a piece of wool cloth and then touch the rod to a small ball made of light wood, such as pith

or balsa wood, suspended by a thread. The small ball is then repelled by the rubber rod, as shown in Fig. 8.1. Presumably something has been transferred from the rubber rod to the ball that causes this force of repulsion. If you rub a glass rod with a silk cloth and then touch it to another small and light ball, as shown in Fig. 8.2, the ball will be repelled by the glass rod. Again we assume that something has been transferred from the glass rod to cause this repulsion. If you now bring the two small balls close together, they attract one another, as illustrated in Fig. 8.3. Let us call the substance that has been transferred from either rod to either ball an *electrical charge.*

From these experiments we see that charges produced in a similar manner repel each other but attract charges produced in a different manner. We can conclude, therefore, that there must be two types of electrical charge. Experiments have failed to discover a third type of charge. The two types of charge could have been labeled *A* and *B*, but in the late 18th century Benjamin Franklin suggested that the charge appearing on the glass rod

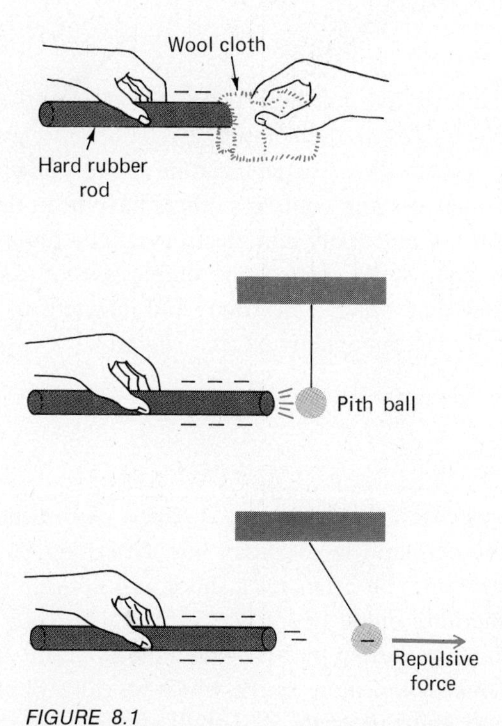

FIGURE 8.1

In the first picture a hard rubber rod is rubbed with a piece of wool. In the second picture the rod is touched to a pith ball, transferring some charge. In the last picture the pith ball is repelled by the rod.

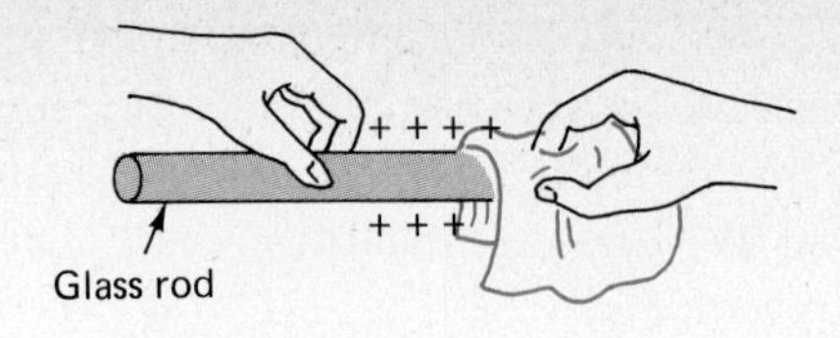

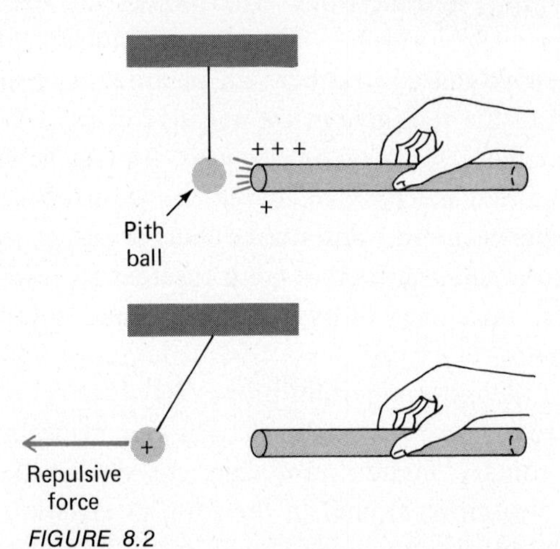

FIGURE 8.2

A similar experiment to Fig. 8.1, using a glass rod and a silk cloth. After the pith ball is charged it again repels the rod.

be called positive (+) and the charge appearing on the rubber rod be called negative (−). This definition is still in use today.

It was not until the beginning of this century that physicists discovered what it was that carried the electrical charge. The ultimate carriers of charge are small particles called *electrons* and *protons*. These particles,

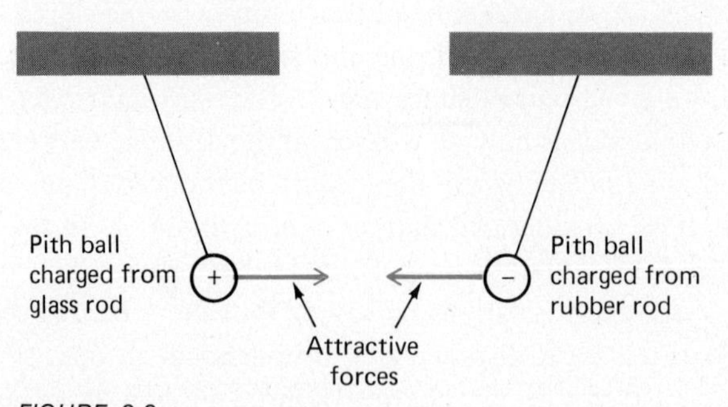

FIGURE 8.3

The two charged pith balls are hung near each other. They attract each other.

along with the *neutron,* a neutral particle, are the building blocks out of which all matter is formed. The electron carries negative charge; the proton, an equal positive charge. We shall have much more to say about these particles in our chapters on modern physics. Here we use them simply to explain electrification.

Ordinary matter is electrically neutral and contains exactly equal numbers of protons and electrons. The protons are tightly bound to the matter that they help form, but the electrons can sometimes be added or removed. When we charge a rubber rod, we are transferring electrons from the wool cloth to the rod, giving it a negative charge. When we charge a glass rod, we remove some of its electrons, leaving behind more protons than electrons and a net positive charge. As we have already mentioned, like charges repel each other and unlike charges attract. It is this attraction between electrons and protons that holds together all solid objects, although we need the full machinery of quantum mechanics to explain exactly how the process works.

One of the most important properties of charge is that it can never be created or destroyed. Opposite charges can neutralize each other, as happens in ordinary matter, producing an electrically neutral object. Charges may also move around in the form of electric currents. The net charge of a closed system, however, is a constant and will never change unless charge is added or removed from the outside. This means that charge obeys a conservation law called *conservation of charge.* This law has been experimentally verified with great precision and is known to hold even under such extreme conditions as a stellar interior or a neutron star.

8.2 ELECTROSTATICS

In 1787, C. A. de Coulomb, a French physicist, measured the forces between small charges, using apparatus similar to that described in Sec. 4.1. He found that for a given pair of charges the force was quadrupled when the distance between them was halved, whereas if the distance between the charges were tripled, the force between them would be reduced to one-ninth of its original value. If we consider two charges of magnitudes q_1 and q_2, the force between them is also found to be proportional to the product $q_1 q_2$. We can summarize both of these observations by saying that the force between charges q_1 and q_2 separated by a distance d is proportional to $q_1 q_2/d^2$. The situation is shown in Fig. 8.4 for two charges of the same sign. If we introduce a constant of proportionality K, we can then write

$$F = K\frac{q_1 q_2}{d^2} \tag{8.1}$$

This equation is known as Coulomb's Law. Notice in Eq. (8.1) that if we halve the distance *d*, the force is increased by a factor of 4, provided that the charges remain the same. Moreover, if the distance of separation is kept constant, we see from Eq. (8.1) that the force between the charges is proportional to the product of the charges. Thus, Eq. (8.1) correctly gives the results of Coulomb's observations. The direction of the force will be an attraction if the charges are of different types, whereas the force will be a repulsion if the two charges are of the same type, as shown in Fig. 8.4.

In order to calculate the value of the constant *K* in Eq. (8.1), we must measure the force between a pair of known charges that are a measured distance apart. The constant *K* is thus an experimental constant. Up to this point, we have a unit for distance (the meter) and a unit for force (the newton). We now need a unit for charge. In practical work the effects of moving charges (electrical currents) are much more important than electrostatic effects; therefore, the fundamental unit in electricity is the *ampere*

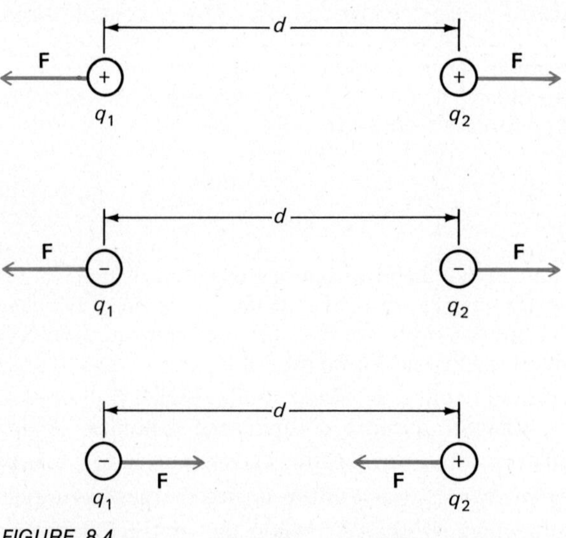

FIGURE 8.4
The force between two charges. The magnitude of the force is given by $F = k\frac{q_1 q_2}{d^2}$. In the top two pictures the charges are alike and the force is repulsive. In the bottom picture they are different and the force is attractive.

(amp), which describes the rate of flow of charge per unit of time. (The ampere is defined in Sec. 9.3.) The ampere is analogous to the rate (perhaps measured in gallons per second) at which a liquid flows through a pipe. If a constant current I (measured in amperes) flows for a time of Δt seconds, the total electrical charge Δq passing a point is then given (in coulombs) by the expression

$$\Delta q = I \, \Delta t \tag{8.2}$$

The unit of charge that we use is the *coulomb,* defined by Eq. (8.2) as the amount of charge flowing past a point when 1 amp of current flows for 1 sec.

Now that we have a unit in which to measure charges, we can determine experimentally the value of the constant K in Eq. (8.1). The value of K turns out to equal very nearly 9×10^9 (when the charges are measured in coulombs, their separation measured in meters, and the force between them measured in newtons). Thus, the coulomb is a very large charge. Two charges, each 1 coulomb in magnitude, separated by a distance of 1 m would exert a force of almost 10 billion newtons on each other, or about 2 billion pounds.

EXAMPLE For example, let us compute the force between charges of 4×10^{-6} coulombs (4 microcoulombs) and -2×10^{-6} coulombs separated a distance of 0.1 m. From Eq. (8.1) we have for this force the value

$$F = 9 \times 10^9 \frac{(4 \times 10^{-6})(2 \times 10^{-6})}{0.1^2} = 9 \times 10^9 \frac{8 \times 10^{-12}}{10^{-2}} = 7.2 \text{ N}$$

Thus, we find that the force between the given charges is 7.2 N and is an attraction, because the charges are unlike. If the separation of the charges is now increased to 0.2 m, we can easily see that the force between them is reduced by the factor 4, so that the force now becomes 1.8 N.

Imagine a more complicated situation. A group of charges are located at various fixed locations in space as in Fig. 8.5. We call these charges the *source charges.* Each of the source charges has a definite magnitude and position, which we shall keep constant in the following discussion. Now let us place another charge q in the vicinity of the source charges. We call q a *test charge.* Each of the source charges will exert a force on the test charge. Because force is a vector quantity, the net force exerted on the test charge could be found by adding the various force arrows, as described in Chapter 2. Each of the forces exerted on the test charge could then be found by using Coulomb's Law. Each force would depend on the size of the particular

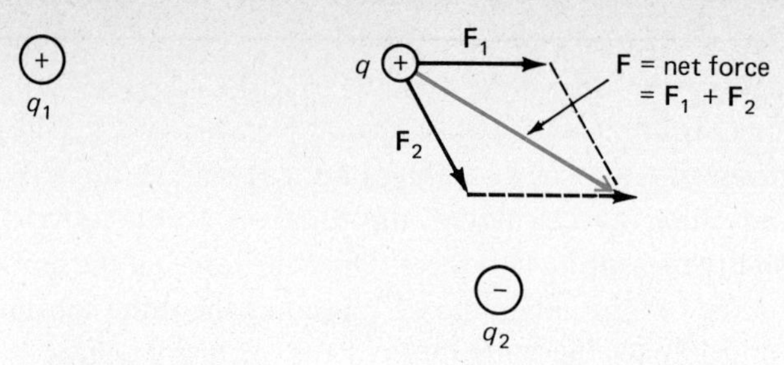

FIGURE 8.5

A test positive charge q is located near two source charges q_1 (positive) and q_2 (negative). The black vectors $\mathbf{F}_1$ and $\mathbf{F}_2$ are the forces exerted on q by the two source charges. The colored vector $\mathbf{F}$ is the vector sum $\mathbf{F}_1 + \mathbf{F}_2$, found graphically, and is the net force on the test charge.

source charge, its distance from the test charge, and, of course, on the size of the test charge itself. In general, this process would be rather complicated; we would have to add together many force arrows, one for each source charge. However, once we had done this calculation for some location of the test charge, we would never have to do it again, because each of the forces on the test charge and thus the sum of them would be proportional to the size of the test charge. If we place another test charge of half the magnitude at the same place as in Fig. 8.6, the force on it would be just half that on our original charge. Suppose that we found the force, perhaps experimentally rather than by calculation, at various points near the source charges by using a test charge of one microcoulomb (10^{-6} coulombs); then the force at each point on a 1-coulomb charge would be 10^6 times bigger. We use a small test charge because 1 coulomb is so large. We could then

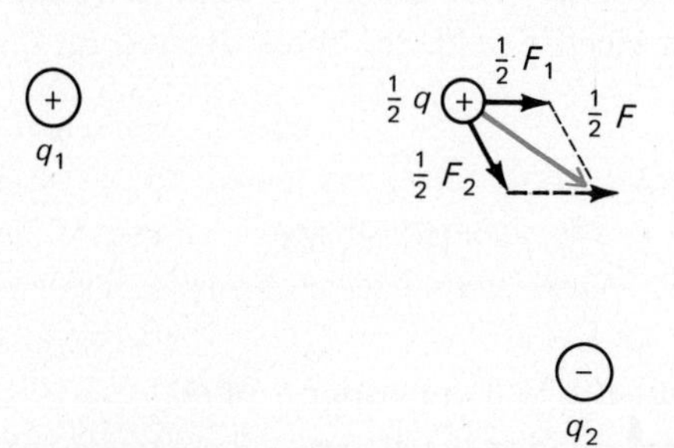

FIGURE 8.6

The same source charges q_1 and q_2 as in Fig. 8.5 with the same geometry and a test charge half as big. All the forces are reduced to half the values they had in Fig. 8.5.

calculate the force on any other test charge by multiplying the force on 1 coulomb by the size of the test charge in coulombs.

In the preceding discussion, you learned that the force exerted by a group of fixed source charges on a given arbitrary test charge could be found either by calculating the force on a unit test charge or by experimentally measuring that force. Once the force on the unit charge was found, the force on any other charge placed at the same location could be found by multiplying the magnitude of the arbitrary charge by q.

We can look at this process in two ways. In one way, we consider that the forces exerted by the source charges are acting directly on the test charge across the distance separating them. This approach is called *action-at-a-distance.* According to the second method, we say that, even in the absence of the test charge, the source charges produce a change in the region of space in which they are located. At each point in space, the source charges produce an *electric field.* When a test charge is placed at a point where there is an electric field, the field exerts a force on the test charge. If the force on a test charge q is **F**, we define the electric field at that point by the equation

$$\mathbf{E} = \frac{\mathbf{F}}{q} \tag{8.3}$$

See Fig. 8.7. From this definition it is evident that the units of electric field are newtons per coulomb. The test charge used in Eq. (8.3) should be small enough so that it does not disturb the source charges and thus change the situation that we seek to measure. It is also obvious from this definition that electric field is a vector quantity and has the same direction as the force that is exerted on a positive test charge. If another charge q' is now placed in a region with an electric field, the force exerted on this charge is given by

$$\mathbf{F}' = q'\mathbf{E} \tag{8.4}$$

which is nothing but Eq. (8.3) rewritten.

We have thus divided the electrical interaction into two parts. First, the source charges produce an electric field in space. Then this electric field exerts a force on any other charge placed in it. The concept of the electric field has proved a tremendously fruitful idea. As we shall see in later chapters, the electric field and its companion, the magnetic field, can carry

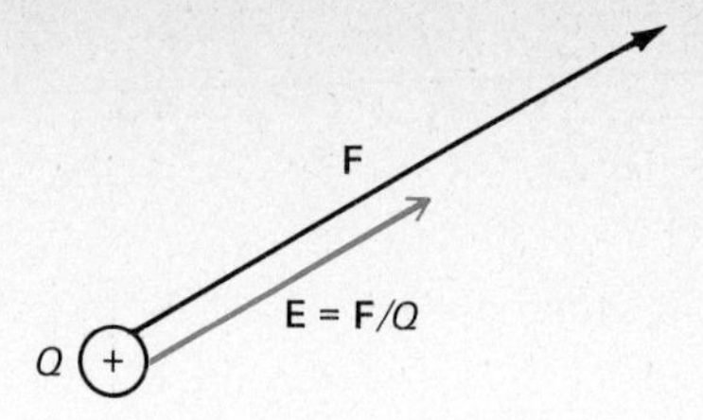

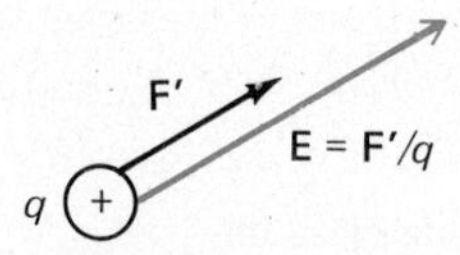

FIGURE 8.7
The force and electric field produced by distant source charges (not shown). In the upper picture the test charge is greater than one coulomb and the force is larger than the electric field. In the lower picture a smaller test charge is used giving a smaller force but the same electric field.

energy and momentum away from a system of moving charges. Ultimately they are responsible for electromagnetic radiation.

A convenient way of illustrating electric fields is by using *lines of flux*. At each point in space the electric field has a definite direction and magnitude. Lines of flux are drawn in such a way that at each point the direction of the flux line gives the direction of the electric field. Furthermore, the spacing between flux lines is chosen so that the number of lines of flux at right angles to a unit area perpendicular to the lines is proportional to the magnitude of the electric field. Both of these ideas are shown in Fig. 8.8. Clearly, such a diagram only gives an approximate value of the magnitude and direction of the electric field at any given point. Fig. 8.9 shows flux lines for some simple situations.

Let us return to the situation shown in Fig. 8.4, in which a charge q_1 exerts a force F on a charge q_2 when they are separated a distance d. From Eq. (8.1), we find that the force on q_2 is $K(q_1q_2/d^2)$. When we substitute this force into Eq. (8.3), which defines the electric field on q_2, we find

$$E = \frac{F}{q_2} = \frac{K\left(\dfrac{q_1q_2}{d^2}\right)}{q_2} = K\frac{q_1}{d^2} \tag{8.5}$$

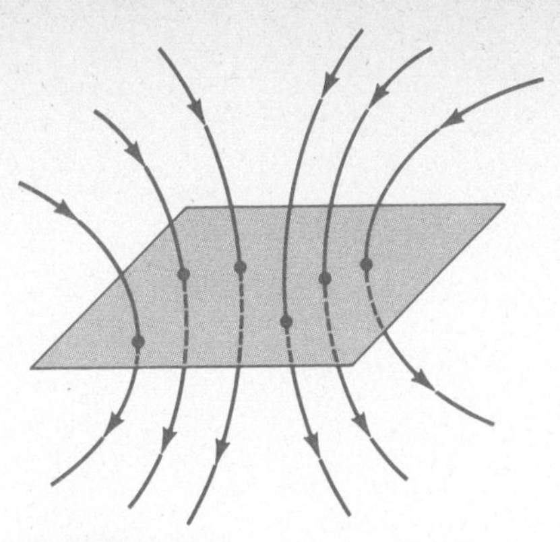

FIGURE 8.8
Electric flux lines. The direction of each line gives the direction of the electric field at that point. The number of them crossing a unit area perpendicular to their direction gives the field intensity. If the area of the shaded square was 1 m^2, then the electric field strength would be 6 N/coulomb.

The magnitude of the test charge q_2 was cancelled from Eq. (8.5), so we could have made it as small as we pleased. Thus, q_2 would not disturb the electric field that we were trying to calculate. The electric field produced by any charge q_1 at a distance d from it is given by Eq. (8.5). If several charges all produce electric fields at a given point, their contributions must be added as vectors, using the method described in Sec. 2.4.

EXAMPLE Let us compute the electric field produced at a distance of 0.2 m from a point charge of 4×10^{-6} coulombs. From Eq. (8.5) we calculate

$$E = 9 \times 10^9 \frac{4 \times 10^{-6}}{0.2^2}$$

$$= 9 \times 10^9 \frac{4 \times 10^{-6}}{4 \times 10^{-2}} = 9 \times 10^5 \text{ N/coulomb}$$

Now if a charge of 5×10^{-6} coulombs is placed at this point, the force on it is given by

$$F = (5 \times 10^{-6})(9 \times 10^5) = 4.5 \text{ N}$$

In this case, it would have been simpler to calculate the force on the charge by the direct use of Eq. (8.1), without bothering to calculate the electric field produced by the first charge.

In the static case, it makes no difference whether we use Coulomb's Law to find the force exerted on a given charge or break the problem into

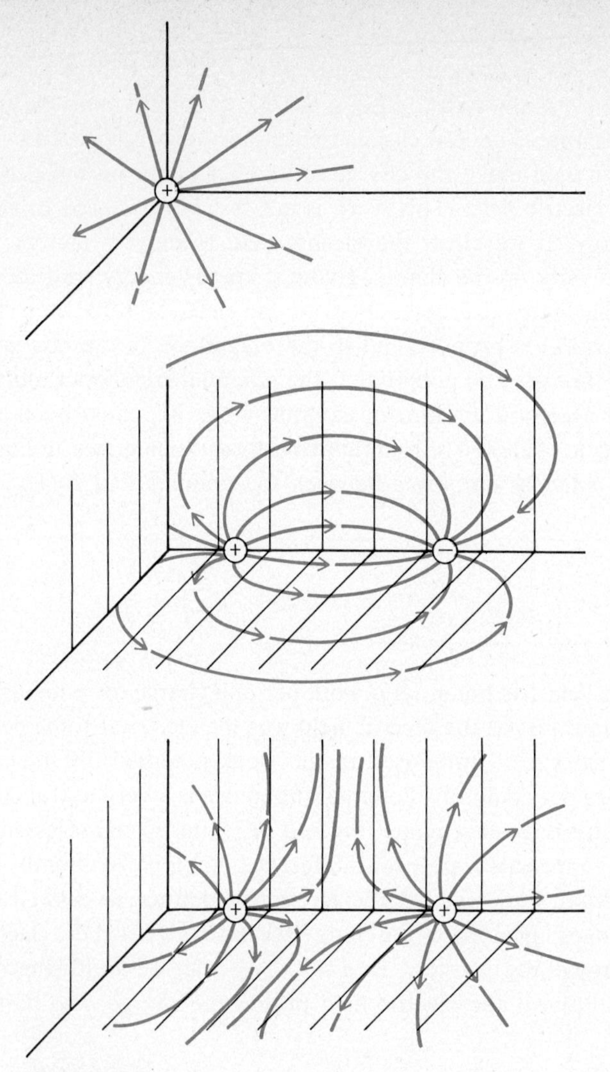

FIGURE 8.9
The electric field, as pictured by lines of flux, near a single positive charge (top), equal and opposite charges (middle), and equal positive charges (bottom).

two parts, in which we first calculate the electric field at the location of the charge and then find the force exerted on the charge by the electric field. If, however, the charge or charges that are establishing the electric field are moving, we *must* use the electric field idea. This procedure must be followed because changes in the force exerted by one moving charge on another are propagated with the speed of light (186,000 mi/sec). If either the charge

exerting the force or the charge experiencing the force has a speed close to the speed of light, changes are not transmitted instantaneously.

If we wish to hold a charge stationary in an electric field, we must exert a force on that charge to balance the force exerted by the electric field. If we then move the charge to another point, we have to do work against the electric field. This work is not lost but is stored in electrical potential energy. If we allow the electric field to move a charge, the electric field does work on the charge, giving it kinetic energy, and the potential energy of the charge decreases. Because the electrical force on a charge (and hence the work) is proportional to the magnitude of the charge, it is customary to define electric potential as the potential energy per unit charge. Suppose that a certain amount of external work W_{ab} must be supplied to move a particle of charge q from point a to point b, as shown in Fig. 8.10. We define the *potential difference* between the points a and b (V_{ab}) by the equation

$$V_{ab} = \frac{W_{ab}}{q} \tag{8.6}$$

Thus, electric potential is work per unit charge (or potential energy per unit charge) just as the electric field was the electrical force per unit charge. In the system of units used in this book potential difference is expressed in joules per coulomb. Potential difference is a very useful concept; therefore, its unit is given a name, the *volt* (V), after Count Alessandro Volta. Thus, 1 V represents a potential difference of 1 joule/coulomb. If we have to do external work against the electric field force in order to accomplish the change in position of the charge, we have increased the electrostatic potential energy of the charge q. By a suitable device we could later recover this work. Similarly, if the electric field pushes the charge q to its new position, we

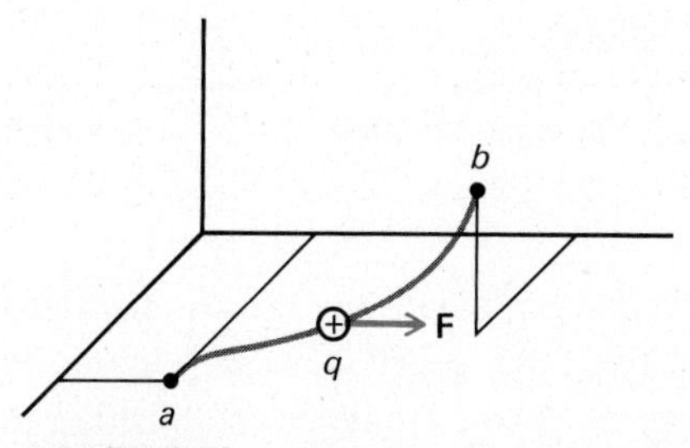

FIGURE 8.10

A charge q is moved from a to b by the force **F** which does work W_{ab} against the electric field. The potential difference is $V_{ab} = W_{ab}/q$.

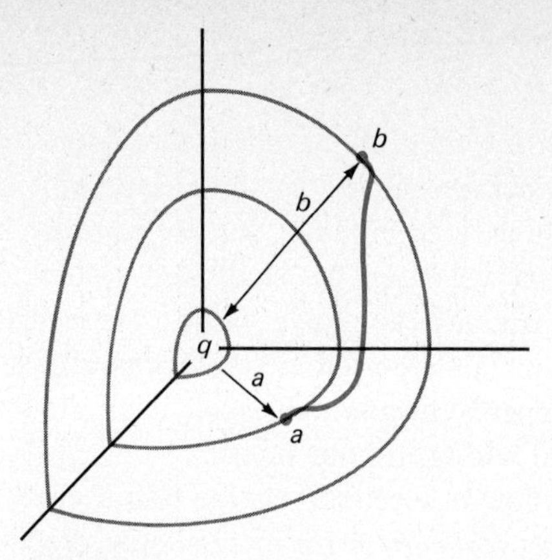

FIGURE 8.11
The potential difference V_{ab} between point a, a distance a from the charge q, and point b, a distance b from the charge q, is given by $V_{ab} = \left(\frac{q}{b} - \frac{q}{a}\right)$.

can extract work during the process and the charge q loses electrostatic potential energy.

Suppose that it requires 12 joules of external work to move a charge of 2 coulombs from terminal a to terminal b of a battery. The potential difference between the terminals of the battery is then given by $V_{ab} = 12/2 = 6$ V. If the battery were allowed to move the same charge from terminal b to terminal a, the work done on the charge would be 12 joules. When we say that the potential difference available in household electrical wiring is 110 V, we mean that 110 joules of work are done on each coulomb of electricity that is moved from one terminal of an outlet to the other. This amount of work is converted into some form of energy, such as heat or light, by the appliance connected between the two terminals of the electric outlet.

Next let us consider the potential difference between a distance a from a charge q to a greater distance b from this charge. The situation is illustrated in Fig. 8.11. The calculation involves mathematics beyond the scope of this book, but the result is as follows:

$$V_{ab} = K\left(\frac{q}{b} - \frac{q}{a}\right) \tag{8.7}$$

Because b is greater than a, V_{ab} is negative, which means that we can extract work from the system by letting a positive charge move from a to b.

8.3 ELECTRIC CURRENTS

As we mentioned in the preceding section, most practical applications of electricity involve moving charges. In some important cases, the charges are accelerated, as in a radio antenna, but we consider here only charges moving with constant velocity. We shall also restrict our discussion to materials in which charges can move easily. These materials are called *conductors*. Usually metals are used as conductors, copper and silver being the best. Because copper is cheaper than silver, it is undoubtedly the material used in the electrical wiring in your house.

The rate of flow of electrical charge past a given point per unit of time is called the *electrical current* (or more simply, the current). The mks* unit of current is the ampere (A), which is defined in Sec. 9.3. Other units of current in common use are the milliampere (10^{-3} A or mA) and the microampere (10^{-6} A). A 100-watt electric light bulb carries a current of about 1 A; an electric stove may use a current of 20 A; and a hearing aid might carry a current of 10 mA.

When we apply a potential difference between the ends of a conductor (perhaps by using a battery), work is done on the charges in the conductor and forces are exerted on these charges. We would expect that the greater the potential difference applied, the greater the force on each charge. Thus, we would assume that a greater applied potential difference would lead to a more rapid flow of a charge and consequently to a greater current. Apparently, about 1800, Henry Cavendish investigated the connection between the potential difference applied between the ends of a conductor and the resulting current. However, he did not publish his results. Therefore, the relation between these two quantities is named after the German physicist, G. S. Ohm. About 1840, Ohm found that for metals (good conductors) the current I was proportional to the applied potential difference V. Ohm's Law can be written as follows (if we introduce a constant of proportionality R):

$$V = IR \tag{8.8}$$

R is found to be almost exactly a constant for pure metals and is called the *electrical resistance* (or more simply, the resistance) of the conductor on which the measurements are made.

When we solve Eq. (8.8) for R, we find

$$R = \frac{V}{I} \tag{8.9}$$

* Meter-kilogram-second system of measurement.

Because potential difference in our system of units (mks) is measured in volts and current in amperes, the unit of resistance is the volts per ampere, which is called the *ohm*. Here you should note that many electrical devices, such as vacuum tubes, transistors, and motors, do not obey Ohm's Law. Those devices that do obey Ohm's Law are often called *ohmic*. Figure 8.12 shows a typical series circuit.

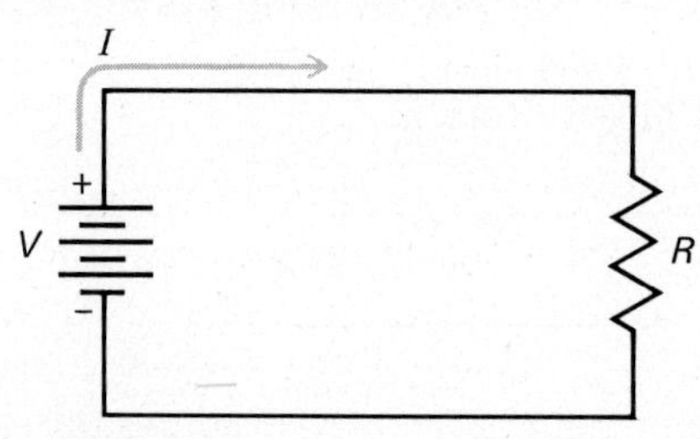

FIGURE 8.12
A simple series circuit with a battery driving current through a resistor. The voltage V of the battery, the current I through the resistor, and the resistance R are related by $V = IR$.

Suppose that a potential difference V is applied between the ends of a conductor and that a current I flows through the conductor. During a time t a charge Q flows through the conductor, according to Eq. (8.2). The work done by the source of potential difference is then given by Eq. (8.6) as

$$W = QV = (It)V \tag{8.10}$$

If we divide both sides of Eq. (8.10) by the time t, we obtain the work per unit time, which we defined in Sec. 3.4 as the power delivered. We have then

$$P = \frac{W}{t} = IV \tag{8.11}$$

The relation given in Eq. (8.11) does not depend on the conductor obeying Ohm's Law; instead, the relation comes directly from the definitions of work, power, potential difference, and current. Equation (8.11) holds therefore for *any* electrical device that conducts a current I when a potential difference V is applied across it. Current is measured in coulombs per second and

potential difference is measured in joules per coulomb; therefore, the power is expressed in watts, as shown below:

$$P = IV = (\text{coulombs/second}) \times (\text{joules/coulomb})$$
$$= \text{joules/second} = \text{watts}$$

If the conductor obeys Ohm's Law, so that it is said to be ohmic, we can combine Eq. (8.8) with Eq. (8.11) to get two additional expressions for the power used in a conductor. If we substitute the value of V from Eq. (8.8) into Eq. (8.11), we find

$$P = I^2R \tag{8.12}$$

Because $I = V/R$ from Eq. (8.8), if we substitute this value for I into Eq. (8.11), we get

$$P = \frac{V^2}{R} \tag{8.13}$$

We must point out that Eq. (8.11) applies to any conductor of electricity, good or bad, ohmic or not, whereas Eq. (8.12) and Eq. (8.13) apply only to conductors that obey Ohm's Law.

EXAMPLE Let us consider an electric light bulb that uses 100 W when connected across 110 V. From Eq. (8.11), we see that the current conducted by this light bulb is given by

$$I = \frac{P}{V} = \frac{100}{110} = 0.909 \text{ A}$$

If we assume that the filament inside the bulb obeys Ohm's Law, we can write

$$R = \frac{V}{I} = \frac{110}{0.909} = 121 \text{ ohms}$$

Note that the resistance of the filament could just as well have been obtained from Eq. (8.13) if we had written

$$R = \frac{V^2}{P} = \frac{110^2}{100} = 121 \text{ ohms}$$

EXAMPLE Suppose that an automobile battery delivers 150 A at a potential difference of 12 V when the car is being started. Under these conditions the power used is given by

$$P = VI = 12 \times 150 = 1800 \text{ W}$$

To start the car, the starter motor and the rest of the circuit connected to the battery must have a resistance given by the following equation:

$$R = \frac{V}{I} = \frac{12}{150} = 0.0800 \text{ ohms}$$

You can see, therefore, that a poor connection at one of the terminals of the battery can keep the battery from delivering enough current and power to start the engine of the car.

The bills that electrical utility companies send their customers are usually expressed in terms of kilowatt-hours (kWhr). We define the kilowatt-hour as the amount of work done when energy is delivered at the rate of 1 kW (1000 W) for 1 hr. Thus, 1 kWhr = 1000 W-hr = 3,600,000 W-sec = 3,600,000 joules. This very large amount of work or energy is delivered to the consumer at a price of approximately 5 cents, depending on his location. If you run a 100-W bulb for 5 hr, you use up $0.1 \times 5 = 0.5$ kWhr of electrical energy, which would cost you several cents. Similarly, if an electric clock draws 2 W of power, in a month of 30 days it would use an amount of electrical energy given by $2 \times (24 \times 30)/1000 = 1.44$ kWhr. Operating this clock would cost about 7 cents. On the other hand, if the broiler of your oven uses 2000 W and you roast a turkey for 5 hr, you would use $(2000 \times 5)/1000 = 10$ kWhr. The cost would be about 50 cents. In all of these cases, however, the cost of the electrical energy used is usually very small compared to the cost of other sources of energy.

DISCUSSION QUESTIONS

1. Electrostatic experiments usually work poorly on days when the humidity is high. Why?

2. When you comb your hair in dry weather, the comb will attract small pieces of paper or lint, even though they are not charged. Explain how this attraction can happen.

3. Is it possible to arrange two electrical charges in such a way that at a minimum of one point the net electrical field produced is zero? If so, sketch the arrangement and state any other necessary conditions.

4. Prove that the electrical field inside a perfect conductor (a material in which the charges are completely free to move) is zero.

5. Prove that the electrical field at the surface of a perfect conductor must be at right angles to the surface of the conductor.

6. Justify the statement that if any charge is put on a perfect conductor, it will reside entirely on the surface of the conductor.

7. The electron is negatively charged. Will it move in the direction of the electrical field or in the opposite direction? Will the electron move towards regions of high or low potential?

8. Suppose that the potential difference is zero between all points in a certain region. Can you say anything about the electrical field in this region?

9. Consider two electric light bulbs rated at 60 W and 100 W, respectively, when operated on 115 V. Which has the higher resistance?

10. On a dry day it is possible to generate a potential difference greater than 10,000 V merely by walking along a rug. Explain why this voltage is not dangerous, whereas an ordinary household electrical outlet furnishing a potential difference of 115 V may be lethal.

PROBLEMS

1. By what factor is the force between two charges changed if the distance between them is doubled? (*Ans.:* $\frac{1}{4}$)

2. If each of two charges is tripled, their separation remaining the same, by what factor is the force between them changed?

3. Compute the charge passing a point when a current of 5 A flows for 2 min. (*Ans.:* 600 coulombs)

4. Compute the force between a charge of -5 microcoulombs and a charge of -10 microcoulombs separated by a distance of 5 cm. Is the force an attraction or a repulsion?

5. Compute the magnitude of a charge that will repel an equal charge with a force of 10 N when the charges are 0.1 m apart. (*Ans.:* 3.33×10^{-6} coulombs)

6. The electric field of the earth is approximately 100 N/coulomb. Compute the force exerted by this electric field on a charge of 5×10^{-6} coulombs.

7. Take the radius of the earth as 7000 km and assume that the earth's electric field is caused by a point charge placed at the earth's center. Compute the magnitude of the charge that would be required to produce an electric field of 100 N/coulomb at the surface of the earth. (*Ans.:* 5.44×10^{5} coulombs)

8. Compute the electric field produced at a distance of 0.2 m from an electric charge with a magnitude of 2×10^{-6} coulombs.

9. If the work required to move a charge of 3 coulombs between two points is 60 joules, what is the potential difference between these two points? (*Ans.:* 20 V)

10. By what factor must the distance between two charges be changed if the force between them is to be tripled? (*Ans.:* 0.577)

11. Suppose that each of two charges is quadrupled, while the distance between them is doubled. By what factor is the force between the charges changed?

12. Compute the time a current of 2 A would have to flow past a given point so that the total charge passing the point is 300 coulombs. (*Ans.:* 150 sec)

13. If a charge of 600 microcoulombs flows out of a capacitor during a time of 10^{-3} sec, compute the average current flowing out of the capacitor during this time.

14. Compute the force between a charge of 4 microcoulombs and a charge of -6 microcoulombs when they are separated by a distance of 8 cm. Is the force an attraction or repulsion? (*Ans.:* 3.37×10^{1} N; attraction)

15. How much work is needed to move a charge of 4 coulombs from one point to another if the potential difference between the two points is 50 V?

16. A 100-W electric light bulb uses a current of approximately 1 A. Compute the charge passing through the light bulb during 20 sec. (*Ans.:* 20 coulombs)

17. If a light bulb uses 1 A of current when it is connected across a potential difference of 115 V, compute the electrical resistance of the light bulb.

    How much power does the light bulb use? (*Ans.:* 115 W)

18. If the oven of an electric stove uses 3000 W at a voltage of 220 V, compute the current used by the oven.

    If your power company charges 3 cents/kWhr, how much would it cost to operate the oven of this stove for half an hour? (*Ans.:* 4.5 cents)

19. How much work is required to move a charge of 5 coulombs between two points if the potential difference between these points is 100 V?

20. The work to move a certain charge between two points where there is a potential difference of 12 V is found to be 24 joules. Compute the magnitude of the charge. (*Ans.:* 2 coulombs)

21. If one burner of an electric stove uses 500 W of power, compute the current drawn by this burner at a voltage of 110 V.

    Compute the resistance of the burner. (*Ans.:* 24.2 ohms)

    If your power company charges 2 cents/kWhr, how much would it cost to operate this burner for 3 hr?

22. A flashlight draws 10 mA from a 3-V battery. How much power is used by the flashlight? (*Ans.:* 3×10^{-2} W)

23. If two equal charges repel one another with a force of 5 N when they are 0.2 m apart, what is the magnitude of each charge? Are both charges of the same sign?

24. How far away from a given point must a charge of 6×10^{-6} coulombs be to produce an electric field of 3 N/coulomb? (*Ans.:* 134 m)

25. At a certain distance from a charge of 5×10^{-6} coulombs, the electric field caused by this charge is 2 N/coulomb. How far is the point in question from the charge?

26. If the work required to move a charge of 5 coulombs from one point to another is 300 joules, what is the potential difference between these points? (*Ans.:* 60 V)

27. Resistors are said to be connected in *series* when they are connected as shown in Fig. 8.13. In this case, exactly the same current flows through each resistor.

    Show that the current delivered by the battery is given by

$$I = \frac{V}{R_1 + R_2 + R_3}$$

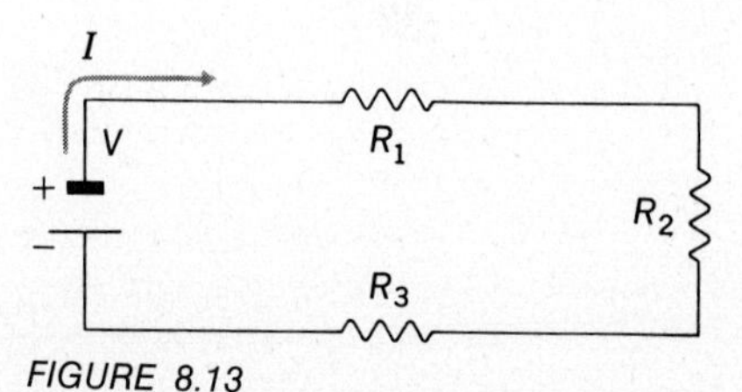

FIGURE 8.13

As far as the battery is concerned, it is delivering current to an *equivalent resistance, R,* given by

$$R = R_1 + R_2 + R_3$$

28. When resistors are connected so that their end points are joined by perfect conductors, we say that they are connected in *parallel.* In this situation, exactly the same potential difference exists across each resistor. The circuit is shown in Fig. 8.14. Show that the current through each resistor is given by one of the expressions:

$$V = I_1R_1 \qquad V = I_2R_2 \qquad V = I_3R_3$$

From the preceding equations, show that the total current delivered by the battery is

$$\begin{aligned} I &= I_1 + I_2 + I_3 \\ &= \frac{V}{R_1} + \frac{V}{R_2} + \frac{V}{R_3} \\ &= V\left(\frac{1}{R_1} + \frac{1}{R_2} + \frac{1}{R_3}\right) \end{aligned}$$

Thus, as far as the battery is concerned, show that it is delivering current to an *equivalent resistance R* given by

$$\frac{1}{R} = \frac{1}{R_1} + \frac{1}{R_2} + \frac{1}{R_3}$$

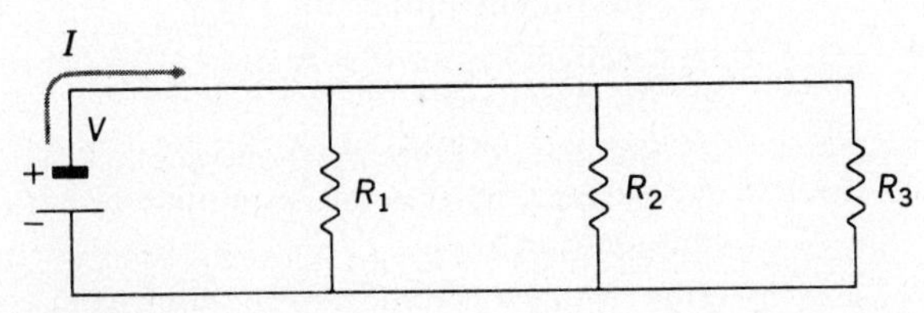

FIGURE 8.14

9

Magnetism

Magnets have been known to man since the dawn of history. At first, they were regarded as magical or mystical objects. Later they were used as navigational tools and then as objects of scientific study. The area of physics that studies and explains magnets and related phenomena is called *magnetism.* In this chapter we shall study magnetism and see how it relates to electricity.

9.1 MAGNETS

We have all played with magnets at one time or another. The forces that they exert on one another are rather complicated. Sometimes they attract each other and sometimes they repel. Most magnets have two ends, which are called north and south because the north end always points to the north pole of the earth when the magnets are hung on a string or floated in water with a piece of cork. These poles behave somewhat like electric charges; that is, like poles repel and unlike poles attract. Furthermore, if the relative orientation of two magnets is kept fixed and they are slowly separated, the force between them will decrease by $1/r^2$ (where r is the separation between them).

But there is no other resemblance. If you were to cut a long magnet such as the one in Fig. 9.1 into two pieces, you would not get one north

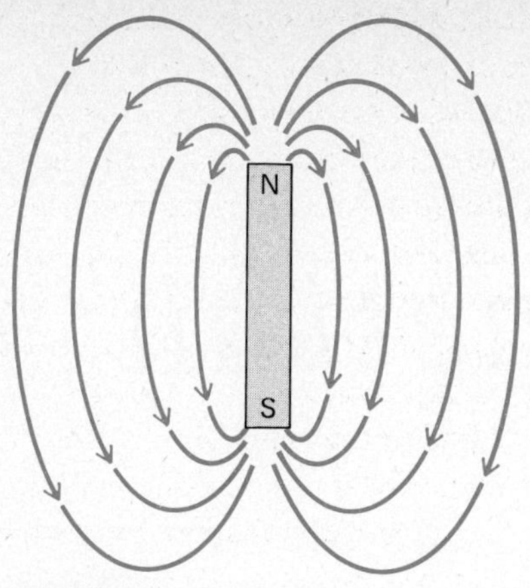

FIGURE 9.1

The magnetic field of a bar magnet. The colored lines represent the magnetic field defined in Sec. 9.2. Notice how the flux lines enter at the south pole and exit at the north pole.

pole and one south pole; instead, you would get two complete magnets as shown in Figs. 9.2 and 9.3. If you were to cut them again, you would get more complete magnets. However small you divided the magnets, you would continue to make smaller magnets until you were left with dust. There is another difference between magnets and charges. Unlike a charge, which is strictly conserved, magnets can be created and destroyed. If you heat a

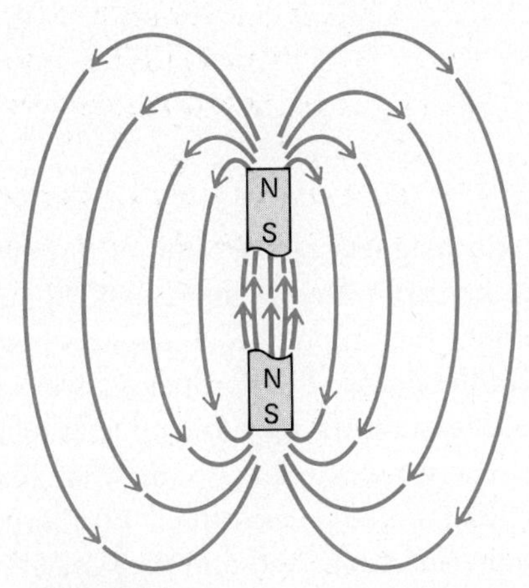

FIGURE 9.2

The magnet of Fig. 9.1 cut into two pieces. The separate parts behave exactly like two smaller magnets. They would attract each other since north poles attract south poles.

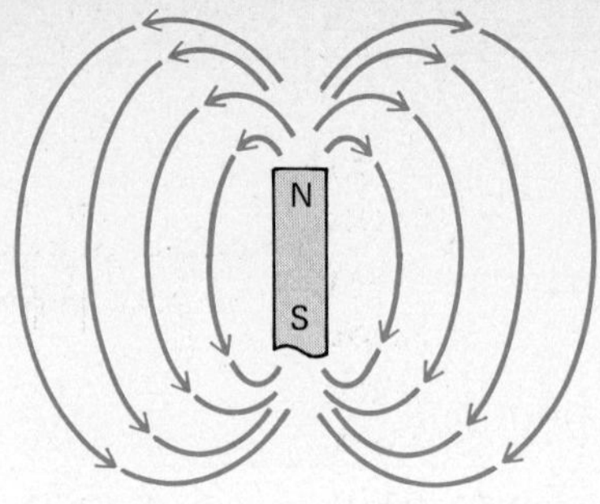

FIGURE 9.3

One of the two pieces from Fig. 9.2 isolated. The magnetic field is exactly the same as Fig. 9.1 for the original magnet.

permanent magnet above a certain critical temperature, called the *Curie temperature,* it loses all of its magnetic properties and behaves like an ordinary piece of iron. If you place a piece of iron in contact with a strong permanent magnet, the iron becomes a weak magnet. This characteristic explains why magnets attract certain materials called *magnetic materials.* If a permanent magnet's north pole is brought near a magnetic material such as iron, it induces a south pole in the part of the iron nearest it and a north pole in the opposite part as shown in Fig. 9.4. The induced south pole then is attracted to the permanent magnet. Similarly, the south pole

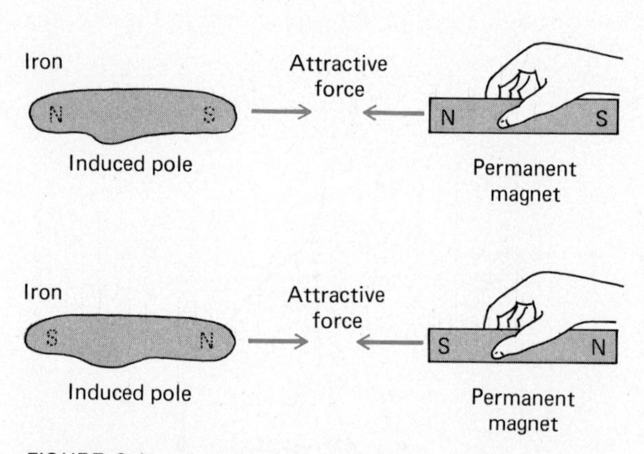

FIGURE 9.4

A permanent magnet is brought near a piece of iron. In the top picture the magnet's north pole induces a south pole in the iron and they attract. In the lower picture the magnet's south pole induces a north pole in the same piece of iron and they again attract.

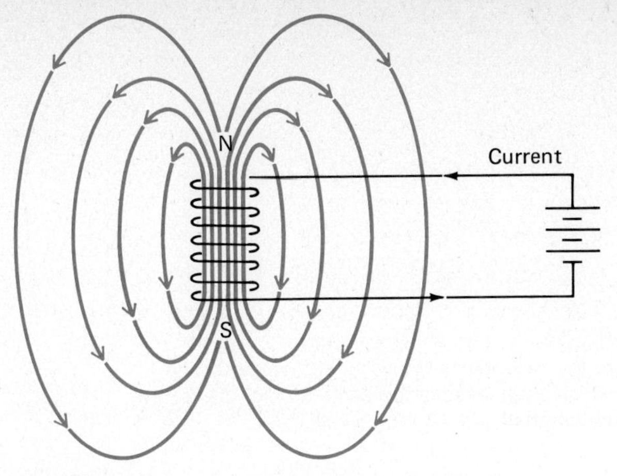

FIGURE 9.5
An artificial magnet formed by passing current through a coil of wire called a solenoid. Compare the magnetic field pattern with that of Fig. 9.1 for a bar magnet.

of a permanent magnet induces a temporary north pole in a magnetic material brought near it and again attracts the material. Note that there is no isolated object called a *magnetic pole.* A magnetic pole is just one end of a magnet.

Magnetism obviously is not an intrinsic property of magnetic materials but is a property that they can acquire and lose. The first real progress in understanding magnetism came during the 19th century when physicists discovered how to make artificial magnets. They found that if electric current were run through a coil of wire, as shown in Fig. 9.5, the coil would behave

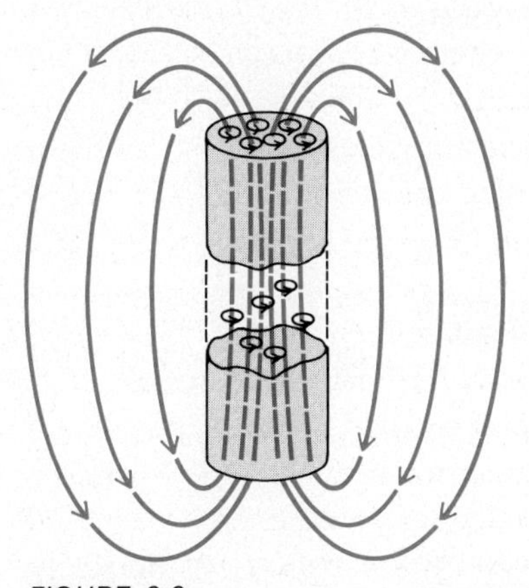

FIGURE 9.6
The magnetic field of a permanent magnet is actually caused by microscopic current loops in the molecules of the magnet.

exactly like a magnet. The ends of the coil would act like magnetic poles. If you had such a coil, you could locate its north pole by wrapping the fingers of your right hand around the coil with your fingers pointing in the direction of the current flow. The end of the coil that your thumb pointed at would always be the north pole of the magnet. Magnets produced in this way are called *electromagnets,* and the coil of wire is called a *solenoid.* We now know that all magnetic effects are produced by currents or, equivalently, by moving charges. Permanent magnets as in Fig. 9.6 are also created by current loops. The loops are microscopic in size and exist in the individual molecules of the material. All the properties of permanent magnets can be understood in terms of these microscopic current loops, although this explanation is beyond the scope of this book.

9.2 MAGNETIC FIELDS

Magnetic forces are caused by currents or moving charges exerting forces on other magnets. As in the case of electricity, we can divide this action-at-a-distance effect into two stages. We say that the electric currents produce a *magnetic field* and that this magnetic field exerts a force on a charge moving in it. Permanent magnets, solenoids, and moving charges all produce magnetic fields. These magnetic fields then exert forces on other objects in a manner we shall define shortly. Clearly, the magnetic field is intimately related to the electric field. A moving charge will produce both an electric field and a magnetic field. However, to an observer moving along with the charge, it will appear to be at rest and will seem to have only a static electric field. Thus, a field that seems to be purely electric to one observer (the one who sees the charge as being at rest) may seem to another observer (the one moving with respect to the charge) to be both electric and magnetic. Electric and magnetic fields are actually different manifestations of the same underlying structure, called the *electromagnetic field.* Whether the electromagnetic field appears to be purely electric, purely magnetic, or a mixture of both depends partly on the nature of the source charges producing the fields and partly on the relative motion of the observer and the sources. However, for a given observer, the splitting of the electromagnetic field into an electric part and a magnetic part is always clear. We shall continue to treat these fields as distinct for the remainder of this chapter.

In the previous chapter we defined the electric field in terms of the force exerted on a *stationary* test charge. We shall retain this definition. Thus, when a stationary charge experiences no force, we say that there is no electric field present. Experimentally, it is found that an uncharged metallic conductor (a wire) with a constant current flowing through it

produces no electric field. This statement is also true for permanent magnets. The field of these systems is purely magnetic and is able to exert a force only on charges in motion. This force is the one that we describe here.

Consider a charge moving in a region of space with a magnetic field and no electric field. The magnetic field might be produced by a permanent magnet or by electric current flowing in a wire, perhaps a solenoid. The charge will, in general, experience a force, called the *magnetic force.* This force depends on several variables, including the magnitude of the charge, its speed, and the direction in which it is moving. At any given point in space there is always one particular direction in which the charge can move so that the force on the charge is zero, regardless of its speed. Motion in any other direction will always produce a force. We define the *direction* of the magnetic field at this point to be in this particular direction, as shown in Fig. 9.7. If we now let the charge move at right angles to the direction in which it experienced zero force (i.e., at right angles to the magnetic field), the force on the charge will become a maximum, which we call F_{max}. This situation is illustrated in Fig. 9.8. The magnitude of F_{max} is found experimentally to be proportional to the magnitude of the charge and to its speed. If the magnitude of the charge is Q and its speed is v, we define the *magnitude* of the magnetic field B by the equation

$$B = \frac{F_{max}}{Qv} \tag{9.1}$$

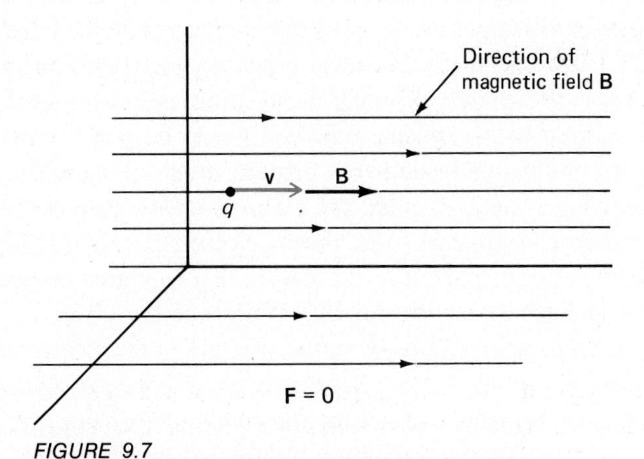

FIGURE 9.7

In this picture the charge **q** is moving with velocity **v** parallel to the magnetic field **B**. There is no force on the charge. Motion in any other direction will produce some force.

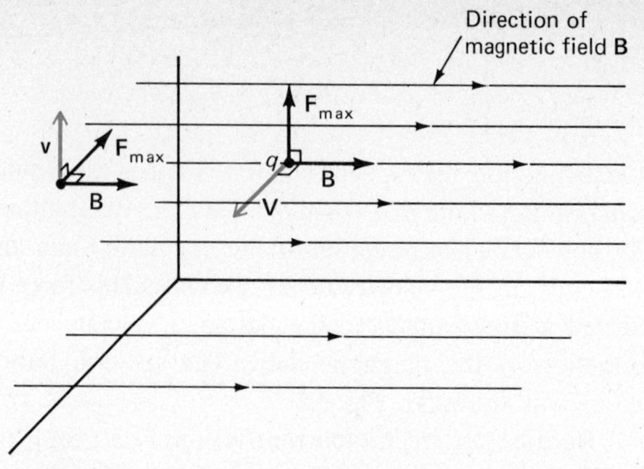

FIGURE 9.8
When the charge moves in a direction perpendicular to the magnetic field—that is, perpendicular to the direction for which the force is zero—the force on it is maximum. Two possible choices for this are shown. Notice that the force is always perpendicular to both the velocity vector and the magnetic field vector.

In the mks system of units used in this book, force is measured in newtons, charge in coulombs, and speed in meters per second. We deduce, therefore, that the unit of *magnetic field strength* (or intensity) is a newton/coulomb-meter/second. Because of the importance of magnetic fields, this unit is given a name, the tesla (T). Sometimes, particularly in older books, this unit is called the weber per square meter. Another common unit for magnetic field strength is the gauss, which equals 10^{-4} tesla. It is found in books using cgs* units. In order to give you some idea about the magnitudes of magnetic fields, the magnetic field of the earth is about 5×10^{-5} T, whereas an electromagnet may produce a magnetic field of 1 T.

We can now find the force on a charge moving at right angles to a known magnetic field by using Eq. (9.1) rewritten as

$$F_{max} = QvB \tag{9.2}$$

This equation is actually a special case of a more general formula that gives the force on a charge moving in any direction in a magnetic field. If a charge Q is moving with speed v in a magnetic field B, and if the angle between the direction of motion of the charge and the magnetic field is θ, then the

*Centimeter-gram-second system of measurement.

magnitude of the force is given by

$$F = QvB \sin \theta$$

Sin θ is a function that is zero when $\theta = 0$ deg, or when $\theta = 180$ deg, and has a maximum value of 1 when $\theta = 90$ deg. We shall not use this formula further however. The direction of the magnetic force on a moving charge also depends on the velocity of the particle. This force is always found to be exerted at right angles to the particle's velocity and at right angles to the direction of the magnetic field. The relation between these various directions was shown in Fig. 9.8.

Because an electric current is simply a group of moving charges, a magnetic field exerts a force on a current. Where a current I is flowing in a straight conductor at right angles to a magnetic field of strength B, as shown in Fig. 9.9, we use the following equation to find the force on a length L of the conductor:

$$F = BLI \tag{9.3}$$

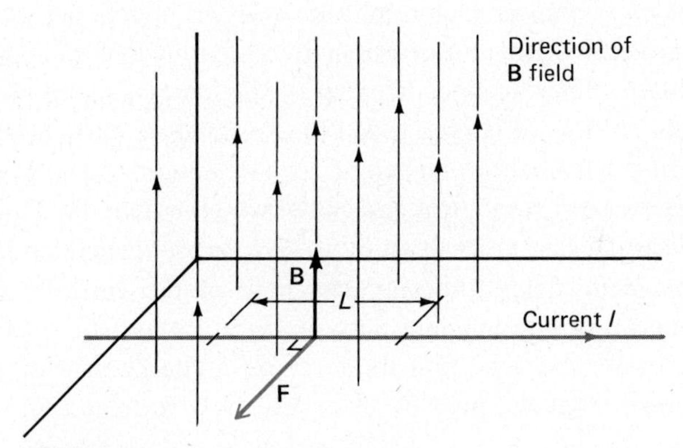

FIGURE 9.9
The force on a length L of a conductor carrying current I at right angles to a magnetic field.

In Eq. (9.3) we must measure the magnetic field in tesla, the length of the conductor in meters, and the current in amperes. Then the force on the conductor comes out in newtons.

EXAMPLE Let us assume that a short circuit in an electric power station produces a momentary current of 10^4 A in a conductor 2 m long at a point where the earth's magnetic field is 5×10^{-5} T. The current, of course, flows at right angles to the conductor.

From Eq. (9.3) we see that the force on the conductor can be calculated as

$$F = 5 \times 10^{-5} \times 2 \times 10^{4} = 1 \text{ N}$$

If the mass of the conductor happened to be 0.5 kg, then the conductor would experience an acceleration given by the equation

$$a = \frac{F}{m} = \frac{1}{0.5} = 2 \text{ m/sec}^2$$

The acceleration of gravity is about 9.8 m/sec^2, so a figure of 2 m/sec^2 is not a negligible acceleration nor is the magnetic force exerted on the conductor small compared to the force of gravity. In cases of short circuits, the magnetic forces may actually rip conductors from their mountings. Here it should be pointed out that forces of the same sort are used in electrical motors.

9.3 MAGNETIC FIELD OF A CURRENT

In the preceding section we discussed the action of magnetic fields and electric currents on moving charges. We now return to the problem of producing magnetic fields by means of electric currents. This process was first done experimentally by H. C. Oersted, J. B. Biot, and F. Savart during the early 19th century. They found that an electric current did in fact produce a magnetic field. In addition, this magnetic field had the same effect on a magnet or another conductor carrying an electric current as would the magnetic field set up by a magnet. For the case of a long, straight conductor carrying a current I, they found that the current established a magnetic field B at a distance d and at right angles to the conductor. The relation is shown in the following equation:

$$B = 2 \times 10^{-7} \frac{I}{d} \tag{9.4}$$

In Eq. (9.4) the magnetic field strength is given in tesla when the current is measured in amperes, and the perpendicular distance from the conductor to the point in question is measured in meters.

The magnetic field of a straight wire runs in circles around the wire. This field is illustrated in Fig. 9.10. The direction of the field is given by another right-hand rule. If you hold the wire in your right hand with your thumb pointing in the direction of the current, your fingers will point in the direction of the field.

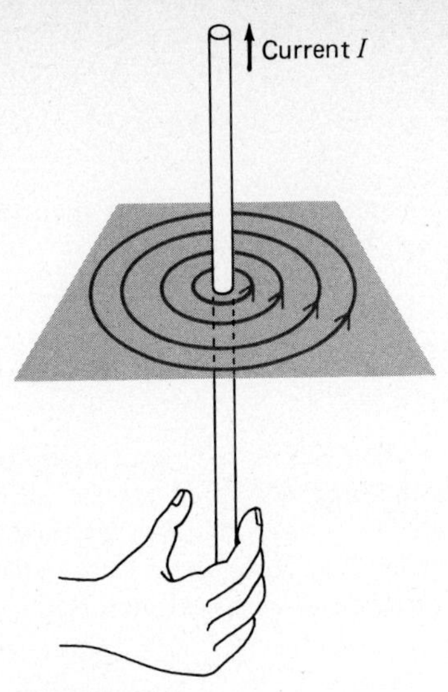

FIGURE 9.10
The magnetic field of a long straight wire.

EXAMPLE In Sec. 9.2 we calculated the force on a conductor that had the large current of 10^4 A. Now let us compute the magnetic field produced by such a current at a perpendicular distance of 1 cm from the conductor. From Eq. (9.4) we have

$$B = 2 \times 10^{-7} \frac{10^4}{10^{-2}} = 0.2 \text{ tesla}$$

A parallel conductor of length L m carrying a current of 10^4 amp also would experience, according to Eq. (9.3), a force of

$$F = 0.2 \times 1 \times 10^4 = 2 \times 10^3 \text{ N}$$

This force is almost certainly very large compared to the weight of the conductor. Thus, if both conductors carry such a large current due to a short circuit, the magnetic force exerted by one conductor on the other may be sufficient to pull one or the other of the conductors from its mountings.

We shall now calculate the force per unit length between a pair of long, parallel conductors separated by a distance d. Let the conductors carry currents I_1 and I_2, as shown in Fig. 9.11. Then the magnetic field produced by the current I_1 at the location of current I_2 is given by

$$B_1 = 2 \times 10^{-7} \frac{I_1}{d} \tag{9.5}$$

Combining Eq. (9.5) with Eq. (9.3), we find for the force per unit length

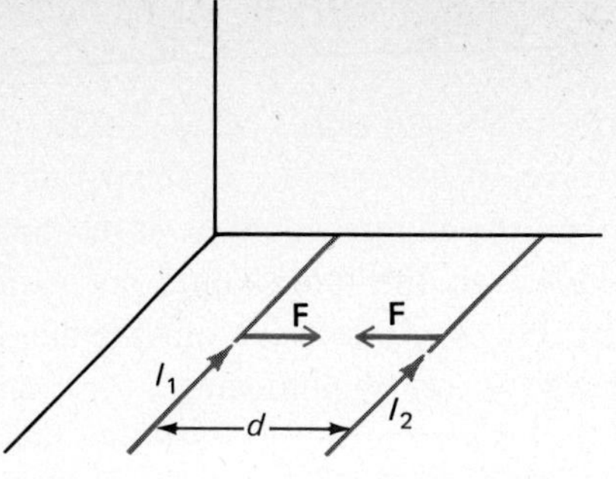

FIGURE 9.11
The attractive force between parallel wires carrying currents I_1 and I_2. If the currents were in opposite directions the force would be repulsive.

on the current I_2 the expression

$$F_2(\text{per unit length}) = B_1 I_2 = 2 \times 10^{-7} \frac{I_1 I_2}{d} \tag{9.6}$$

If the two currents, I_1 and I_2, are the same and are equal, say, to I, Eq. (9.6) takes the form

$$F(\text{per unit length}) = 2 \times 10^{-7} \frac{I^2}{d} \tag{9.7}$$

Equation (9.7) provides the definition of the ampere, from which the coulomb and other electromagnetic units are defined. We define the *ampere* as the current that causes a force of 2×10^{-7} N/m when flowing in long, parallel conductors spaced 1 m apart. As is often the case, practical measurements of currents measured in amperes use indirect means derived from Eq. (9.7).

EXAMPLE Let us calculate, for instance, the force per meter between two long, parallel conductors spaced 1 cm apart when each carries a current of 10^4 A. We have then

$$F(\text{per meter}) = 2 \times 10^{-7} \frac{10^4 \times 10^4}{10^{-2}} = 2 \times 10^3 \text{ N}$$

This result agrees with the earlier calculation in which we first found the magnetic field produced by one conductor at the location of a second conductor and then found the force exerted on the second conductor by this magnetic field. Naturally, the two different calculations concerning the same situation must give the same result.

9.4 INDUCED ELECTROMOTIVE FORCE

In the preceding sections we said that a charge moving through a magnetic field experienced a force. If we consider this experience from the point of view of the charge, then the charge would be at rest and the magnetic field would be moving. However, the force would not vanish just because we change our viewpoint. We conclude that a moving magnet must have some sort of effect on a charge at rest by producing a force on it, and thus setting up an electric field. In this section we shall see that this inference is correct and can be generalized.

About 1831, Michael Faraday noticed that a current would flow in the coil during the time that the nearby magnet was being moved about (as shown in Fig. 9.12). A year earlier an American, Joseph Henry, had observed that changing a current in one coil produced a momentary current nearby in a second coil. The experimental arrangement is shown in Fig. 9.13. We see that the common feature in these two experiments is the production of a current in a coil whenever the coil is subjected to a *changing* magnetic field. In Faraday's experiment, the motion of the magnet made the magnetic field at the location of the coil change; whereas according to Henry's observations, the varying current in the first coil produced a changing magnetic field in the vicinity of the second coil. Although Henry published his results earlier than Faraday did, this discovery of the relation between a changing magnetic field and its effect on charges is credited to Faraday, an Englishman, because at this time the United States commanded very little attention from the world's scientists.

Before continuing the discussion of these experiments, we must digress somewhat. As we said in Sec. 8.2, work is required to move a charge. We defined the potential difference between two points as the work needed to move a unit positive charge from one point to the other. Algebraically, the definition of potential difference (or voltage) is given by Eq. (8.6). If we do work to move a charge through a conductor, the work is converted

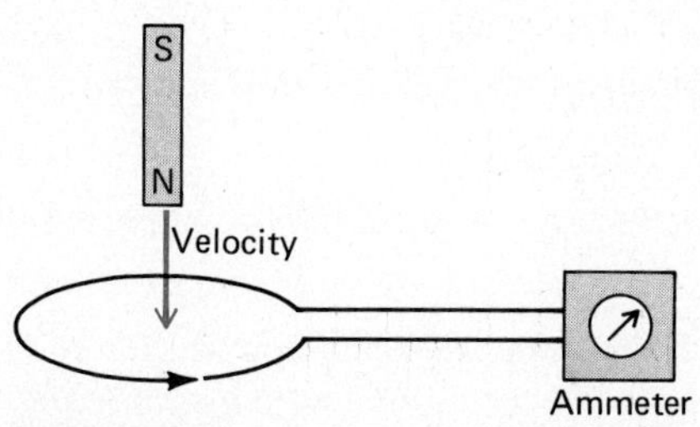

FIGURE 9.12
When a magnet is moved through a wire loop it induces a current which is measured by the ammeter.

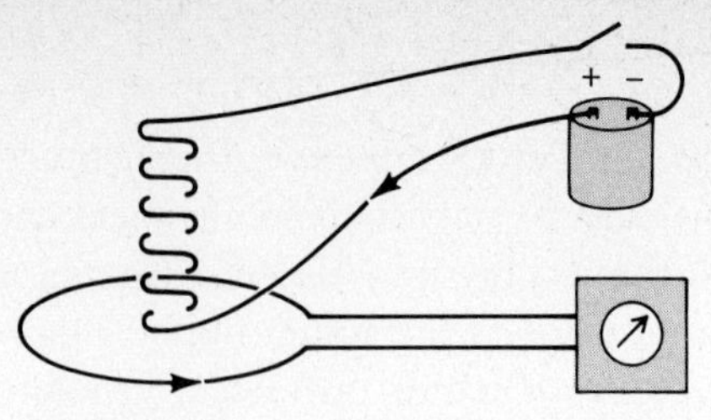

FIGURE 9.13
When the switch is closed current will flow through the solenoid. An induced current then will appear in the loop.

into heat and other forms of energy, such as mechanical energy. According to the Second Law of Thermodynamics (treated in Sec. 7.2), not all of the heat energy thus produced can be converted back into electrical or other forms of energy. We say therefore that such a process is *irreversible.* In some cases, at least ideally, no heat is involved, and we can completely transform electrical energy into some other form (or forms) of energy. Furthermore, we can entirely transform the other forms of energy into electrical energy, at least in principle, if heat energy plays no part. Processes in which a mutual conversion is possible between electrical energy and some other form of energy are said to be *reversible.* Because of such effects as electrical heating and mechanical friction, no process in nature is truly reversible, but we can often approach this ideal situation very closely.

When a device produces a potential difference between its terminals, it is said to be a source of *electromotive force,* which is often abbreviated emf. (Note that the word "force" is not used correctly. The device must be able to reverse work or energy, not force. However, this misuse of the word is the common practice.) If a device transforms an amount of energy W into electrical energy when a charge Q passes through the device, or if the device transforms an amount of electrical energy W into some other form of energy (not heat energy) when a charge Q passes through the device, we say that the device is a source of electromotive force $\mathcal{E}$ of a magnitude given by the equation

$$\mathcal{E} = \frac{W}{Q} \tag{9.8}$$

If the work or energy is measured in joules and the charge is measured in coulombs, then the magnitude of the emf is measured in volts. We see from the definition that a source of electromotive force always produces a

potential difference, but a potential difference is not always due to a source of electromotive force.

Perhaps the simplest example of an emf source is the car battery. When the battery is used to start the car, chemical energy in the battery is converted (almost) reversibly into electrical energy, which is then converted into mechanical work by the starter motor. Once the car is running, mechanical energy from the motor operates a generator, which produces electrical energy. This electrical energy is then converted into chemical energy and stored in the battery, so that the battery is ready to start the car another time. A second example of an emf source is the motor or generator. When electrical energy is supplied to a motor, the motor converts this energy into mechanical energy. In reverse, if mechanical energy is supplied to the same device, electrical energy is produced. This procedure takes place in a generator. In both the battery and the motor, the processes are not completely reversible because some of the energy is converted into heat by electrical resistance or mechanical friction. Nevertheless, the concept of a reversible device can be very useful.

Although the experiments of Faraday and Henry seem superficially different, all of them can be explained in the same way if we introduce one more idea. Suppose that we have an area A in a plane and the area is bounded by a conductor. This area is shown in Fig. 9.14. Let the magnetic field B be uniform and at right angles to the area. Then we define the *magnetic flux* f through the area bounded by the conductor in the equation

$$f = BA \tag{9.9}$$

Because the magnetic field is measured in tesla and the area in square meters, magnetic flux is measured in tesla–meters squared or webers. Although it will not concern us in this book, if the magnetic field is not at right angles to the area bounded by the conductor, we use the component (part) of the magnetic field that is at right angles to the area in calculating the flux from Eq. (9.9).

EXAMPLE The vertical part of the earth's magnetic field is about 5×10^{-5} tesla. A typical house of rectangular shape might be 5 by 10 m, so that it has an area of 50 m^2. The magnetic flux through a horizontal floor of this house would then be given by

$$f = (5 \times 10^{-5})50 = 2.5 \times 10^{-3} \text{ webers}$$

Similarly, a large electromagnet might produce a magnetic field of 1.5 webers/m^2 throughout an area of 20 m^2; therefore, the total magnetic flux through this area would be

$$f = 1.5 \times 20 = 30 \text{ webers}$$

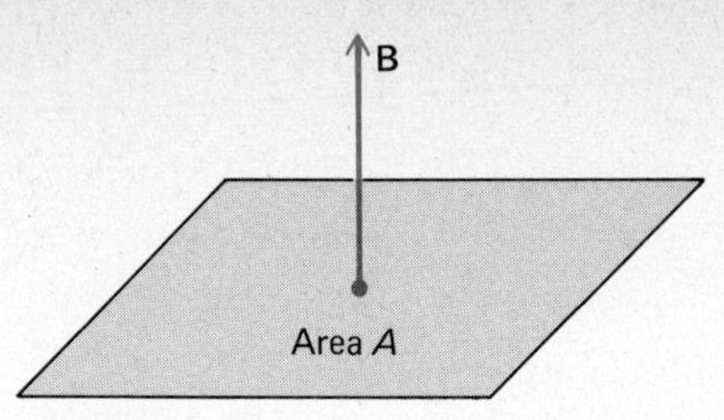

FIGURE 9.14

In a region of constant magnetic field, the flux through an area A oriented at right angles to the field is $f = AB$.

Now that we have defined electromotive force and magnetic flux, we can return to the discussion of the effect of a changing magnetic situation on electric charges. Suppose that the magnetic flux through a plane area A bounded by a conductor changes from f_1 to f_2 during a time Δt. The experiments of both Faraday and Henry can be combined into a single law stating that the average induced electromotive force $\mathcal{E}$ in the conductor is given by

$$\mathcal{E} = \frac{f_1 - f_2}{\Delta t} = -\frac{\Delta f}{\Delta t} \qquad (9.10)$$

In Faraday's experiment, the magnetic flux changed because the magnet was moved; whereas in Henry's experiment, the magnetic flux changed because the current in the first coil was varied. Generally, the magnetic flux through a closed conducting circuit may change whenever the magnitude of the magnetic field, the direction of the magnetic field, or the magnitude of the area enclosed changes. To summarize, whenever there is a change in the magnetic flux through a plane area, there is an electromotive force induced around the boundary of the region.

If we use the subscript 1 to designate quantities at the initial instant and the subscript 2 to designate quantities at an instant Δt later, Eq. (9.10) takes the form

$$\mathcal{E} = \frac{(BA)_1 - (BA)_2}{\Delta t} = -\frac{\Delta(BA)}{\Delta t} \qquad (9.11)$$

We see from Eq. (9.11) that the unit of electromotive force must be a unit of magnetic field strength multiplied by an area unit and divided by a time unit. In Sec. 9.2 we noted that 1 tesla = 1 N/coulomb–m/sec according to the definition of magnetic field strength given in Eq. (9.1). Thus, in the

system of units used in this book, electromotive force is measured in the following units:

$$\mathcal{E} = \frac{(\text{N/coulomb–m/sec})(\text{m}^2)}{\text{sec}} = \text{N–m/coulomb}$$
$$= \text{joules/coulomb} = \text{V}$$

We also see that the units of magnetic flux divided by time turn out to be volts, in agreement with our earlier discussion. This example shows the internal consistency of physical units.

EXAMPLE Let us now discover what happens when a coil consisting of 1000 turns, each having an area of 0.1 m^2, is rotated through 90 deg in a region in which there is a uniform magnetic field of 1.5 tesla. We assume that the initial orientation of the coil is at right angles to the magnetic field and that the final position of the coil has the plane of the coil parallel to the magnetic field. Figure 9.15 illustrates these ideas. Thus, there is no magnetic flux through the coil in the final position, because no part of the magnetic field has been at right angles to the plane of the coil. If the time required for this rotation is 0.1 sec, the average electromotive force induced in the coil can be calculated from Eq. (9.11). It has the following value:

$$\mathcal{E} = 1000 \frac{1.5 \times 0.1}{0.1} = 1500 \text{ V}$$

The factor 1000 appears in this calculation because there are 1000 turns in the coil and each turn has induced in it the same emf.

You might consider how different our whole world would be if induced electromotive forces could not be produced. Generators of electrical energy are devices in which large coils are rotated in magnetic fields, so

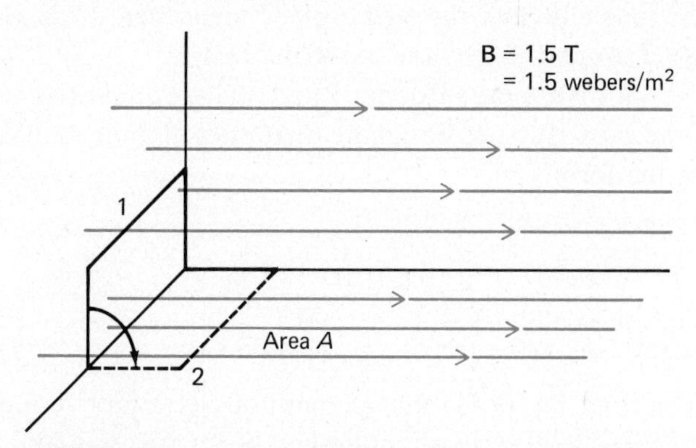

FIGURE 9.15
When the wire loop rotates from position 1 to position 2 the flux changes from *AB* to zero.

that electromotive forces are induced in the coils. Because energy is conserved in such a process, mechanical energy is transformed into electrical energy, although there are small losses due to electrical resistance and mechanical friction. This electrical energy is then sent for many miles through electrical conductors and converted back into various useful forms of energy. If this process were not possible, the sources of mechanical energy, such as coal, flowing water, and moving air, would have to be transported to the point at which this energy was needed. Certainly, ours is a society based upon electrical energy.

DISCUSSION QUESTIONS

1. Explain the difference between the terms "potential difference" and "electromotive force." Is it possible to have one without the other?

2. A positively charged particle moves through a region without being deflected from its straight line path. Can we say with confidence that in this region either the electric field or the magnetic field is zero, or both are zero?

3. A charged particle enters a region and is deflected sideways. What can we say about the magnetic and electric fields in this region?

4. Circular rings made of copper and wood are placed in the same varying magnetic field. Are the electromotive forces induced in the rings the same? Are the currents induced in the rings the same?

5. Two iron bars attract one another, no matter which ends are brought near. Do you conclude that both are magnetized? If you conclude that only one is a magnet, how would you tell which one was the magnet?

PROBLEMS

1. Calculate the magnetic field at a distance of 10 m from a power line carrying a current of 2000 A. What effect might this current have on a compass used below this power line?

2. Find the force per meter between two parallel conductors, each carrying a current of 1000 A and separated by a distance of 0.05 m.
(*Ans.:* 80 N/m)

3. If the earth's magnetic field of 5×10^{-5} T is at right angles to a coil with an area of 2 m^2, compute the magnetic flux through the coil.

If the coil is turned 90 deg to a position where the earth's magnetic field is parallel to the plane of the coil during a time of $\frac{1}{20}$ sec, compute the electromotive force induced in the coil. (*Ans.:* 2×10^{-3} V)

4. A coil with an area of 0.2 m^2 is to be rotated from a position in which the magnetic field is 2 T at right angles to the plane of the coil to a second position in which the magnetic field is parallel to the plane of the coil. Compute the time for this rotation if an electromotive force of 5 V is to be induced in the coil.

5. If the earth is thought of as a very large magnet, is the north pole of the earth a north or south magnetic pole?

6. A charge of 3 coulombs moves at a speed of 10 m/sec in a direction at right angles to a magnetic field of 0.5 T. Compute the force on the charge. (*Ans.:* 15 N)

7. Assume that the earth's magnetic field is 5×10^{-5} T. What current must flow in a conductor 10 m long at right angles to the earth's magnetic field if the conductor is to experience a force of 2 N?

8. A current of 10 A flows in a conductor 8 cm long that is at right angles to a magnetic field of 1.5 T. Compute the magnetic force on the conductor. (*Ans.:* 1.2 N)

9. Compute the magnitude of the current flowing in a conductor 5 m long and directed at right angles to the earth's magnetic field of 6×10^{-6} T so that the conductor will experience a force of 1.5 N.

10. A conductor of length 2 m carries a current of 5 A at right angles to a uniform magnetic field of 0.5 T. Compute the force on this conductor. (*Ans.:* 5 N)

11. A conductor with a mass per unit length of 5 gm/m is at right angles to a uniform magnetic field having a strength of 1.5 T. Compute the current that this conductor must carry so that the magnetic force will equal its weight.

12. Compute the speed at which a particle with a charge of 1.6×10^{-19} coulombs must travel at right angles to a magnetic field having a strength of 5×10^{-5} T so that the force on the particle will be 10^{-32} N.

13. A charge of 3.2×10^{-19} coulombs moves with a speed of 10^7 m/sec at right angles to a uniform magnetic field having a strength of 2 T. Compute the force on this charge. (*Ans.:* 6.4×10^{-12} N)

    If the mass of the particle is 6.8×10^{-27} kg, compute the acceleration of the particle.

14. Compute the magnetic field strength such that a particle with a charge of 1.6×10^{-19} coulombs will experience a force of 10^{-12} N when it moves with a speed of 10^9 cm/sec in a direction at right angles to the magnetic field. (*Ans.:* 6.25×10^{-1} T)

15. A conductor carries a current of 100 A. How far from this conductor will the magnetic field of the conductor be no more than 10^{-4} T? (*Ans.:* 0.2 m)

16. Two parallel conductors each carry a current of 2000 A. How far apart must they be for the force per unit length between them to be less than 0.5 N/m?

17. A square coil of 50 turns, each 10 cm on a side, is at right angles to the earth's magnetic field of 5×10^{-5} T. What is the total flux through the coil? (*Ans.:* 2.5×10^{-5} webers)

    If the coil is rotated 90 deg during 0.10 sec, what is the emf induced in the coil?

    What is the average current in the coil while it is being rotated if its resistance is 5 ohms? (*Ans.:* 5×10^{-5} A)

10 Wave Motion

Nearly everyone has sat on the seashore and watched waves roll in. A less evident fact is that the sound we hear and the light we see are also carried by waves. In this chapter we shall discuss the various aspects of waves. We begin with those properties of waves that are essentially the same for all types of wave motion.

10.1 GENERAL PROPERTIES OF WAVES

Even the most casual observation of water waves shows that these waves carry *energy*. Energy is a general characteristic of all waves and can be measured in terms of the energy transferred per second. In the metric units used in this book, the rate at which a wave delivers energy from a source to a receiver would be measured in joules/second or watts.

Here you should note that the material carrying the disturbance (for example, water) merely acts as a device for transferring energy from one point to another; the material itself is not permanently changed. Similarly, the noise and energy of a thunderclap are transmitted to our ears by the intervening air, without the air itself being much affected. In a less obvious way, energy is transferred from a radio transmitter to our receiver by means of radio waves. Neither device is changed by the waves.

If you watch water waves first on a calm day and then on a day on which the water is quite rough, you would decide that the higher the

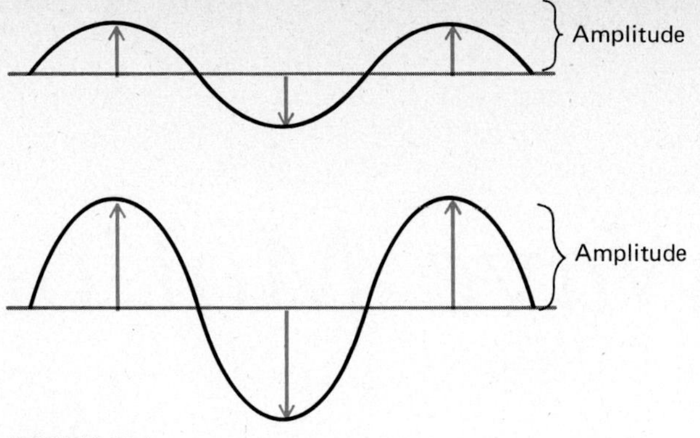

FIGURE 10.1

The amplitude of a wave is the distance from the average position to the crest of the wave as shown by the colored arrows.

waves, the more energy they brought in to shore. Suppose that you measure the disturbance of the water compared to its average or completely still level. The height of the wave's crest above the average water level or the depth of the wave's trough below the average water level is called the *amplitude* of the wave. Waves of two different amplitudes are pictured in Fig. 10.1. For simple waves, the energy carried by the wave per unit of time is proportional to the square of the amplitude of the wave. Thus, doubling the amplitude increases the energy carried by the wave by a factor of 4.

If you observe water waves from the edge of a dock and have a friend to help you make measurements, you can measure the distance between two successive crests or between two successive troughs, as shown in Fig. 10.2. This distance is called the *wavelength* λ of these waves and is measured in meters. Wavelengths of water waves may range from a very small fraction of a meter for ripples up to many meters for ocean waves.

Another obvious feature about water waves is the speed at which a given crest or trough travels past a fixed point, as illustrated in Fig. 10.3. By standing on a dock or an anchored boat, you can easily observe this

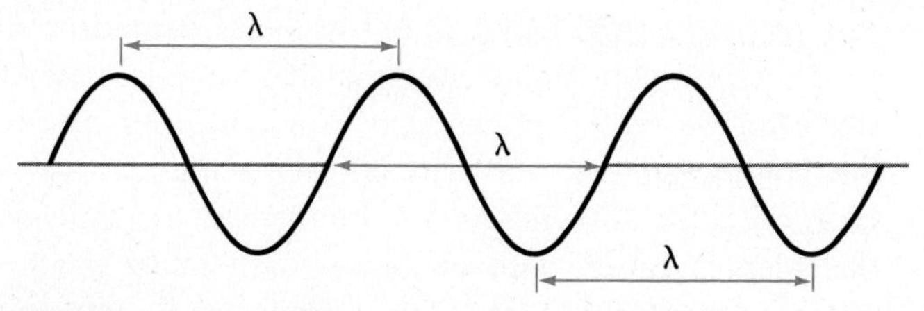

FIGURE 10.2

The wavelength of a wave is the distance between crests or the distance between troughs.

speed, which is known as the *wave velocity v*. It is measured in meters per second. As examples, water waves typically have velocities of a few meters per second; sound travels in air at a velocity of about 330 meters per second; and the velocity of a light wave is about 300 million meters per second. The velocity of a given type of wave depends primarily upon the material that transmits the wave, but it may also depend on other factors, such as the temperature.

If you continue to observe water waves from a fixed point, you can also measure the number of crests or troughs that pass per second. This quantity is known as the *frequency f* of the wave. Frequency is variously referred to in terms of waves per second, vibrations per second, or cycles per second (which is often abbreviated as cps). A recently adopted unit of frequency is the hertz (1 Hz = 1 cps). All three of these terms mean the same thing, namely, the number of waves passing a given point per second.

In the same situation, you can also measure the time interval

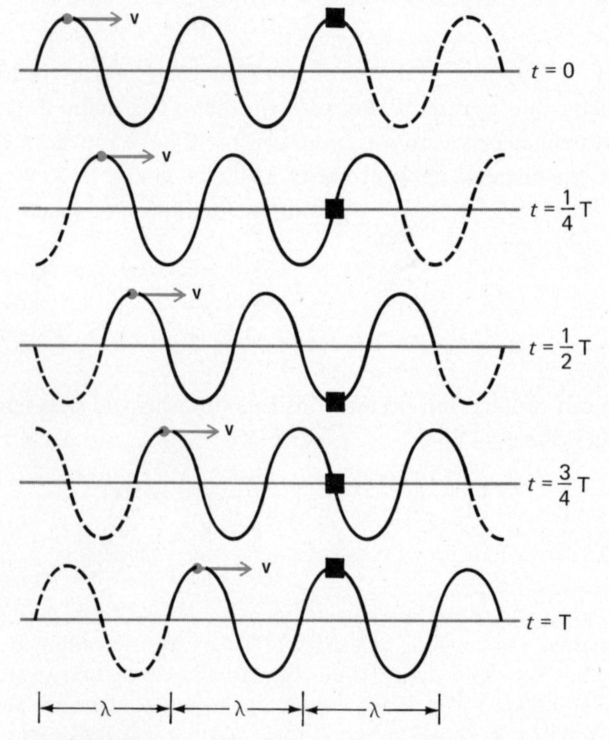

FIGURE 10.3

One complete period of a wave shown in quarter period intervals. Notice that in the time *T* the crest of the wave, moving at speed *v*, travels a distance of one wavelength. During this time the black cork makes one complete round trip down and back up.

between two successive crests or troughs. This time is known as the *period* T of the wave and is measured in seconds. Clearly, if f waves pass a given point per second, then the time interval between successive waves is $1/f$ seconds. Figure 10.3 shows a wave over a time of one complete period. This time interval has been defined as the period T of these waves, so we have the useful relation:

$$T = \frac{1}{f} \tag{10.1}$$

EXAMPLE Suppose that water waves strike a beach every 10 sec. Then the period of these waves is 10 sec. Their frequency is $f = \frac{1}{10}$ per sec. A sound wave might transmit a sound wave having a frequency of 250 vibrations/sec, which is just below the frequency of middle C. The period of such a wave is then given by $T = \frac{1}{250} = 4 \times 10^{-3}$ sec. As a last example, a frequency in the FM radio band is 100 megacycles or 10^8 cps. The period of these waves is then calculated as $T = 1/10^8 = 10^{-8}$ sec.

Suppose that you observe the crest of a wave for a time equal to exactly one period T. Because the period is defined as the time interval between successive crests, your crest will have moved a distance just equal to λ, the distance between crests, as shown in Fig. 10.3. We can then compute the speed v of the wave by using the definition of speed as distance divided by time, giving

$$v = \frac{\lambda}{T} \tag{10.2}$$

We can replace the period T in Eq. (10.2) by the frequency f if we use Eq. (10.1). We find that

$$v = f\lambda \tag{10.3}$$

EXAMPLE Let us compute the wavelength of a sound wave that has a frequency of 300 vibrations/sec traveling at a speed of 330 m/sec. According to Eq. 10.3, this wavelength is $\lambda = \frac{330}{300} = 1.1$ m. If the distance between successive crests of a water wave is 20 m and if waves strike the beach every 5 sec, then the speed of these waves is given by $v = \frac{20}{5} = 4$ m/sec. If radio waves travel with a speed of 3×10^8 m/sec, the frequency of radio waves that have a wavelength of 1 cm is given by $f = 3 \times 10^8/10^{-2} = 3 \times 10^{10}$ cps = 30 gigacycles. From the preceding examples, it should be clear that values of velocity, frequency, wavelength, and period cover an enormous range.

If you watch a cork floating on water as waves pass, you see the cork bob up and down as the waves move horizontally. The situation was shown in Fig. 10.3. In this case, the vibrations of the material are at right angles to the direction of motion of the wave itself, so we call this type of wave a *transverse wave*. Similarly, the waves on a stretched wire, such as a violin string, travel along the length of wire, while the parts of the wire move at right angles to the wire's length. Thus, waves on stretched wires are also transverse waves. A less obvious example of a transverse wave is a light or radio wave, in which variations in electric and magnetic fields take place in a direction at right angles to the direction in which the disturbance itself moves.

Let us consider another type of wave motion. Suppose that you suddenly move a piston along the length of a tube, as shown in Fig. 10.4. The rapid motion of the piston compresses the air in front of it, so that a pulse of high pressure travels ahead of the piston. On the other hand, if you move the piston sharply in the opposite direction, the pressure near the face of the piston is lower than normal air pressure, so that a region of low pressure travels down the tube. If a drumhead vibrates back and forth, alternating regions of high and low pressure are transmitted to the air. When these variations in pressure reach your ears, you experience the sensation of sound. In this example, the direction of the motion of the drumhead and the direction in which the sound wave moves are parallel; therefore, we call this type of wave a *longitudinal wave*. If you hammer on one end of a metal rod, this impulse will be transmitted to the other end

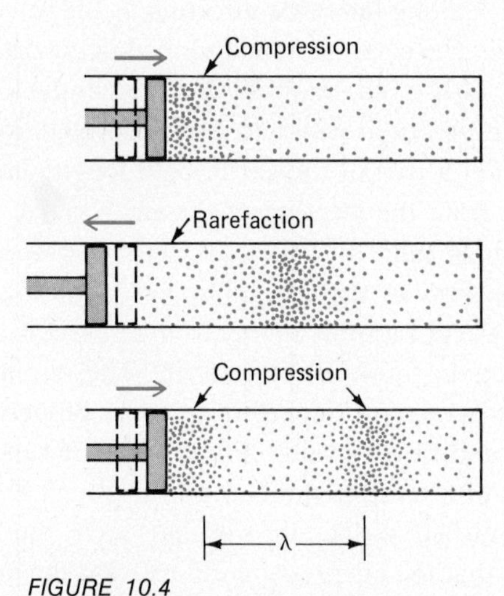

FIGURE 10.4

A longitudinal sound wave in air generated by moving the piston back and forth rapidly.

of the rod as longitudinal waves. Similarly, if you take a spring and attach one end to a wall while you hold the other end in your hand, you can easily see a longitudinal wave. After the spring has been stretched a bit, suddenly move your hand towards the wall. A compression will travel down the spring, as shown in Fig. 10.5. This compression is an example of a longitudinal wave.

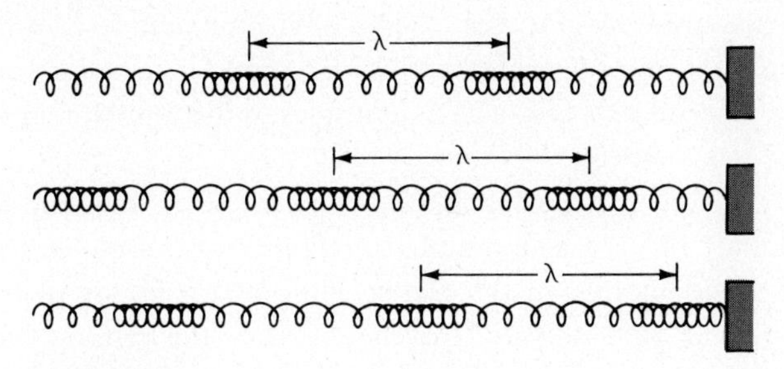

FIGURE 10.5
A longitudinal compression wave traveling down a spring.

We shall now try to make clear the difference between transverse and longitudinal waves. In the case of a transverse wave, the particles of the material move back and forth at right angles to the direction in which the wave and the energy carried by it travel. The waves that we see on water provide an example of such a wave. If the particles of the material move back and forth along the same direction as the wave travels, we have a longitudinal wave. The compression described in the preceding paragraph and shown in Fig. 10.4 is an example of a longitudinal wave. In a solid it is possible for both types of waves to be transmitted. For instance, when an earthquake shakes a part of the earth, both longitudinal and transverse waves travel away from the location of the earthquake.

Here we must point out that the general properties of all waves are the same, regardless of the type of wave or the material in which it is transmitted. All waves transmit energy from a source to a receiver without appreciably affecting the transmitting material. The amplitude of any simple wave is always the maximum departure of some property of the material from its average or undisturbed value. For instance, the amplitude of a water wave is the height of a crest or the depth of a trough as compared to the level of the undisturbed water. In a sound wave the amplitude is the maximum amount that the air pressure differs from the pressure of still air. Similarly, the velocity of any wave is the speed at which some characteristic of the wave travels through the material, and the frequency of any wave is the number of complete waves or vibrations that pass a given point in

1 sec. Finally, the wavelength is the distance between two successive points at which the disturbance caused by the wave in the material is the same.

10.2 STANDING WAVES

When a string quartet is playing in a room, the sound from each instrument fills the entire room, yet the parts played on the individual instruments can be easily distinguished. This observation, that sound from several different sources can travel through the air simultaneously and still retain their individual character, is a result of one of the fundamental laws of wave motion, the *Principle of Superposition.* According to this principle, when two waves cross each other, they retain their individual motions. Their only effect on each other is the addition of their amplitudes as they cross. Two examples of this behavior are illustrated in Figs. 10.6 and 10.7. In Fig. 10.6, two pulses are moving towards each other. Both pulses are in the same direction. As they cross, the net amplitude becomes the sum of the two, as shown. After they have passed each other, they regain their original shapes and continue on their way. In Fig. 10.7, the pulses are in opposite

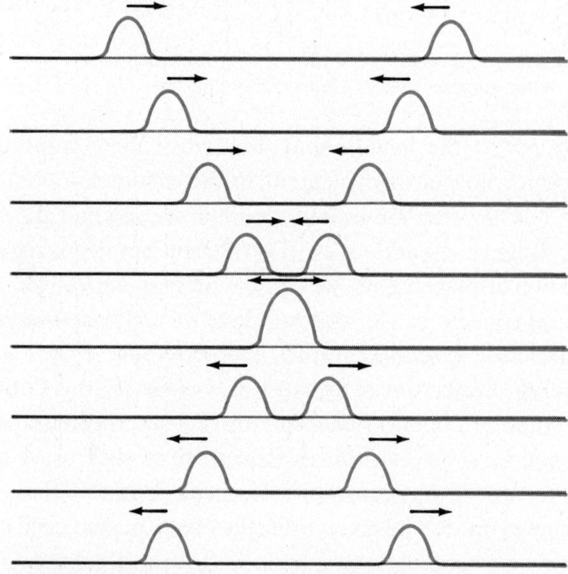

FIGURE 10.6

Two pulses moving along a string towards each other. When they cross the net amplitude is the sum of the individual amplitudes. After they cross they continue with their original shape.

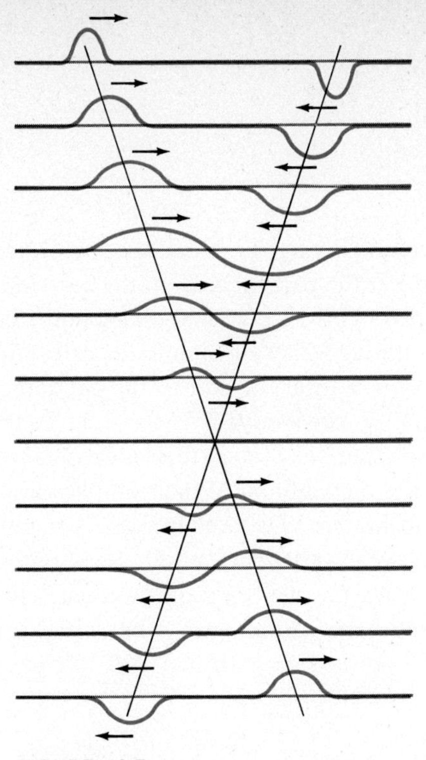

FIGURE 10.7
Two pulses of opposite amplitude moving towards each other. When they cross their amplitudes cancel.

directions. At the instant that they cross, their amplitudes exactly cancel to produce no net displacement in the medium.

When two waves add together, we say that they *interfere* with each other. When they add so as to produce a net displacement that is the sum of the two displacements, we call the process *constructive interference.* When they add so as to cancel one another, we call it *destructive interference.* We shall describe several examples of interference effects later in this chapter. Interference effects arise because waves satisfy the Principle of Superposition. Another common example of the superposition of two waves occurs when the bow waves from two boats cross each other on a calm lake. At the point where they cross, interference effects can be seen as the individual amplitudes unite. However, after they have passed each other, they continue in their original shape and leave calm water behind. The only limitation to the Principle of Superposition is that the individual amplitudes must not be too large. If they are too large, then, as the two waves cross each other, new waves will be created.

One very interesting example of the interference of two waves occurs when two waves traveling with the same wavelength pass each other going

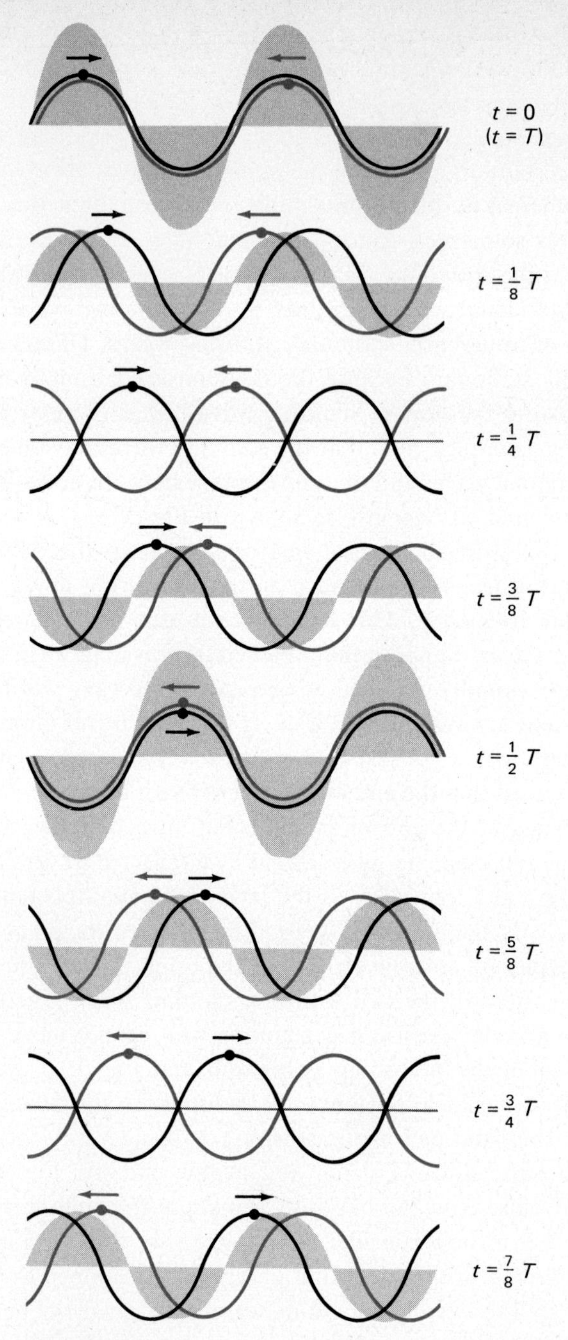

FIGURE 10.8

Two identical waves, one black and moving to the right, the other colored and moving to the left, cross each other. The shaded shape is the sum of their amplitudes and is the actual amplitude present. Notice that it has no apparent motion, thus it is called a standing wave.

in opposite directions. This situation is illustrated in Fig. 10.8, where each wave is drawn along with the sum of the two amplitudes. In the diagram the total time elapsed is one complete period of the waves. Each of the individual pictures is drawn one-eighth of a period apart in time. Notice that at certain points, called the *nodes,* the waves always interfere destructively, whereas at other points midway between the nodes, called *antinodes,* the waves sometimes cancel and sometimes add constructively. The net motion at the antinodes is a periodic displacement with twice the amplitude of the individual waves. The net result of the two waves as they cross, is a series of apparently stationary *standing waves.* Of course, the motion is not really stationary because the medium is continually moving at all the points except the nodes. Standing waves occur in many situations. In the following examples, note that the standing waves have the same frequency as the original waves and that the distance between nodes is exactly one-half of the original wavelength, as shown in Fig. 10.9.

All stringed musical instruments, from the piano to the guitar, produce musical notes because a stretched string always vibrates with a particular frequency. This characteristic can be understood in terms of standing waves. Suppose that you send a traveling wave down a stretched wire that is rigidly attached at one end. This wave will have a frequency of f cps and a wavelength of λ m. It will then travel towards the fixed end of the wire with a velocity v, given by $v = f\lambda$, carrying energy with it. We have assumed that the end of the wire is fixed; therefore, this energy cannot be absorbed at the end of the wire but must be reflected back along the wire. The reflected energy is carried by a reflected wave of exactly the same wavelength and frequency as the incoming wave; this reflected wave must interfere with the incoming wave. The reflected wave has the same amplitude, frequency, and wavelength as the original wave; therefore, the interference pattern of the two will be a standing wave. The end of the string must be a node because it is clamped so it cannot move. In terms of the discussion in the preceding paragraph and Fig. 10.8, we mean that the reflected wave has to have exactly the opposite amplitude of the incoming wave at the fixed end of the string. You could verify this fact by sending a single pulse down a string tied at one end. As Fig. 10.10 shows, the reflected pulse is of the opposite amplitude from the incident pulse.

Let us now consider a complete wire with both ends fixed. If the wire described in the preceding paragraph is clamped at one of the nodes of the standing wave, the standing-wave pattern will not be affected because the nodes do not move. However, if you try to clamp the wire at a point other than a node, you will destroy the pattern. Instead, you will produce a situation in which the incident and reflected waves do not interfere constructively, so no standing-wave pattern will result. You can clamp the

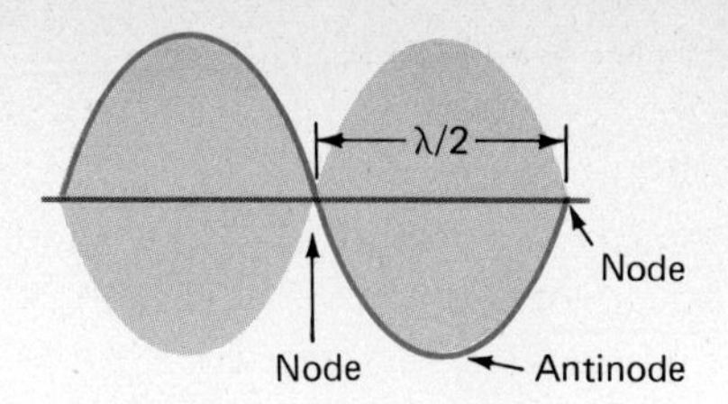

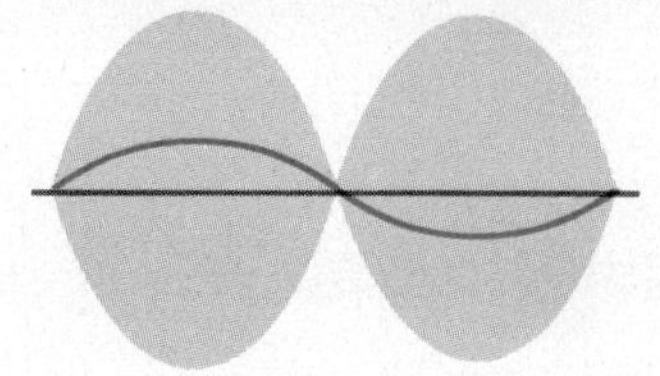

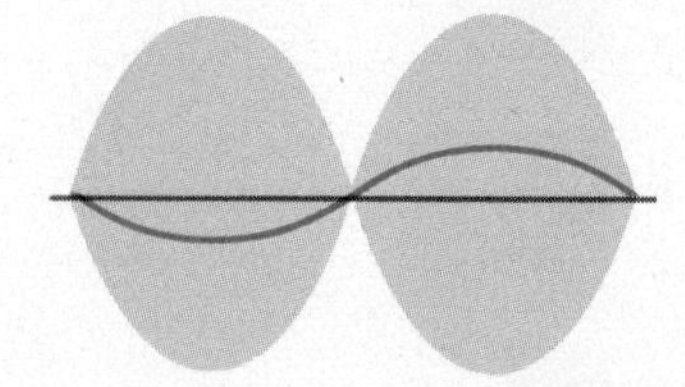

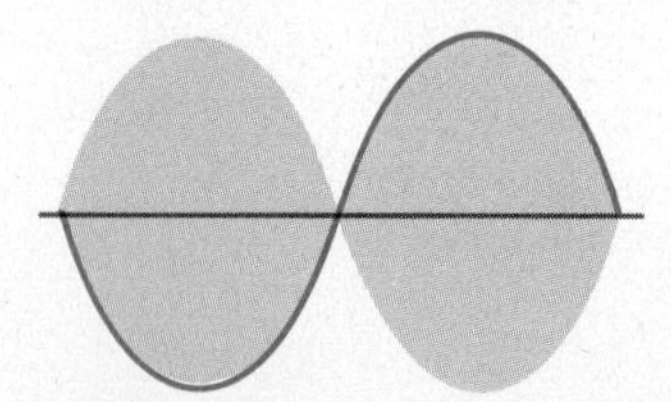

FIGURE 10.9

A standing wave showing only the motion of the string. The points where the string is stationary are called nodes. The distance between nodes is $\lambda/2$.

wave at the first node, the second node, or any other node that you wish. Because the distance between nodes is $\lambda/2$, the length of the clamped wire can be any integral multiple of $\lambda/2$. Therefore, to have a standing wave of wavelength λ on a string of length L, we must satisfy the following relation:

$$L = n\frac{\lambda}{2} \qquad (n = 1, 2, 3, \cdots) \tag{10.4}$$

For a given wire of fixed length L, Eq. (10.4) tells us that the possible wavelengths producing standing waves are given by

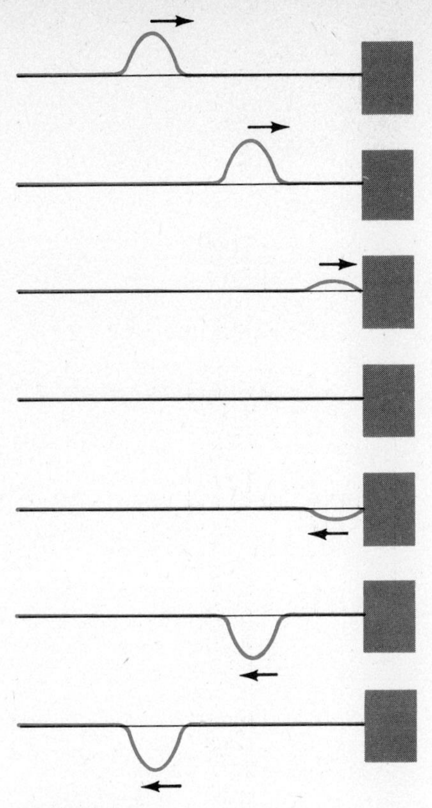

FIGURE 10.10

A single pulse is sent down a string and reflected at the fixed end. The reflected pulse has the opposite amplitude of the incoming pulse.

$$\lambda_n = \frac{2L}{n} \qquad (n = 1, 2, 3, \cdots) \tag{10.5}$$

as shown in Fig. 10.11. If the speed of the waves in your string is v, Eq. (10.5) can be used to find the frequencies f_n associated with the various possible wavelengths. We find that

$$f_n = \frac{nv}{2L} \tag{10.6}$$

The lowest frequency, f_1, that can produce a standing wave is given by $n = 1$; thus, $f_1 = v/2L$. All the frequencies of a stretched wire, therefore, are related to the lowest frequency by the equation

$$f_n = nf_1 \qquad (n = 1, 2, 3, \cdots) \tag{10.7}$$

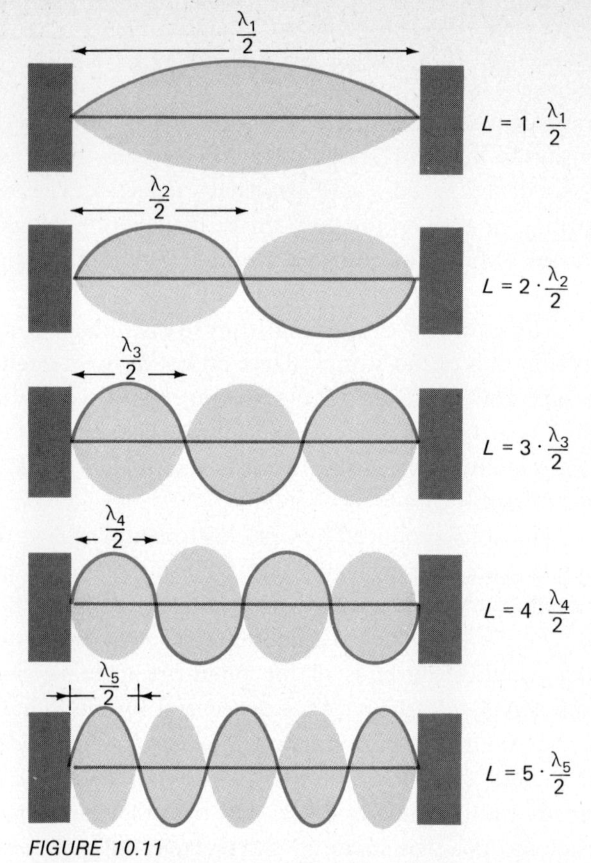

FIGURE 10.11

The fundamental and first four harmonics of a stretched string clamped at both ends. Notice that the distance between nodes is always $\lambda/2$.

The lowest frequency of a stretched wire is called its *fundamental frequency*. Its higher frequencies are called *harmonics* by scientists and *overtones* by musicians. If $n = 2$, the frequency is $2f_1$, which is called either the second harmonic or the first overtone. Similarly, if $n = 3$, we obtain the third harmonic or the second overtone. In the case of a stringed instrument, such as the violin, guitar, and piano, all harmonics are possible.

The musical sensation that we refer to as *pitch* is determined by the frequency of the sound. The higher the frequency, the higher the pitch. A pitch difference of one octave represents a doubling of the frequency of the sound. Thus, the first harmonic of a string is always one octave above its fundamental.

The fundamental frequency of a string (and thus its pitch) can be changed in two ways. The length of the string can be altered, or the speed of the wave can be changed. If the string is shortened, the fundamental

frequency rises, because Eq. (10.6) says that the fundamental frequency is proportional to $1/L$. This method of altering the fundamental frequency is used in such instruments as the piano and the autoharp, which have strings of various lengths. The speed of a wave can be altered either by changing the tension in the string or by changing its mass. If the mass of the string is increased, the speed v and frequency f decreases. For this reason, the bass strings on a piano are very thick. If the string is tightened, the speed v increases. Musicians use this method for the fine tuning of all string instruments.

The sound produced by a stringed instrument is not caused directly by the vibration of the strings. They do not move enough air. Instead, the strings are always linked to a sounding board which they vibrate. The sounding board then generates the sound. The main difference between a pawnshop violin and a Stradivarius is the quality of the sounding board, not the strings.

The notes produced by wind instruments, such as the trumpet, flute, and organ, can also be explained by using standing waves. For these instruments, the standing waves are actual sound waves in the pipes forming the body of the instruments. For standing sound waves in a pipe, we must consider whether the ends of the pipes are open or closed when fitting standing waves into them. At a closed end the standing wave must have a *node,* but at an open end the wave must have an *antinode*. Thus, for pipes closed at both ends, the situation is the same as it was for a string, and we find that all harmonics occur. On the other hand, for a pipe open at one end, the possible standing waves have different wavelengths, so we discover that all the even harmonics are missing. These two cases are shown in Fig. 10.12.

If we study instruments like the drum and the cymbal, we see that they also vibrate with a fundamental frequency determined by a standing wave. However, the higher harmonics are not related to the fundamental frequency by an integer. Therefore, except for the kettle drum whose higher harmonics are less important, they do not give the sense of having a given pitch. Actually, no real instrument produces pure harmonics. For instance, the sounding board of a violin, which actually produces the sound, is driven by the vibrations of the strings. It has its own set of higher harmonics, which, like the drum, are not related in any simple way to the fundamental harmonic. No two instruments have the same set of higher harmonics. Thus, the timbre of a violin is very different from that of a piano, even though a vibrating string is the ultimate cause of the note produced.

The note C three octaves below middle C has a frequency of about 32 vibrations/sec, whereas the C octaves above middle C has a frequency

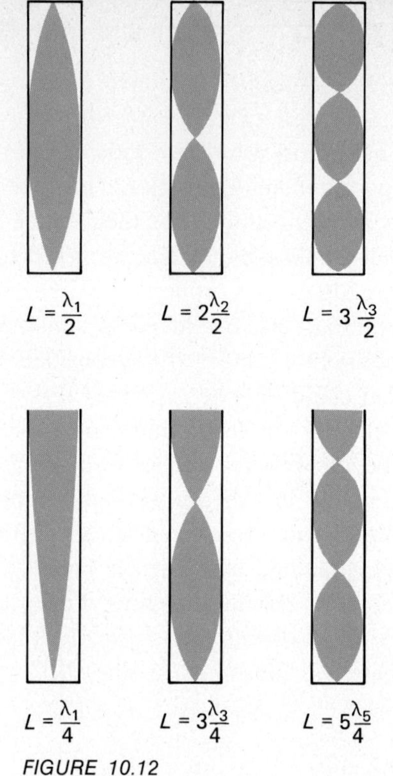

FIGURE 10.12

Standing wave patterns in a pipe. For the pipe closed at both ends (or open at both ends) all of the higher harmonics are possible. For a pipe open at one end only the odd harmonics are possible.

of about 2048 vibrations/sec. (For this example, we assume middle C to have a frequency of 256 vibrations/sec; in practice, musicians use a frequency of 260–265 vibrations/sec.) At this point, you might wonder why sound reproduction equipment is commonly designed to produce frequencies up to 5000 or 20,000 vibrations/sec. The reason is that a musical instrument does not sound realistic unless the harmonics that it produces are also heard. Because the normal human ear is sensitive to frequencies above 10,000 vibrations/sec, it is necessary to reproduce quite high frequencies if a musical sound at a fundamental frequency of 1000 vibrations/sec is to sound natural. If the relative energies present among the harmonics is changed by the reproducing apparatus, the sound is unreal and is said to be distorted. A basic problem in the reproduction of music is trying to present the original harmonics of each instrument in their proper relationship to one another.

10.3 DIFFRACTION AND INTERFERENCE

At the beginning of the 18th century, when the properties of light were first seriously investigated, there were two schools of thought about the nature of light. One school, whose most prominent member was Newton himself, felt that light was composed of small particles. This model explained such phenomena as shadows and reflection. The other school, led by Christiaan Huygens, a Dutch physicist, believed that light was a wave. This assumption enabled physicists to explain all the properties of light that had been observed at that time. Gradually more evidence was gathered for the wave theory. By the middle of the 19th century, it was considered to be conclusive: Light had properties that could only be explained if it was a wave. As we shall see in the next chapter on light, this conclusion was incomplete. Nevertheless, in this section we shall discuss two effects, diffraction and interference, that are observed in light and strongly support the conclusion that light is a wave phenomena. Interference and diffraction occur only in waves and cannot be explained by a particle model.

Suppose that a set of parallel traveling waves, perhaps in water, are incident on a wall with an opening much smaller than the wavelength of the waves. This situation is illustrated in Fig. 10.13. The water on the

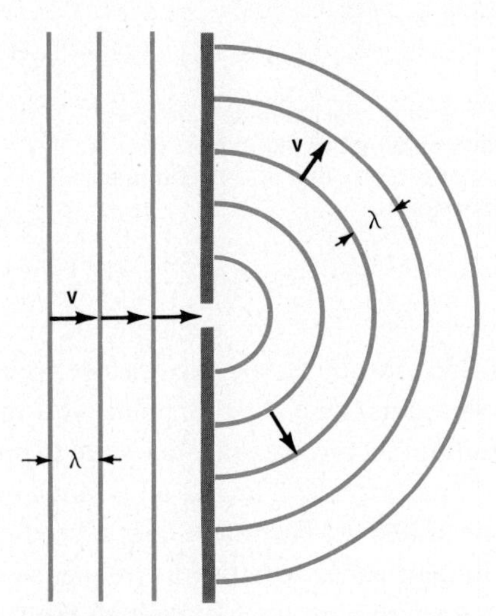

FIGURE 10.13

A plane wave in water is incident on a wall with an opening very small compared to the wavelength of the wave. The transmitted wave is circular and has the same wavelength frequency, and velocity, as the original. The lines represent the crests of the wave and are moving in the direction indicated by the velocity vectors.

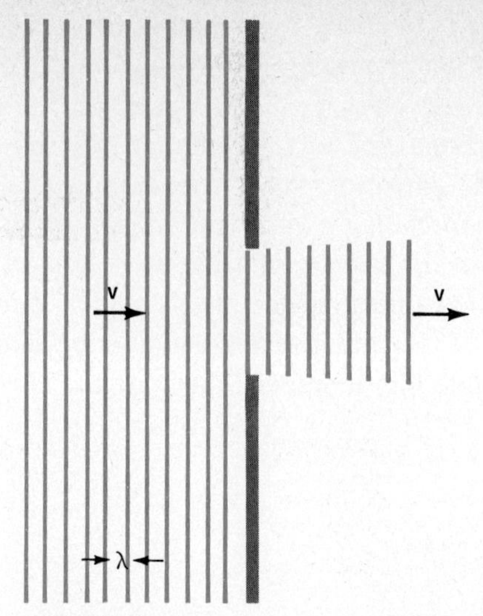

FIGURE 10.14

A plane wave is incident on an opening much wider than the wavelength of the wave. The part of the wave incident on the opening is transmitted almost unchanged and the part of the wave incident on the wall is blocked.

other side of the wall is disturbed by the wave arriving at the hole. This disturbance acts like a point vibrating with the frequency of the incident wave. Anyone who has thrown a small stone into a still lake knows that the wave produced is a circular one. The wave arriving at the small hole acts like a vibrating point and produces a circular wave on the far side of the wall. This wave has the same frequency as the incident wave. The speed of the wave is assumed to be the same on both sides of the wall, so the wavelength is also the same. Here is an extreme case of *diffraction* of a wave.

Another extreme case occurs when a set of parallel traveling waves are incident on an opening in a wall much larger than the wavelength, as shown in Fig. 10.14. In this case, the wall blocks all of the wave except for the part hitting the opening. Thus, the wave on the far side is still essentially a set of parallel waves, but they have been cut off at the sides by the wall. (Actually the wave is not cut off cleanly, but the effect is small as long as the hole is much larger than the wavelength.) In everyday language we would say the wall casts a shadow. In this case, there is almost no diffraction.

What about the intermediate cases where the hole is neither much larger or much smaller than the wavelength of the incident wave? We would obviously expect something intermediate between the two examples just given. Suppose that we have an incident wave hitting a wall with an

adjustable opening in it. When the opening is much larger than the wavelength, the situation will be as shown in Fig. 10.14. Now imagine slowly making the opening narrower. Several different possibilities are illustrated in Fig. 10.15. Notice that the wave on the far side starts to spread out as the opening is made narrower. When the opening is roughly the size of the wavelength, this spreading is quite appreciable. By the time the opening is much smaller than the wavelength, the wave has spread so much that it has lost its original direction completely and is circular. Although a detailed discussion of the causes of diffraction is beyond the scope of this book, several available film loops well illustrate the effect for water waves in a ripple tank.

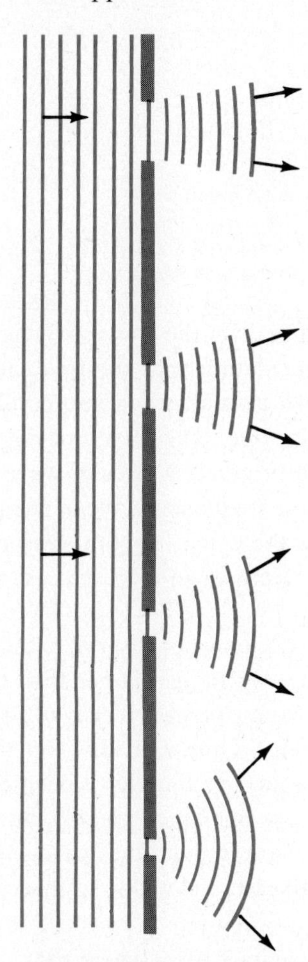

FIGURE 10.15

A plane wave incident on a wall with openings of various sizes. Notice that the smaller the opening the more the transmitted wave spreads.

Diffraction occurs for any wave motion, including light and sound. When diffraction effects were observed in light, it was taken as very strong evidence that light was a wave. If light were composed of particles, the only possible effect of a wall with a hole in it would be to stop those particles that struck the wall and leave those that did not, thus casting a shadow.

Although waves on water or a stretched string can be seen, it is not directly evident that sound and light are effects transmitted by waves. The phenomenon of interference, however, provides another crucial test of whether a given effect is caused by waves or particles. We have already seen one example of an effect produced by interference, standing waves. Let us now consider another example.

Suppose that particles from two sources strike the same point. The effect of two particles hitting the same point will be greater than if only a single particle struck that point. Similarly, if crests of two waves reach a given point at the same time, the Principle of Superposition implies that their combined effect will be greater than if only a single wave had arrived. When two waves having different sources but the same frequency, period, wavelength, and velocity arrive at the same point, their crests may arrive at the same time. In this case, their troughs will also arrive simultaneously, as will all other corresponding points on the two waves. When corresponding points of two or more waves arrive at the same time, we say that the waves arrive *in phase.* The sum of the two waves is then greater than either wave separately. We call this situation *constructive interference.* Constructive interference of two waves was responsible for the antinodes of the standing waves discussed in Sec. 10.2.

On the other hand, when two waves strike a point, it may happen that their crests do not arrive at exactly the same instant. The waves are then said to be *out of phase.* In the extreme case, the crest of one wave occurs half a period later than the rest of the other wave, so that the crest of one wave falls on the trough of the other. Then the Principle of Superposition implies that the waves subtract from one another. The resultant effect is less than would be caused by either wave alone. Here we have an example of *destructive interference.* This partial cancellation of one wave by the other can only happen continually if the waves have the same velocity, frequency, wavelength, and a constant phase relation. One way of achieving a constant phase relation is to have parts of a given wave travel from the source to the detector by different but constant paths. Because the times needed for the two waves will be different, one wave will arrive later than the other. If the situation is arranged properly, a crest of one wave can be made to fall on the trough of the other, so destructive interference results. If, in addition, the amplitudes of the two waves are the same, then complete cancellation can be achieved. When complete cancellation occurs, no effect

is noticed by the detector. At the nodes of the standing waves described in Sec. 10.2, destructive interference caused the two waves to cancel completely.

When two particles strike a detector, the effect is always greater than if only one particle had been used. Thus, cancellation never occurs when particles are involved. However, as we have just seen, partial or complete cancellation can occur with waves. Whenever we observe that two effects cancel one another, if only partially, we know that the phenomenon is caused by a wave.

On the other hand, if two effects striking a detector produce a greater result than one alone would, we cannot tell if we are dealing with a stream of particles or a wave. Thus, the crucial distinction between an effect caused by particles and one caused by waves is that only a wave can produce destructive interference or cancellation. Let us illustrate these effects by describing a simple experiment that can be performed with many different types of waves.

Suppose that a series of parallel waves is incident on a wall with two small slits, *A* and *B*, in it. Imagine that these waves are in water. None of our conclusions, however, will depend on that assumption. The size of the individual slits *A* and *B* is small compared to the wavelength of the incident wave, and so the waves leaving the slits on the far side of the wall are circular because of the diffraction effects. Figure 10.16 shows the wave on both sides of the wall when one of the slits is closed. When both slits

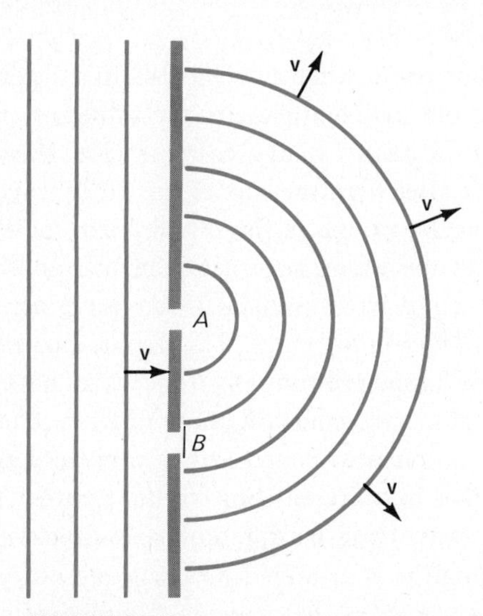

FIGURE 10.16
With one slit closed the transmitted wave is circular as in Fig. 10.13.

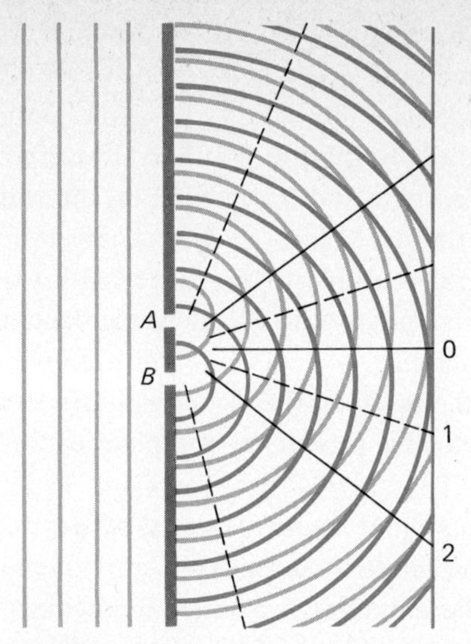

FIGURE 10.17

A plane wave incident on two small slits. The crests of the transmitted circular waves are drawn as if they were not interfering. The solid lines leading to points 0 and 2 are regions where the waves are in phase and interfering constructively. The dotted line leading to point 1 is a region where destructive interference is occurring.

are open, two sets of circular waves are present, as shown in Fig. 10.17. Both of the circular waves are generated by the same parallel wave. Therefore, they send out crests at exactly the same time, and they leave points A and B exactly in phase. The resultant amplitude of the two waves at a given point can be found by using the Principle of Superposition. Point O in Fig. 10.17 is equidistant from points A and B; thus, the crests from each slit require exactly the same time to travel to this point. The crests and troughs of the two waves arrive simultaneously (i.e., in phase), and the waves interfere constructively, producing a large net amplitude at point O. The point marked *1* in Fig. 10.17 is not equidistant from points A and B; point B is closer. We have chosen point *1* so that the distance from A to *1* is exactly one-half of a wavelength longer than the distance from B to *1*. Because it takes a wave exactly 1 period T to travel a distance of 1 wavelength λ, it takes the wave from A exactly $T/2$ sec longer to reach point *1* than it does the wave from B. Consequently, when a crest arrives at *1* from point A, a trough will be arriving from point B, and vice versa. The two waves arrive exactly out of phase and interfere destructively. The amplitudes of the two waves cancel, and there is no net amplitude at point *1*. Point *2* is 1 wavelength farther from A than from B, and so waves from

A require 1 period *T* longer to reach *2* than do those from *B*. Thus, crests arrive from *A* in phase with crests emitted one period earlier from *B*, and they interfere constructively, producing a large net amplitude. In general, if we wish to find the net amplitude at an arbitrary point, we must know the difference between the distance traveled by the wave from *A* and the distance traveled by the wave from *B* in reaching the point. If this path difference is an integral multiple of the wavelength (λ, 2λ, 3λ, etc.), then the waves will arrive in phase and interfere constructively. On the other hand, if the path difference is a half-integral multiple of the wavelength ($\lambda/2$, $3\lambda/2$, $5\lambda/2$, $7\lambda/2$, etc.) then the waves will arrive exactly out of phase and interfere destructively, producing complete cancellation and no net amplitude.

In order to show that sound is caused by waves, let us consider the common situation in which two loudspeakers are driven by the same amplifier. If the input to the amplifier is a single frequency (say, from a tuning fork), you can find places in front of the loudspeakers at which the sound intensity will be nearly zero. The situation is depicted in Fig. 10.18. If you move from point *O* to points *1*, *2*, and so on, you will pass places where the path difference causes alternating loud and soft sound. Experiments of this sort show that destructive interference can be produced with sound; therefore, sound is a wave phenomenon.

About the year 1800, England's Thomas Young performed an experiment to see if light was caused by waves. In his experiment a narrow beam of light struck two slits, and the result was observed on a screen, such as the one pictured in Fig. 10.19. At the point marked *2*, the wave from slit *A* has traveled a distance of one wavelength farther than the wave from slit *B*. Thus, crests still fall on crests and troughs on troughs, so there is

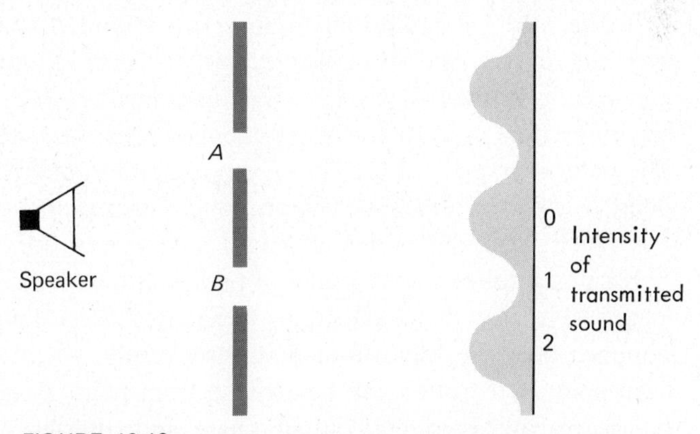

FIGURE 10.18

A speaker sends sound waves towards two slits. The intensity of the transmitted sound is shown at the right.

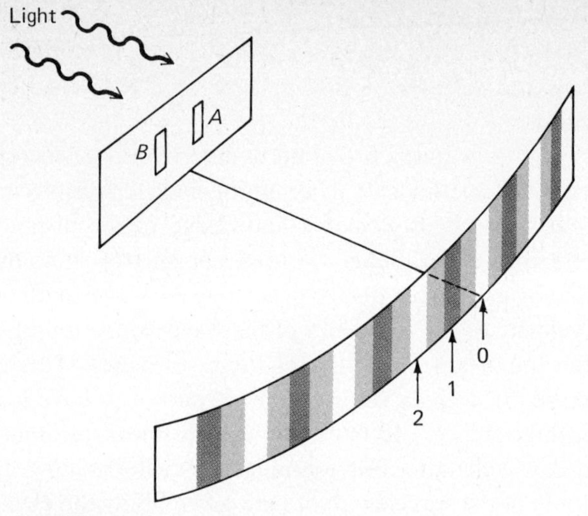

FIGURE 10.19

A schematic diagram of Young's twin-slit experiment with light. Interference causes alternate light and dark bands on the screen.

brightness at point *2*. At point *1*, the wave from slit *A* travels a distance only half a wavelength greater than that traveled by the wave from slit *B*. In this case, a crest of one wave falls on a trough of the other wave, or vice versa, so here there is darkness. A photograph of this effect is shown in Fig. 10.20, in which the center of the pattern and point *1* and *2* are

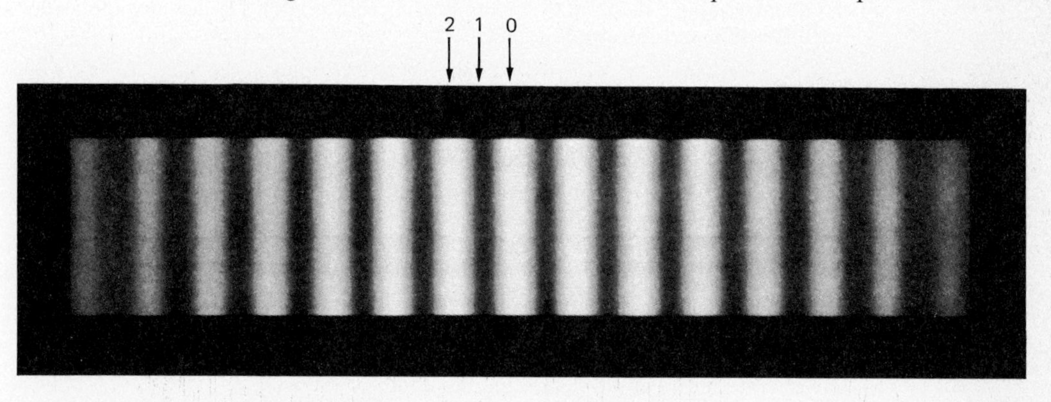

*FIGURE 10.20**

A photograph of a twin-slit interference pattern taken using an apparatus similar to the one shown in Fig. 10.19.

*Reprinted from Cagnet, Françon, and Thrierr, *Atlas of Optical Phenomena* (Springer-Verlag/Prentice-Hall, 1962) with the permission of the authors and Springer-Verlag.

marked. Each dark line is a place where the path difference is a half-integral multiple of the wavelength, causing destructive interference. In this way, Young and scientists following him concluded that light is a wave phenomena.

The patterns produced by interference effects can be used in many ways. Most methods of determining wavelengths accurately are based on an interference effect and a knowledge of the distances involved. Thus, interference experiments are used not only to show that a given effect is caused by a wave but also to determine the wavelength of the wave. If either the velocity or the frequency of the wave can be found, then we know most of the important properties of the disturbance. This process can also be reversed. If a given color of light is known to have a certain wavelength, then interference effects can be used to measure other distances in terms of this wavelength. For example, the meter is now defined as a certain multiple of the wavelength of light given off by the element krypton. When precision distance measurements are made, they use interference effects to compare distances to the known wavelength of this light.

DISCUSSION QUESTIONS

1. The human ear responds only to a range of frequencies between approximately 20 cycles per second and 15,000 cycles per second. List some sources of sound which produce frequencies below the limit of human hearing (infrasonic) and frequencies above the limit of human hearing (ultrasonic). Discuss any applications of infrasonic and ultrasonic sound.

2. Discuss any evidence that the speed of sound is the same for all wavelengths.

3. Discuss how a bugler can produce different frequencies, even though the length of the air column of a bugle is fixed.

4. The speed of sound in air is about 1100 ft/sec and the speed of light is about 186,000 miles/sec. If you divide the time in seconds between the instant you see the lightning bolt and the instant you hear the thunder by five you will learn the approximate distance away of the lightning bolt in miles. Explain why this is so.

5. A stone is dropped down a deep well. The time between releasing the stone and hearing it splash in the water is measured. Discuss how from this measurement you could determine the depth of the well.

6. Since light is a wave, why do we not observe interference effects when two lights are turned on in the same room? When two persons are talking in a room?

7. Is it possible to observe interference effects when two loudspeakers operated from the same amplifier are used in a room?

8. Explain qualitatively why an approaching train whistle sounds higher in frequency than the same whistle when it is at rest with respect to you. (This is known as the *Doppler effect*.)

PROBLEMS

1. Standing on a dock you observe wave crests passing you on the water every 2 sec. If the distance from one crest to the next following is 100 ft, what is the velocity of these waves? (*Ans.*: 50 ft/sec)
2. If the velocity of sound in air is 1100 ft/sec, what is the wavelength of sound waves of frequency 500 cycles per second?

   Repeat for a frequency of 50 cycles per second. (*Ans.*: 22 ft)
3. Compute the value of the longest wavelength which can exist as a standing wave on a violin string 2 ft long.

   If the string is bowed properly, the third harmonic may be excited. What is its wavelength? (*Ans.*: 1.33 ft)

   If the velocity of waves along the violin string is 2000 ft/sec, what frequency is associated with this wavelength?

   What wavelength does this violin string produce in air, if the velocity of sound in air is 1100 ft/sec? (*Ans.*: 2.2 ft)
4. How long must an organ pipe closed at one end be if its fundamental wavelength is to be 10 ft?

   If the velocity of sound in air is 1100 ft/sec, what is the fundamental frequency of this organ pipe? (*Ans.*: 110 cps)

   What is the frequency of the second overtone of this organ pipe?
5. Take the velocity of sound in air as 1100 ft/sec. Compute the length of an organ pipe closed at one end which will have a fundamental frequency of 20 cycles per second. (*Ans.*: 13.8 ft)
6. An organ pipe produces a wavelength of 5 m in air. Take the speed of sound in air as 330 m/sec and compute the frequency of these waves. (*Ans.*: 66 cps)
7. Compute the wavelength of a sound wave of frequency 10,000 cycles per second.
8. Compute the value of the longest wavelength which can exist as a standing wave on a cello string 1 m long. (*Ans.*: 2 m)
9. A fundamental frequency produced by a piccolo is 2000 vibrations per second. What is the fourth harmonic of this note?
10. Thunder is heard 10 sec after the lightning flash is seen. How far away did the lightning strike? (*Ans.*: 3300 m)
11. If the amplitude of a simple wave is cut in half, by what factor is the energy carried by the wave changed? (*Ans.*: $\frac{1}{4}$)
12. If we wish to increase the energy carried by a wave by a factor of 100, by what factor should we change the amplitude of the wave?
13. Calculate the period of a sound wave of frequency 40 vibrations per second. (*Ans.*: 2.5×10^{-2} sec)
14. Water waves strike a beach every 5 sec. What are the period and frequency of these waves?
15. By what factor must the amplitude of a wave be changed for the energy carried by the wave to be increased by a factor of 9? (*Ans.*: 3)
16. If the amplitude of a wave is quadrupled, by what factor is the energy carried by the wave changed?

17. Calculate the period of a sound wave of frequency 15,000 vibrations per second. (*Ans.*: 6.67×10^{-5} sec)

18. Calculate the frequency of a sound wave which has a period of 1 millisecond.

19. Two loudspeakers are 3 ft apart. When you stand 4 ft directly in front of one of the speakers, the sound intensity is very low. Calculate the probable wavelength of the sound from the loudspeakers.
(*Ans.*: 2 ft, $\frac{2}{3}$ ft, $\frac{2}{5}$ ft, etc.)

20. You stand 20 ft from the vertical brick wall of a building. A point source of sound of wavelength 10 ft is moved slowly from the building directly toward you. Describe the changes in sound intensity received by your ear as the source of sound approaches you.

Calculate a location of the source such that would expect to receive maximum intensity at your ear. (*Ans.*: Source 5, 10, or 15 ft from wall)

Calculate a location of the source such that you would expect to receive minimum intensity at your ear.

21. Longitudinal earthquake waves have a speed of 7 km/sec, while transverse earthquake waves have a speed of 4 km/sec. If the time interval between the arrivals of the two waves at your laboratory is 10 min, how far away did the earthquake occur? (*Ans.*: 5600 km)

22. Intense laser light is incident on a pair of parallel slits that are separated by 1 mm. A piece of photographic film 1 m away from the slits records the interference pattern. If the distance between adjacent maxima is 0.5 mm, find the wavelength of the light. (*Ans.*: 5000 A)

23. Sound from a single source is transmitted to a second location by two long pipes, one of which is 4 cm longer than the other. What is the longest wavelength for which destructive interference will occur? If the speed of sound is 350 m/sec, find the frequency of this sound.

11 Light

One of the most challenging and satisfying problems in physics concerns light and its interactions with matter. The problem is challenging because the phenomena associated with light are extremely varied and complex; it is satisfying because physicists now understand the nature of light and the forces controlling it better than any other phenomenon of nature. In this chapter we shall discuss light as it is presently understood. You will see that neither Sir Isaac Newton, who felt light was composed of particles, nor Christiaan Huygens, who thought light was a wave, were completely correct. Light is both a wave and a particle, behaving sometimes like one, sometimes like the other. By complementing each other, these two aspects account for all the interactions of light.

Electrodynamics is the theory that predicts light to be a wave. It was formulated by J. C. Maxwell in 1873 and is one of the major triumphs of 19th-century physics. With this theory, scientists could explain all of the data then available, and it seemed that Huygens had triumphed. However, during the first two decades of the 20th century, new data began to appear and new effects were observed to indicate that electrodynamics was incomplete. This data spawned two new theories, *relativity* and *quantum mechanics,* which completely revolutionized physics. As a result, the scope of physics was extended down to the atomic and the subatomic level and out to the cosmological sphere. It is the subject of the fourth part of this book. In this chapter, however, we shall see that among this wealth of new data were

effects that could be explained only if light were composed of particles, which have named *photons*. In fact, the complete picture of light involves both waves and photons.

11.1 ELECTROMAGNETIC WAVES

Light waves, like sound waves, are a continual part of our everyday experience. The speed of these waves is so great that our senses are unable to detect any time delay in their propagation. Whereas the speed of sound waves in air is about 330 m/sec, the speed of light in empty space is 300 million m/sec. Thus, we see a lightning flash almost instantaneously, but seconds pass before we hear the accompanying clap of thunder. The speed of light in empty space, denoted by c, is a universal constant of nature; that is, whenever c is measured, the same value is always found. Neither the observer's motion nor the motion of the light source affect it. This fact is somewhat surprising. With sound or water waves, when an observer, moving through the medium in which a wave propagates, measures the speed of the wave, he finds it to be the vector sum of his velocity plus the velocity of the wave in the medium. No such effect is observed in light. The speed of light, c, is so fast that it was not until the very end of the 19th century that A. A. Michelson and E. W. Morley were able to measure it with sufficient accuracy to show that light is, in fact, a universal constant, the same for all observers. As an example of just how large c really is, it takes just under 3 seconds for light to make the round trip from the earth to the moon (a very noticeable effect when listening to a conversation between an astronaut on the moon and his ground control), and the light reaching us from the sun starts its trip about $8\frac{1}{2}$ minutes before we sense it.

Most of the energy in the sunlight that we see lies in the range of wavelengths between about 4×10^{-7} and 7×10^{-7} meters. Therefore, through the process of natural selection the human eye responds chiefly to wavelengths in this range, which is known as the *visible spectrum*. The shortest wavelengths correspond to the color violet. As we look at longer and longer wavelengths, we get the sensation of blue, green, yellow, orange, and red. A chart showing the relation between approximate wavelengths and corresponding colors is shown in Fig. 11.1. Here we must point out that certain colors, such as brown, do not correspond to a single wavelength. The sensations we name by these colors are produced by mixtures of wavelengths. Producing various colors by mixing is used in color photographs, color television, and color printing.

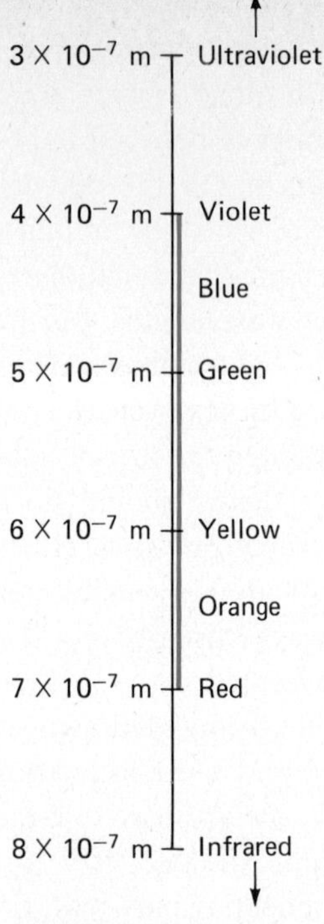

FIGURE 11.1
The visible part of the spectrum. The colored line shows the portion the human eye is sensitive to.

EXAMPLE Suppose that we take a wavelength of 6×10^{-7} m (yellow) as a typical wavelength of visible light. Because the velocity of light in empty space (essentially the same as that in air) is 3×10^8 m/sec, the frequency of such a wave is given by

$$f = \frac{3 \times 10^8}{6 \times 10^{-7}} = 0.5 \times 10^{15} = 5 \times 10^{14} \text{ cps}$$

To show how large this frequency is, we might note that a typical standard radio frequency is about 10^6 cps; a typical television broadcasting frequency is about 10^8 cps. Thus, showing that light is a wave is difficult indeed, so short is the interval between the impact of successive waves on the eye.

In the case of sound waves, the disturbance that is propagated as a wave is a series of variations of pressure above and below the average pressure in the material. Light waves, however, consist of a propagation through space of variations in electric and magnetic fields. Thus, light waves are called *electromagnetic waves.* The electric and magnetic fields occur in directions at right angles to the direction in which the effect travels, as shown

in Fig. 11.2. Light waves, therefore, are transverse, in contrast to sound waves, where the particles of the transmitting material (air, for example) move in a direction parallel to the direction in which the disturbance moves. This behavior is one important difference between light and sound waves.

At this point, you might guess that there could exist electromagnetic waves of frequencies or wavelengths to which the human eye is not sensitive. About a century ago, James Maxwell predicted that there should be electromagnetic waves of all frequencies and wavelengths. Furthermore, all of these waves should travel at the speed of light, namely, 300 million m/sec. At the time that the theory was proposed by Maxwell, the only electromagnetic waves known were light waves. Although the theory adequately accounted for the properties of light waves, a theory should do more than merely explain known facts. In 1887, a German physicist, H. R. Hertz, succeeded in producing electromagnetic waves with frequencies greatly different from optical frequencies. Because these waves also obeyed Maxwell's theory, the electromagnetic wave theory was accepted.

During the 20th century, electromagnetic waves of a wide range of frequencies have been observed and used for various purposes. Frequencies in the range of 10^4 to 10^9 cps are used for radio and television communications. Frequencies of approximately 10^{12} to 10^{14} cps lie in the *infrared* range, which is often used for heating purposes. Frequencies ranging roughly from 10^{15} to 10^{17} cps lie in the ultraviolet region. Frequencies greater than about 10^{18} cps are known as *X rays*. Figure 11.3 is a chart showing some of the commonly used frequencies.

The variations in electric and magnetic fields are called *electromagnetic waves*. Let us see how they are produced for frequencies used in

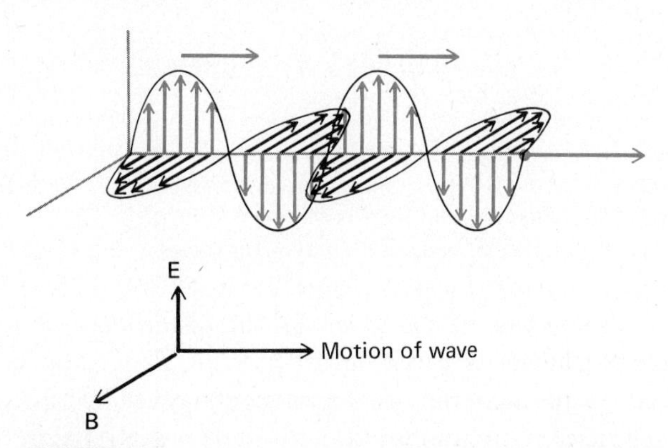

FIGURE 11.2

An electromagnetic wave. This is a transverse wave with the electric and magnetic fields perpendicular to each other and to the direction of propagation.

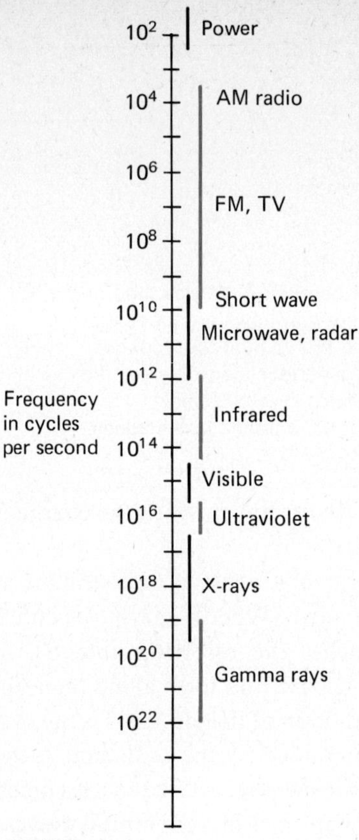

FIGURE 11.3

The electromagnetic spectrum with some of its uses and special regions.

radio and television. A different mechanism is needed to explain electromagnetic waves of much greater frequency.

Suppose that an electric charge (usually an electron) is made to move back and forth in a regular manner along a metal rod (antenna), as depicted in Fig. 11.4. The electric field produced by the charge at a point such as *P* will therefore change as the charge moves along the length of the rod. In addition, a moving charge is equivalent to an electric current, so that a changing magnetic field is also produced at the point *P*. We saw in Sec. 9.4 that a changing magnetic field could produce an electric field. Maxwell's equations not only predicted this behavior but also explained how a changing electric field would produce a magnetic field. The triumph of Maxwell's theory was in showing that these variations of electric and magnetic fields would be propagated through space with a constant velocity equal to the known velocity of light. Furthermore, the velocity did not

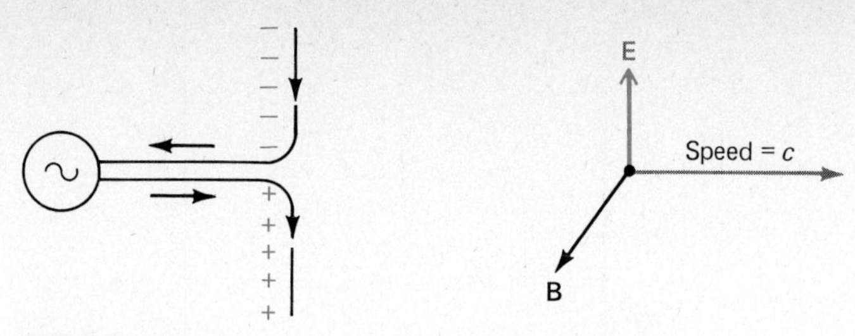

FIGURE 11.4

A dipole antenna. The signal generator charges the antenna as shown. The current that flows produces the magnetic field and the charge that is built up produces the electric field. The signal generator then reverses the current and charge and the fields reverse. This disturbance moves away from the antenna at the speed of light, producing a wave like the one in Fig. 11.2.

depend on the particular frequency at which the charge oscillated back and forth along the metal rod.

A simple radio antenna consists of a straight metal rod in which many electrons are free to move. When a varying electric field from a distant transmitting antenna reaches the receiving antenna, forces play on the electrons, which then move back and forth along the length of the receiving antenna. The oscillatory motion of the electrons is the same as an oscillating current. This current is amplified by the radio and eventually is converted into sound waves by the loudspeaker. At frequencies much higher than those of radio the reception mechanism is more complex. Nevertheless, in all cases, the source emits electromagnetic waves because of the motions of charged particles, and the detector responds to the energy carried by these waves in one fashion or another.

Information may be transmitted by radio waves if some property of the waves varies in a way proportional to the information. For instance, a variation in the amplitude of the wave or in the frequency of the wave can be used. In either case, we say that the wave is *modulated,* so that we speak of *amplitude modulation* (AM) or *frequency modulation* (FM). The most common example of AM is the broadcast band in the vicinity of 1 megacycle; FM is broadcast at a frequency near 100 megacycles. Television combines both forms of modulation by using AM for the picture and FM for the sound.

11.2 PHOTONS

When you hold your hand in front of a stove or a fire (both of which emit large amounts of infrared radiation), you can readily see that light carries energy. Although the fact is less apparent, all wavelengths of light carry

energy. Maxwell's equations for light correctly made this prediction by indicating that light of all wavelengths could transmit both energy and momentum. They also correctly described the overall features of this transmission. However, around the turn of the century, physicists performed several experiments that were very sensitive to the details of just how this energy was transmitted. The crucial discovery among these experiments was that of the *photoelectric effect*.

When light strikes certain metals, such as zinc, the metal emits a small electric current. This effect was noticed by H. R. Hertz in 1887. A typical arrangement for studying the photoelectric effect is shown in Fig. 11.5. The glass bulb is evacuated to a high vacuum, so that nothing interferes with the electric current between the two electrodes. During the last decade of the 19th century, experimenters showed that the electric current traveling from the negative to the positive electrode was carried by electrons, called *photoelectrons*. When light strikes the zinc, it delivers energy to the electrons, enabling them to escape from the surface of the zinc. These negative electrons are then drawn to the positive electrode and collected there. The ammeter indicates the flow of electrons per second, or the *photoelectric current*. Not surprisingly, the photoelectric current in amperes is proportional to the intensity (power delivered by the light per square meter or watts per square meter) of the incident light over a very wide range of intensities. This calculation is important in applications of photoelectric tubes, which are used in such devices as light meters to measure the intensity of light. The output of the device (current) should be proportional to the input (light

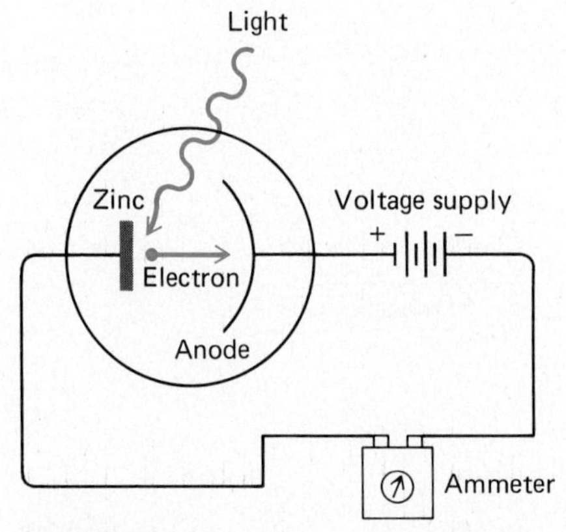

FIGURE 11.5

An apparatus for studying the photoelectric effect. Light strikes the zinc, delivering enough energy to the electrons in the surface to enable them to escape and be collected by the anode. A current then flows which is measured by the ammeter.

intensity). This proportionality is what we might expect by using Maxwell's equations.

However, if we examine the details of the photoelectric effect, we find several very surprising results. Suppose that we measure the maximum energy of the various photoelectrons produced from a given surface by light of a given wavelength. We first find that this maximum energy does not depend on the intensity of the incident light. Normally, we would expect that if we raise the light intensity and thus hit the surface with more energy per second, at least some of the photoelectrons would get more energy. Clearly, the result that we do get has not been predicted. Furthermore, if we vary the frequency of the incident light, keeping the intensity the same, we find that the maximum energy of the photoelectrons is directly proportional to the frequency of the light, as shown in Fig. 11.6. Below a certain critical or threshold frequency for a given surface, no photoelectrons are emitted, regardless of how large the intensity of the incident light might be. Classically, the maximum energy of the photoelectrons should depend on the energy hitting the surface and not depend at all on the frequency of the incident light, but here we find just the reverse.

As another example of the failure of the classical predictions, suppose that we illuminate the surface of the metal with a very low-intensity blue light. The intensity is sufficiently low that it would take several seconds for it to deliver the necessary energy to a given electron for that electron to escape the metal. When we turn on this light, we would expect a delay of several seconds before we detect the first photoelectrons. However, such

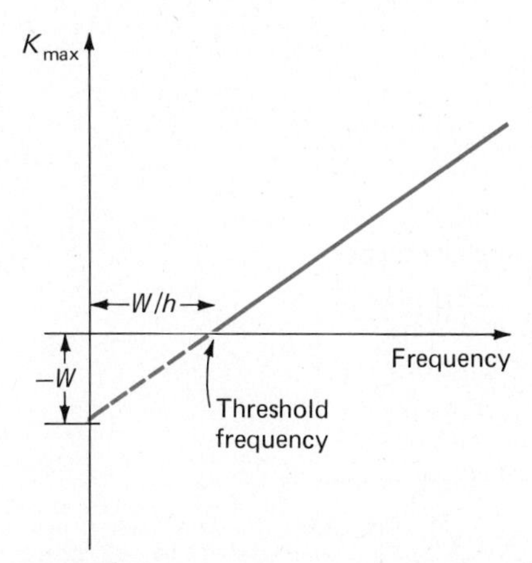

FIGURE 11.6

The maximum kinetic energy of the photoelectrons plotted against the frequency of the light illuminating the zinc. The plot is a straight line showing that K_{max} is directly proportional to f, the frequency.

a delay is not observed. When the light is turned on, photoelectrons are detected immediately. Clearly, the classical theory cannot account for the experimental facts of the photoelectric effect. The problem lies in the precise method by which light delivers energy to the surface of the metal.

In 1905, Einstein solved this puzzle by assuming that the energy delivered to the surface by electromagnetic radiation is transmitted in discrete units of energy called *quanta*. In his model, a beam of light is composed not of continuous waves but of individual particles called *photons*. Each of these photons travels with speed c (the speed of light), has a frequency f equal to the frequency of the beam of light, and carries a discrete quanta of energy given by the equation.

$$E = hf \tag{11.1}$$

where h is Planck's constant. (This constant will be explained in the last part of the book.) When the light shines on a surface, the energy is not delivered continuously to the entire surface; instead, all the energy of a given photon is given to a single electron. If this amount of energy is large enough, the electron might escape the metal and eventually be collected as a photoelectron.

Einstein's ideas concerning the nature of photoelectric emission are shown schematically in Fig. 11.7. The incident photon of frequency f brings in an amount of energy given by $E = hf$. All of this energy is then given to a single electron near the surface of the metal. Part of this energy, W',

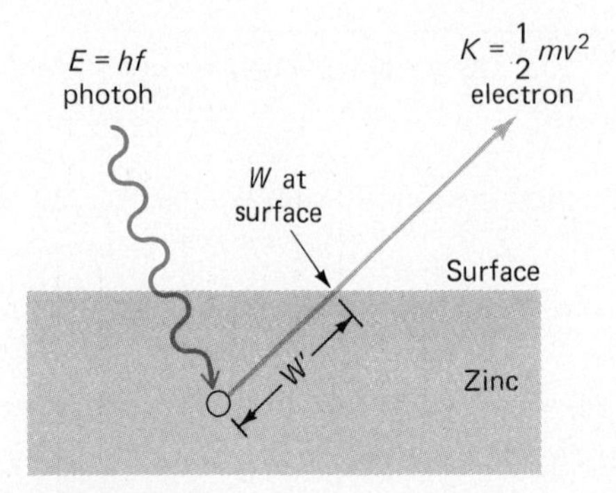

FIGURE 11.7
The processes involved in emission of a photoelectron. A photon with energy $E = hf$ is absorbed by the electron. The electron moves to the surface, losing energy W'. It then leaves the surface losing energy W and is free with kinetic energy $K = \frac{1}{2}mv^2 = hf - W - W'$ left.

is used up in the electron's trip to the surface of the metal, and additional energy W is used up when the electron passes through the forces existing at the surface of the metal. (If it were not for these forces at the surface of the metal, electrons would leak out spontaneously.) The remaining energy is then retained by the electron in the form of kinetic energy after it has emerged from the metal. If we apply the idea of conservation of energy to this situation, we have the following equation:

$$E = hf = W' + W + \tfrac{1}{2}mv^2 \tag{11.2}$$

Unfortunately, Eq. (11.2) cannot be directly verified, because the amount of energy W' that an electron loses in reaching the surface of the metal varies from one electron to another. Some electrons may not be aimed directly at the surface or many start too far from the surface to reach it at all. Other electrons may begin at the surface or may lose no energy in reaching the surface, so that W' equals zero for these electrons. The values of W' for other electrons will lie between these two extremes. However, there is no way to tell what the various values of W' will be for a given metallic surface.

On the other hand, the energy W lost in passing through the surface is the same for each photoelectron, although it will differ from metal to metal. W is known as the *work function* of the metal.

The situation is simpler if we consider only those electrons for which W' equals zero. According to Eq. (11.2), these electrons emerge from the surface with maximum kinetic energy for the particular frequency and metal used. We can write then

$$hf = W + (\tfrac{1}{2}mv^2)_{\text{max}} \tag{11.3}$$

If we solve Eq. (11.2) for the kinetic energy $(\tfrac{1}{2}mv^2)_{\text{max}}$ of the most energetic electrons, we find

$$(\tfrac{1}{2}mv^2)_{\text{max}} = hf - W \tag{11.4}$$

Examination of Eq. (11.4) will reveal why the photoelectrons depend on frequency for their maximum energy. Light intensity does not appear in Eq. (11.4); therefore, the maximum energy of the photoelectrons should be independent of the light intensity, as is observed. If the energy hf is less than W, there should be no emission of photoelectrons at all, because negative kinetic energy has no physical meaning. On the other hand, if hf is greater than W, Eq. (11.4) predicts a straight-line relation between the

frequency f and the maximum kinetic energy of the photoelectrons. Finally, because photons travel at the speed of light, when weak blue light is turned on the surface, some of the photons reach the surface almost instantly and are absorbed, giving rise to a few photoelectrons with negligible time delay. Thus, all of the features of the photoelectric effect are explained by Einstein's theory. From the success of this theory, we conclude that the energy of an electromagnetic wave is transmitted by photons, each carrying energy in accord with the equation $E = hf$. In 1921, Einstein received the Nobel prize for his explanation of the photoelectric effect.

At this point, you might wonder why the discovery of photons did not occur until 1905. The answer lies in the extremely small energy carried by an individual photon. Planck's constant h has the value of 6.6×10^{-34} joule-sec. Thus, a photon of violet light, having frequency $f = c/\lambda = 7.5 \times 10^{14}$ cps, has an energy of only 5×10^{-19} joules. This energy is far too small for Einstein or anyone preceding him to have detected. The existence of photons was predicted on the type of indirect evidence given by the photoelectric effect. Only in recent years, with the development of highly sensitive photomultiplier tubes, has it been possible to detect individual photons of visible light. Photons of radio frequencies (around 10^7 cps) have energies on the order of 10^{-28} joules and remain completely unobservable even today. At the other end of the electromagnetic spectrum lie the very high frequencies (10^{20} cps) of gamma rays; the photons associated with these rays are easily detectable.

We have seen that under certain circumstances electromagnetic radiation (and light) seems to be composed of photons rather than waves. The larger the frequency of the radiation, the more obvious this fact becomes. Therefore, was Newton correct and Huygens wrong? The answer is no. Even at gamma-ray frequencies, where the quantization of light into photons is most evident, diffraction and interference effects are still apparent. Both the wave theory and the particle theory of light are needed to explain all the observed data. During the first three decades of the 20th century, these two ideas remained distinct. Between 1930 and 1950, however, a new theory was developed. This theory, *quantum electrodynamics,* predicted both photons and waves. It incorporated Maxwell's equations and Einstein's photons into a complete picture of electromagnetic radiation. In this picture, both waves and photons appear as complementary aspects of a single phenomenon. Quantum electrodynamics predicts the electromagnetic interactions of matter and light with a precision that is unrivaled by any other physical theory. The principle architects of quantum electrodynamics, Richard P. Feynman, Julian Schwinger, and S. Tomonaga, received the Nobel prize in 1965.

11.3 REFLECTION AND REFRACTION

In this section we shall discuss two of the laws governing the propagation of light, the Laws of Reflection and Refraction. We assume that the objects involved have dimensions that are large compared to the wavelengths of visible light, which are about 6×10^{-7} meters. We therefore assume that beams of light do not bend or spread and that they move in straight lines called *light rays,* because diffraction and interference phenomena are unimportant.

The law concerning a ray of light reflected from a polished, plane surface was known by the Greeks. As shown in Fig. 11.8, it is customary to measure the angle of incidence i and the angle of reflection r with respect to a line normal to the reflecting surface. We can then state the *Law of Reflection* from a plane surface in the following form:

> *1. The angle of incidence equals the angle of reflection, or* $i = r$.
> *2. The incident ray, the normal, and the reflected ray all lie in the same plane.*

This law can be most easily derived by assuming that the photons are reflected elastically from the surface. The particle aspect of light is not

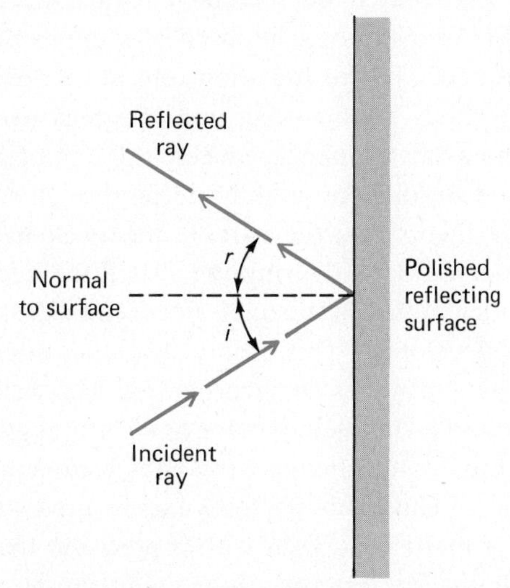

FIGURE 11.8

A light ray is reflected from a polished surface. The angles of incidence and reflection are measured with respect to the normal to the surface.

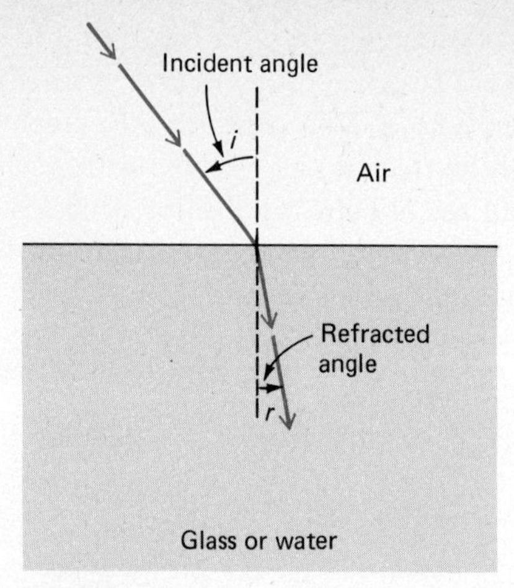

FIGURE 11.9

As light enters the transparent medium it is bent or refracted towards the normal.

necessary, however, and all other waves, such as sound and water waves, obey the same law.

Suppose that a beam of light travels from air into a transparent material, like glass or water. The light ray's direction is changed when it enters the material. This change is known as *refraction* and is illustrated in Fig. 11.9. The direction of the light ray is customarily indicated by measuring its angle with respect to the normal or perpendicular line to the surface, as was done for reflection. In the example shown in the diagram, the light ray is bent closer to the normal as it passes from air into the glass. Refraction is an effect common to all waves. It is most easily seen in light and in water when waves cross from deep into shallow water and are also bent towards the normal. In both cases, the reason for the refraction is the same: The speed of the wave changes when it enters the new medium. This fact was first shown for light by J. B. L. Foucault in 1850 when he conclusively demonstrated that the velocity of light in water is less than it is in air.

As we mentioned in Sec. 11.1, the velocity of light in a vacuum is conventionally given the symbol c and has a value of approximately 3×10^8 m/sec. Let us call the velocity of light in a material medium v. Then the *index of refraction* of the material n is defined by the equation

$$n = \frac{c}{v} \tag{11.5}$$

Because the light always slows down when entering a material medium, n is always larger than unity. Typical values of n for some common substances are given in the table below. Note that for all except the most precise measurements, air can be treated as a vacuum with regard to its index of refraction. As we shall see, a light ray is more deflected when entering a material with a high index of refraction than it would be when entering a material with a low index of refraction.

INDEX OF REFRACTION FOR COMMON MATERIALS

Material	n	Material	n
Dry air	1.0003	Glycerine	1.470
Water	1.333	Benzene	1.501
Ethyl alcohol	1.354	Crown glass	1.517
Acetone	1.359	Flint glass	1.627
Chloroform	1.446	Carbon disulphide	1.628
Quartz	1.458	Diamond	2.419

To see how the change in the velocity of light produces refraction, we must look in some detail at the wave as it crosses the interface between two mediums. Consider a light ray incident normal (that is, perpendicular) to the surface of a piece of glass, as shown in Fig. 11.10. The wave fronts or crests are parallel to the interface and are also shown in the diagram. As the waves cross the interface, each wave front passes in turn from one medium into the other. The number of wave fronts striking the interface per second is exactly equal to the number leaving per second; that is, the

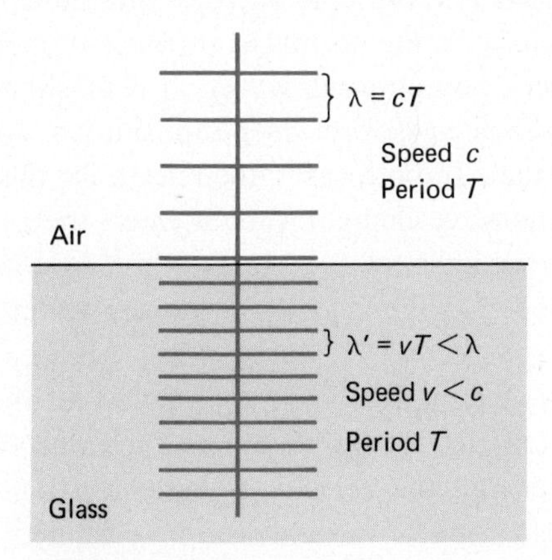

FIGURE 11.10

As a light ray with parallel wavefronts enters glass the speed of the wave decreases and its wavelength, the distance it travels in one period, also decreases. The period and frequency are the same on both sides.

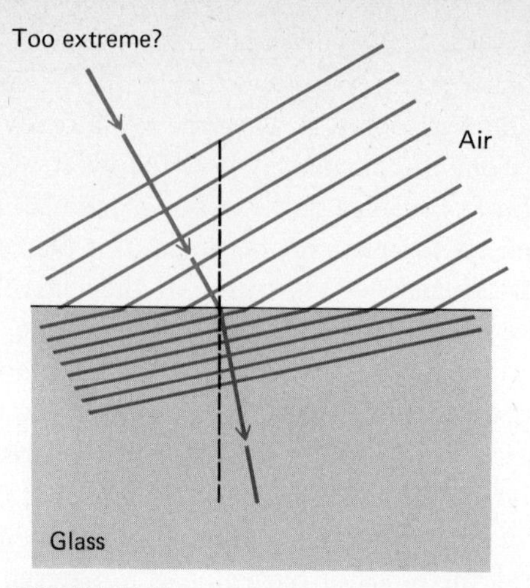

FIGURE 11.11
A parallel wavefront light wave strikes a glass surface at an angle. In the glass the waves are closer together and they connect to the waves in the air forcing the ray to bend.

frequency and thus the period of the waves is identical on both sides. But the wavelength of a wave is the distance that each crest travels in one period. Because the wave is traveling slower in the glass, its wavelength is decreased. We have indicated this behavior in Fig. 11.10 by drawing the wave fronts closer together in the glass. A simple calculation shows that since $\lambda = vT$, the wavelength λ' of the light in the glass is related to the wavelength λ of the light in air by the equation $\lambda' = \lambda/n$. It is this shift in wavelength that produces refraction.

Consider now a ray of light incident at an angle to a piece of glass, as pictured in Fig. 11.11. The light must satisfy two constraints as it passes into the glass. First, the wavelength must be decreased to λ/n, for the reasons just given. Second, each wave front must be continuous across the interface. Because the incident wave produces the refracted wave, each wave front in the glass must connect to one in the air. The only way to meet both of these conditions is for the wave fronts in the glass to meet the interface at a smaller angle than the wave fronts in the air do. Thus, the ray is bent towards the normal. The amount that the light ray is bent is given by Snell's Law, which we shall mention only here. This law states that

$$\sin i = n \sin r$$

where the angles are defined in Fig. 11.9.

You should note one of the important properties of light rays: If

they are reflected back on their original path, they retrace it exactly. Thus, we do not have to consider separately what happens when a light ray leaves a material medium and enters a vacuum. We already have the solution: It is bent away from the normal, just as before it was bent towards the normal. For example, consider light entering a prism, as shown in Fig. 11.12. When the light enters the glass of the prism, it is bent towards the normal and thus towards the base of the prism. It then travels to the other side of the prism where it leaves the glass. At the second side, however, the angles have changed; although it is now bent away from the normal at the surface, this normal is tipped with respect to the normal at the first surface. As the net result, the ray is bent towards the base of the prism, as illustrated in the figure. The larger the top angle of the prism and the larger the index of refraction of the glass, the more pronounced this effect is.

In general, the index of refraction of a material varies slightly with the wavelength of the light used; that is, different wavelengths have slightly different velocities in a material medium. This property is known as *dispersion.* If a beam of light is passed through a prism, as shown in Fig. 11.13, different wavelengths will be refracted through different angles. For glass, the long wavelengths (red) are refracted least and the short wavelengths (violet) are refracted most. The array of colors that emerges from the prism is known as a *spectrum.* Probably the most common example of a spectrum is the rainbow, which is caused by the dispersion of light by water droplets. From such simple observations, we can conclude that white light (sunlight) is a mixture of colors (wavelengths). In the last part of the book we shall discuss the origin of these various wavelengths.

We have all looked through a plate glass window and seen our reflection superimposed upon the scene beyond. Obviously, when light is incident on an optically dense medium such as glass, part of the ray is reflected and part is refracted and thus transmitted. It is almost impossible to eliminate the reflected ray completely, although the coatings on lenses reduce it greatly.

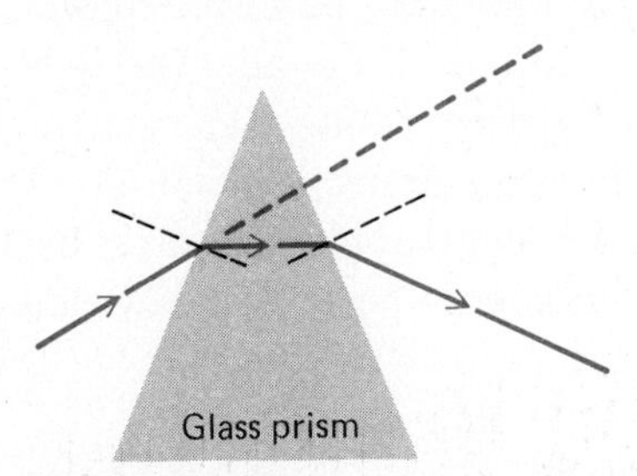

FIGURE 11.12
When a light ray passes through a prism it is bent towards the base of the prism.

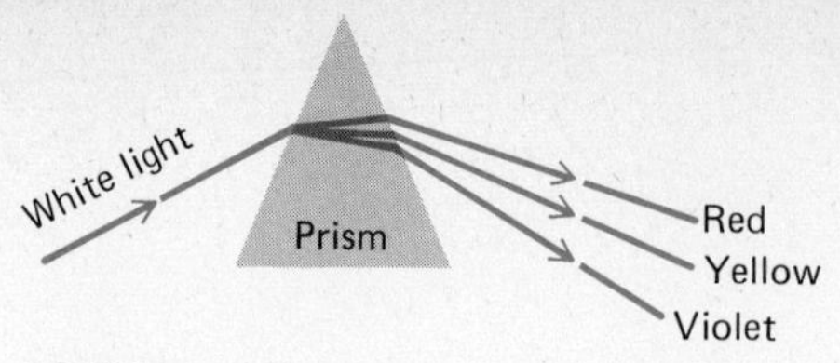

FIGURE 11.13

When white light passes through a prism the various wavelengths are bent by different amounts; the short wavelengths bend the most.

Under certain circumstances, it is possible to have no transmitted ray. Consider a ray leaving a piece of glass, as shown in Fig. 11.14. We have chosen a ray that has a rather large angle of incidence and is thus refracted almost parallel to the surface. We have also drawn the reflected ray in the diagram. Imagine that we further increase the angle of incidence of the ray. Eventually we shall reach an angle such that the refracted ray is bent parallel to the surface. At this angle, called the *critical angle,* the energy carried by the photons in the refracted ray is moving parallel to the surface and no net energy leaves into the air. Consequently, *all* of the energy goes into the reflected ray. As a result, no light is transmitted. This property is called *total internal reflection.* The critical angle in glass is about 42 deg. If light strikes the surface of glass from the inside with a greater angle of incidence, it will be totally reflected.

Figure 11.15 depicts the type of prism used to "fold the optics" in an ordinary pair of binoculars. The rays enter perpendicular to the glass and strike the second surface at an angle of 45 deg. At this surface they are reflected without the need for any silvering and exit through the third

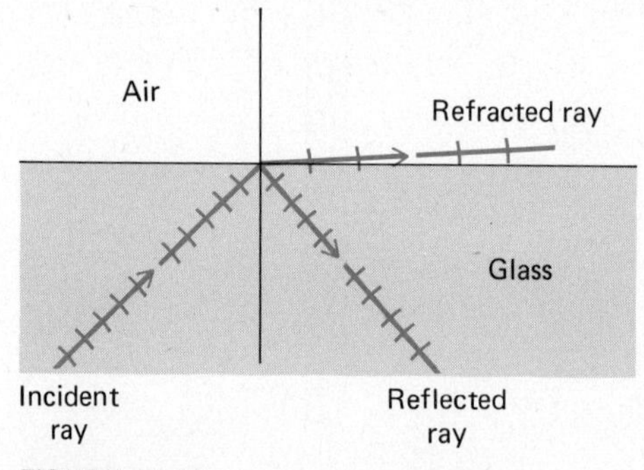

FIGURE 11.14

In this picture the angle of incidence is so large that the refracted ray is bent almost parallel to the surface. If the angle of incidence were increased slightly, total internal reflection would occur.

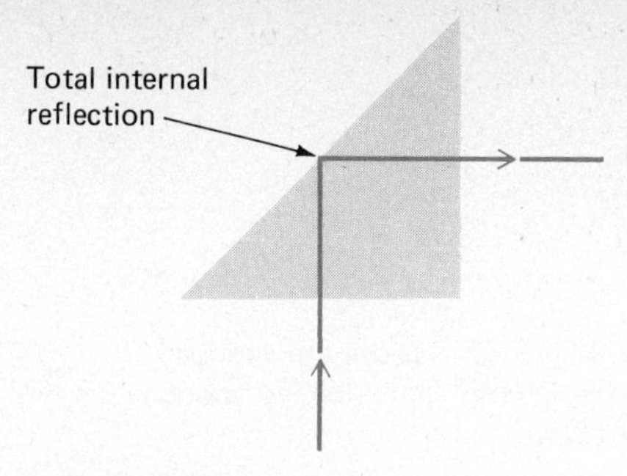

FIGURE 11.15
Total internal reflection in a prism. Compare this light path with the one shown in Fig. 11.12.

face. Four such prisms are used in each tube of a pair of binoculars to shorten the overall length of the instrument.

Another important application of total internal reflection is the field of fiber optics. Figure 11.16 shows a tube of solid lucite. A light is shining into the end of the tube. Some of the light rays are lost at the opening of the tube. But some of them enter at an angle that is small enough to give them an angle of incidence larger than the critical angle by the time they strike the side of the tube. As a result, they are totally reflected. As long as the tube is not bent into too sharp a corner, the light rays will continue to bounce along the inside of the tube, always being reflected at

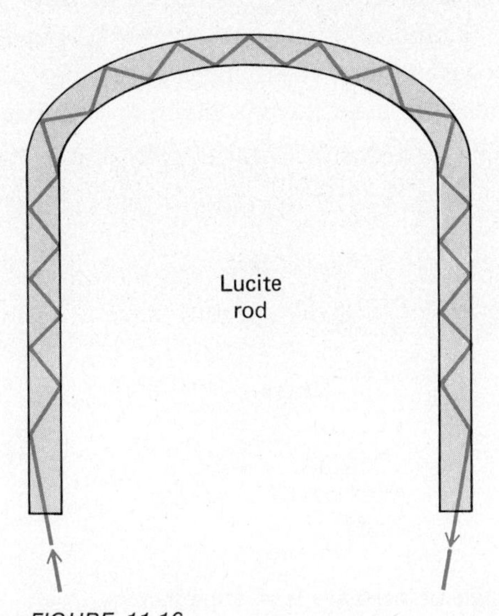

FIGURE 11.16
A lucite rod bent into a U shape carries a beam of light around inside it by many successive internal reflections.

the surface. The beam of light will be led along the tube and will emerge at the other end essentially undiminished. In fiber optics, millions of such tubes, each a thread of plastic or glass, are bound together into a bundle. When an image is placed in front of one end, it illuminates each of the fibers with differing amounts of light. If the bundle were perfectly combed so that the fibers were parallel, this image would emerge undistorted at the other end, even if the bundle were tied in a knot. In practice, the fibers are tangled, and the image must be sorted out with a computer. The principle, however, remains the same. Fiber optics have important applications in such areas as medicine where they can be used to obtain photographs of places like the inside of a person's stomach.

11.4 POLARIZATION

Earlier in this chapter we stated that sound waves are longitudinal, with the vibrations taking place in the same direction as the wave itself moves; whereas light waves are transverse, with the variations in electric and magnetic fields occurring at right angles to the direction in which the disturbance moves. In this section we shall learn how to discriminate between the two types of wave motion by using a method called *polarization*. Only transverse waves can be polarized.

Let us consider first a fence with water waves striking it, as shown in Fig. 11.17. If the rails are horizontal obstacles, the up-and-down motion of the water will be greatly reduced, and only a fraction of the energy carried by the waves will pass through the fence. On the other hand, if the fence has vertical rails, as illustrated in Fig. 11.18, the vertical motion will be only slightly affected, and most of the energy of the waves will get past the fence.

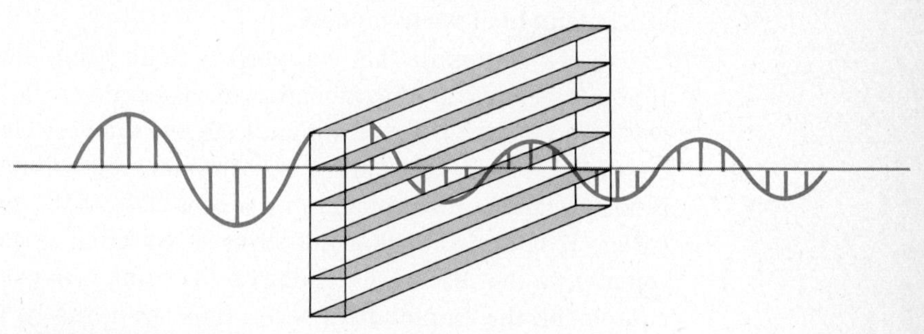

FIGURE 11.17

A water wave incident on a horizontal fence is strongly decreased in amplitude and intensity. If the horizontal slats were wider and closer together the wave would be almost completely blocked.

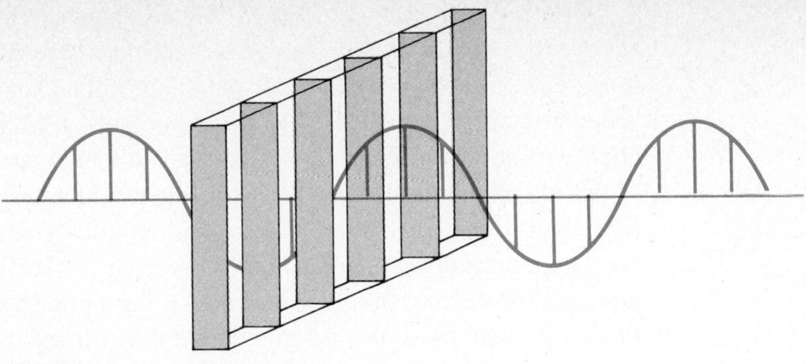

FIGURE 11.18

A water wave incident on a vertical fence is hardly affected. If the slats were wider and closer together the wave would still pass through with almost its original amplitude and intensity.

Thus, the waves are transmitted if directions in the obstacle do not interfere with the motion of the water. If there are no directions of easy transmission, then the waves will be almost completely stopped. We say then that water waves are *polarized* in a vertical direction; that is, they can pass by vertical obstacles but not horizontal obstacles.

Now let sound waves strike a set of parallel obstacles similar to those pictured in Figs. 11.17 and 11.18. Here the vibrations of the material (air) are along the direction of the motion of the waves themselves. If we change the direction of the obstacles, the waves are not affected, because they can pass between obstacles that are oriented in any way whatever. Thus, no matter how obstructions are rotated, a longitudinal wave is always transmitted in the same way through a set of parallel obstacles. A longitudinal wave does not show any polarization effects. In the case of sound, we find experimentally that there is no polarization effect, so we conclude that sound is a longitudinal wave motion.

Now suppose that you pluck a violin string in a vertical direction. If you hold a piece of cardboard with its edge vertical to the string, there will be no effect on the string until the two are very close. If you turn the edge of the cardboard until it is horizontal, you will notice that the string is hitting the cardboard before its edge is close to the visible postion of the string. It is apparent that the string is vibrating vertically but not horizontally, so that the wave existing on the string is transverse. With suitable equipment, the amplitude of waves on a string can be made large enough for the eye to see the direction in which the vibration takes place. Such an experiment is the most convincing evidence that waves in a string are indeed transverse.

In a solid there may be both transverse and longitudinal waves. For instance, earthquake waves in the earth consist of both types. Because the two types of waves travel at different speeds, they take different times to travel from the center of an earthquake to the detecting equipment. The speeds of both wave types are known; therefore, a measurement of the time difference allows scientists to calculate the distance covered by the waves. If several observing stations determine these distances, the location of the center of the earthquake can easily be found. From such observations, scientists can also obtain considerable information about the interior structure of the earth.

To find out if an electromagnetic wave is transverse or longitudinal, we can perform a simple experiment. If we use wavelengths of the order of centimeters, we can easily construct a set of parallel conducting wires similar to the fences shown in Figs. 11.17 and 11.18. The oscillating electric field can pass through such a grid only when it is parallel to wires or parallel to the spaces between the wires. If the electric field is at right angles to the wires, the energy of the field is used up in heating the wires. Waves from a straight antenna have their electric fields parallel to the back-and-forth motions of the electrons along the length of the antenna. We observe transmission of the electric field when the grid is parallel to the antenna. After the grid has been rotated through 90 deg, very little energy is found to pass through the grid. From these observations, we decide that radio (electromagnetic) waves are transverse.

When electromagnetic radiation of visible wavelengths is emitted from a gas or solid, the "antennas" are usually individual atoms or molecules. We shall discuss this topic in Chap. 13. These sources are oriented at random, so we must first find out how to sort out those waves that have their electric fields all in the same direction. This selection is done with a device similar to a set of parallel conducting wires. For instance, a material made by the Polaroid Corporation and called Polaroid consists of a set of parallel hydroquinone crystals. When light passes through a piece of Polaroid, only those waves or parts of waves that have their electric fields parallel to the crystals are transmitted. The waves emerging from the piece of Polaroid are then said to be *polarized* parallel to the crystals in the material. To test that these waves are polarized, we use a second piece of Polaroid as an analyzer. When the crystals of the analyzer are parallel to the crystals of the first Polaroid sheet, we observe transmission of light through the analyzing sheet of Polaroid. If the analyzer is rotated 90 deg, no light passes through it. These experiments are illustrated in Fig. 11.19. Because these experiments show light as the property of polarization, we conclude that it is a transverse wave. In fact, all electromagnetic waves are transverse.

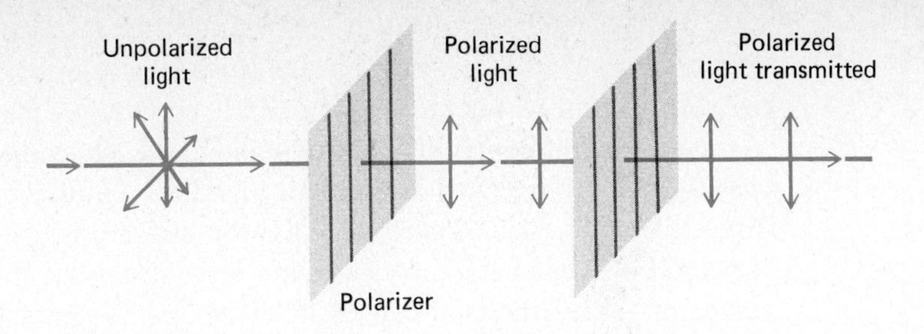

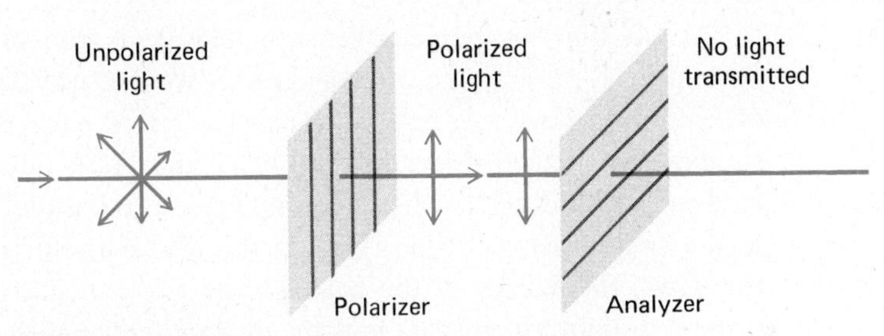

FIGURE 11.19
In these two pictures unpolarized light is incident on a piece of polaroid called the polarizer. The colored arrows represent the transverse electric field of the incident and polarized beams. In the upper picture the polarized light is transmitted through a second piece of polaroid called the analyzer and oriented parallel to the polarizer. In the lower picture the analyzer is rotated 90° and no light is transmitted.

However, scientists need widely differing measuring devices to cover the wide range of wavelengths and frequencies for this type of wave.

11.5 OPTICS

Everyone has had some experience with optical devices, ranging from simple mirrors to complex cameras and binoculars. When you look into a mirror, you see a clear replica of the objects behind you (as well as yourself). When you look into a pair of binoculars, you see a distant scene brought closer. This process of *image formation* is common to all optical instruments. The light coming from some object is bent and focused in such a way that it appears to come from a different place, called the *image*. The image may be larger or smaller than the object, and it will generally be in a different location. In this section we shall utilize the Laws of Reflection and Refraction to explain image formation by a lens or a mirror. Because a sufficiently

small part of any mirror, plane or not, can always be considered flat, images produced by mirrors can be found geometrically if we use the Law of Reflection.

For example, let us consider the image formed by a large, plane mirror. Each ray of light traveling from an object to the eye obeys the Laws of Reflection stated in Sec. 11.3. (Naturally, only rays of light directed so that they strike your eye contribute to the image that you see.) As illustrated in Fig. 11.20, the eye sees light rays appearing to emerge from an object as far behind the mirror as the actual object is in front of the mirror. This type of image is known as a *virtual image,* because the rays of light do not actually pass through the image points but only appear to do so. A careful drawing of rays from an extended object, such as your face, will show that the right and left sides are interchanged. We say then that the image is *reversed.* Through experience, you adjust to this fact when you use a mirror to shave or fix your hair. As can be seen from the diagram, the image formed by a plane mirror is upright and the same size as the object.

Frequently, mirrors shaped like part of a sphere are used. An example is the magnifying shaving mirror. Suppose that light from a distance source strikes such a mirror. In this case, the rays of light will all be parallel. Approximately all of the parallel rays are reflected in such a way that they pass through a single point, known as the *focus* of the mirror. This fact is shown for a *concave mirror* in Fig. 11.21. Furthermore, the focal length f is one-half the radius of curvature R of the mirror. In a similar way, parallel rays of light are reflected from a *convex mirror* as if they came from a focal point, as illustrated in Fig. 11.22. Again, the focal distance is one-half the radius of curvature of the mirror.

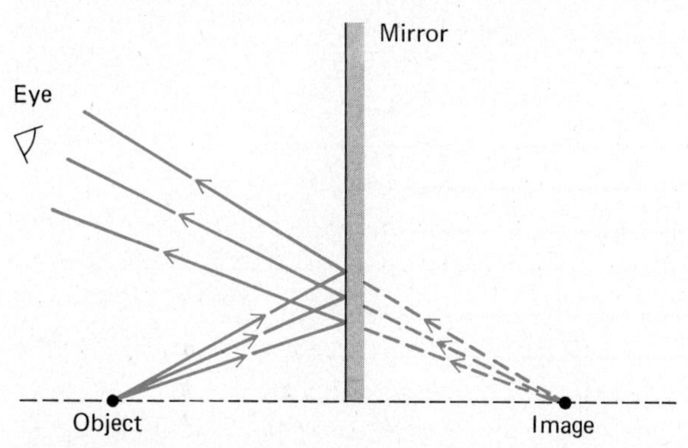

FIGURE 11.20

Image formation by a plane mirror. The light rays leave the object and are reflected by the mirror. The eye, which assumes that they have travelled in a straight line, sees them as coming from the object.

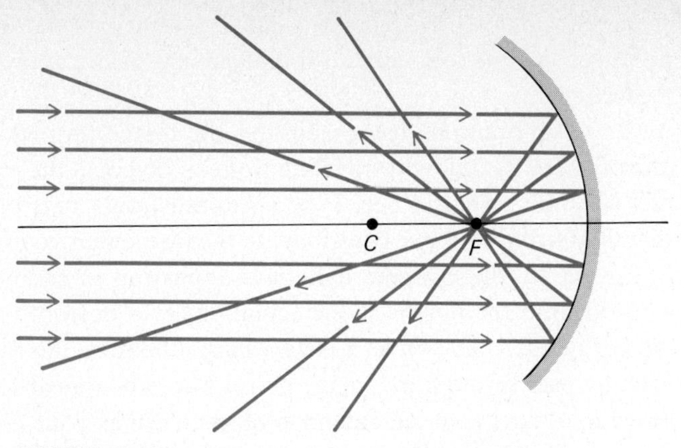

FIGURE 11.21

Light rays entering parallel to the axis of a spherical mirror seem to emerge from a point midway between the center of curvature, *C*, and the mirror. These light rays are reversible; if a light source were placed at *F*, its rays would leave parallel to the axis. This is the principle of a flashlight reflector.

Suppose that an object is not a great distance from the mirror and we would like to construct graphically the image produced by the mirror. A ray from the object is parallel to the axis of the mirror and is reflected as if that ray had come from a very distant object. A ray striking the center of the mirror is reflected, by symmetry, just as it would be from a small plane mirror located at the center of the actual curved mirror. The point

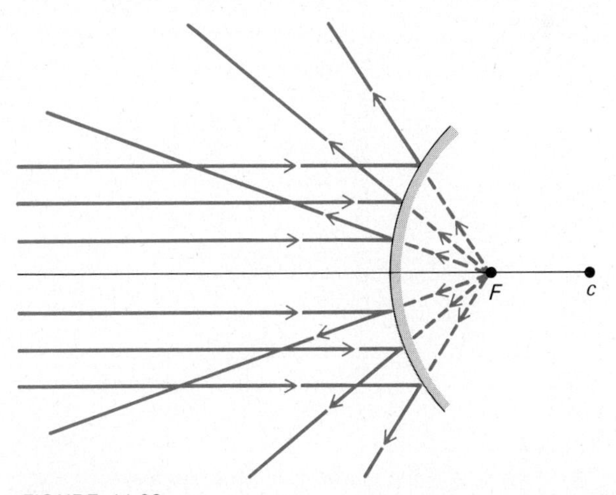

FIGURE 11.22

Light from a distant source is reflected from a convex mirror as if it came from a point behind the mirror.

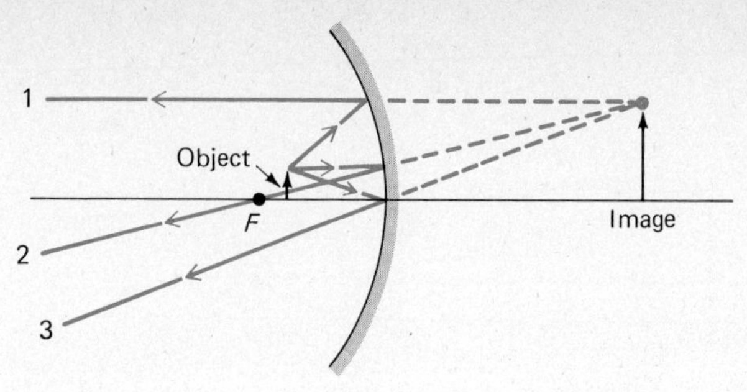

FIGURE 11.23

Image formation by a concave mirror. The object is inside the focal point and the image is virtual. Ray 1 leaves as if from *F* and is reflected parallel to the axis. Ray 2 leaves parallel to the axis and is reflected through the focal point *F*. Ray 3 is reflected symmetrically. This is the way a shaving mirror magnifies.

at which these two rays intersect is then the image point corresponding to the original object point. (It can be shown that all rays from a given object point pass through approximately the same image point.) The two special rays just described are shown for both a concave and a convex mirror in Figs. 11.23, 11.24, and 11.25. In the case of the example with the concave mirror (Fig. 11.24) the image is *real,* because the light rays actually pass through the image points. Furthermore, the image is *inverted, reversed,* and *magnified.* This is not always the case, however, with a concave mirror. When a convex mirror is used, the image is always *virtual, erect, perverted* (turned left for right), and *smaller,* as the diagram shows.

It is possible for a concave mirror to give a magnified, erect, and perverted image, as illustrated in Fig. 11.23. A shaving mirror would be

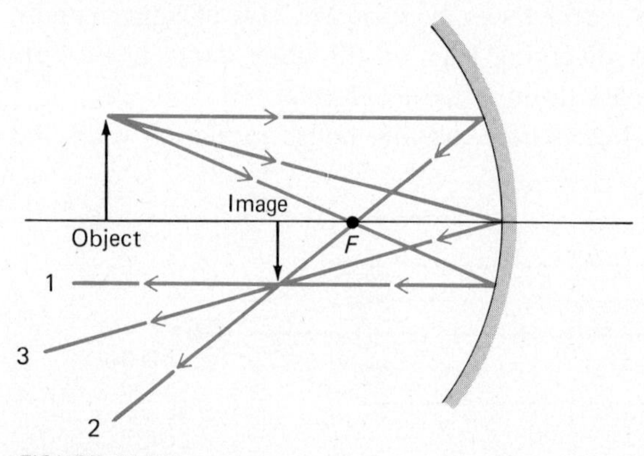

FIGURE 11.24

The same mirror as in Fig. 11.23 with the object outside the focal point. The same three rays are shown.

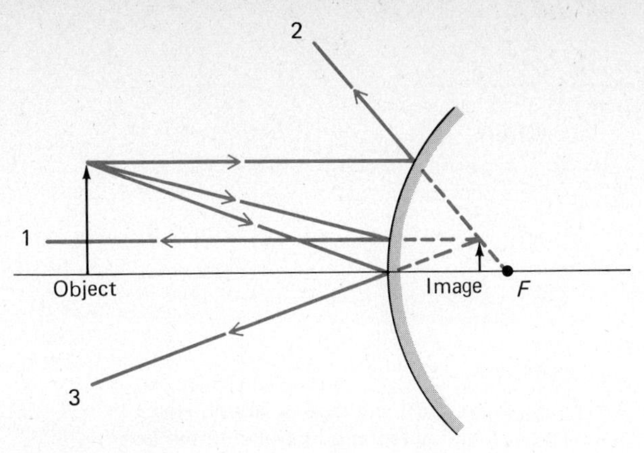

FIGURE 11.25
Image formation by a convex mirror. This is the way a Christmas tree ornament reflects light.

such an example. The only requirement is that the object be closer than its focal length to the mirror.

In Sec. 11.3 we discussed refraction of a light ray by a prism. A simple lens usually has surfaces that are portions of a sphere rather than the flat surfaces of a prism. If light from a distant object, such as the sun, strikes a lens, the rays of light will be parallel to the axis of the lens. In Fig. 11.26 various parallel rays are shown striking a *converging lens*. All the rays go through almost a single point after passing through the lens. This point is called the *focus* of the lens.

In a similar way, when parallel rays of light strike a *diverging lens,* after passing through the lens they all appear to head for a focal point. This situation is shown in Fig. 11.27. You should note that a converging lens has a real focus, because the rays of light actually pass through this point. A diverging lens, on the other hand, has a virtual focus; the rays do not pass through the focal point.

Let us now consider image formation when the object is not very

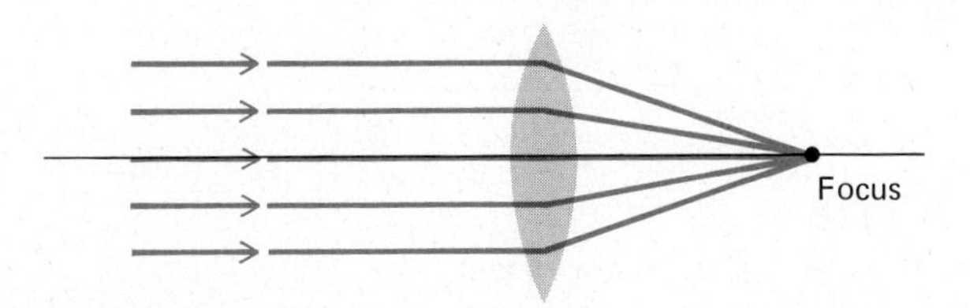

FIGURE 11.26
Parallel rays enter a converging lens and are all bent to the focus.

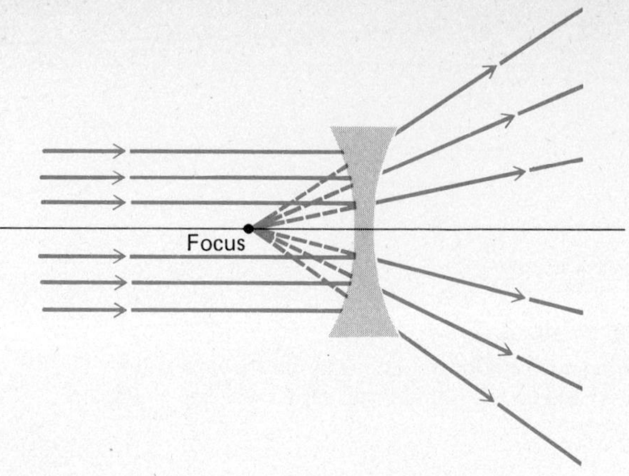

FIGURE 11.27

Parallel rays of light enter a diverging lens. They exit from the lens as if they came from the focus.

far from the lens. For either type of lens, we can draw a line from a point on the object parallel to the axis of the lens. On its outward passage this line will pass through the focus of a converging lens or will appear to come from the focus of a diverging lens. A second ray that we can draw from the point on the object is one passing through the center of the lens. Because of symmetry, this ray will not be deviated by the lens. The point at which these two special rays intersect is then the image point corresponding to the original object point. We shall now show that other lines from the same point on the object also pass through approximately the same image point.

Suppose that an object is situated with respect to the lens as shown in Fig. 11.28. By constructing the two special rays, we can determine the

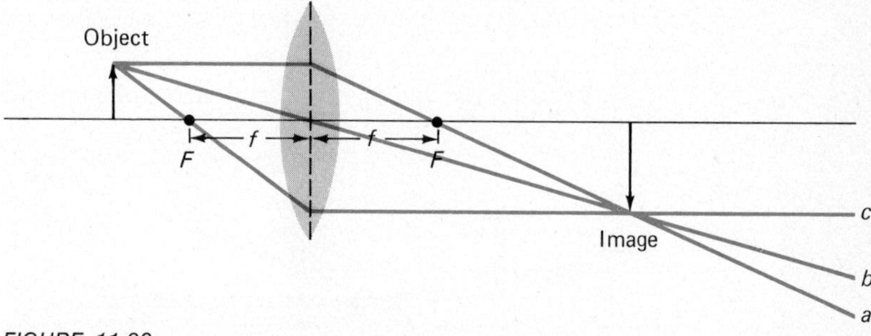

FIGURE 11.28

Image formation by a converging lens. Ray *a* enters parallel to the axis and is bent through the focus. Ray *b* enters through the center of the lens and is not bent. Ray *c* is the reverse of ray *a*, entering as if from the focal point and leaving parallel to the axis.

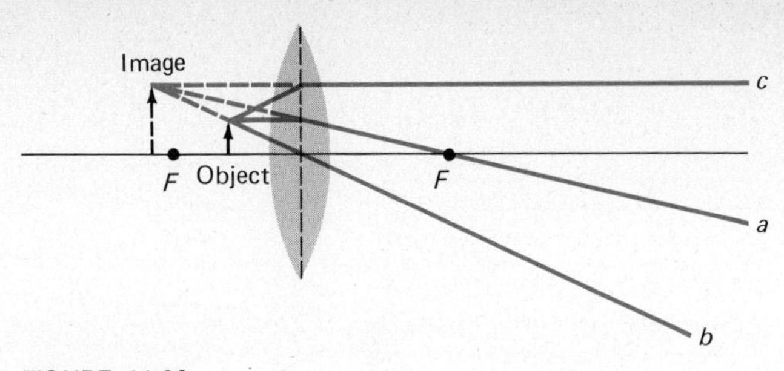

FIGURE 11.29
In this example the object is inside the focus and the image is virtual. The same three rays as in Fig. 11.28 are shown.

location of the image. We see that the image is inverted, real, and larger than the object. If we made a three-dimensional diagram, we would also find that the image is perverted.

If the object is moved closer to the lens than in the preceding example, the ray diagram becomes the one shown in Fig. 11.29. Here we find that the image is virtual, upright, and larger than the object. In this case, the image is not perverted. When we use a converging lens to magnify an object in this way, we have a simple *magnifying glass.* These are commonly used by coin and stamp collectors, for instance. They also help in reading scientific instruments.

In studying the image formed by a diverging lens, we again use the two special rays that allow us to locate the image easily. These lines and the image formed are shown in Fig. 11.30. In this case, we see that the image is virtual, upright, and smaller than the object. Moreover, the image is not perverted. If the object were moved a different distance from the lens, the size and position of image would change, but the image would still be virtual, erect, smaller, and not perverted. Thus, a diverging lens never gives a real image or magnification.

Because the index of refraction of all materials varies somewhat

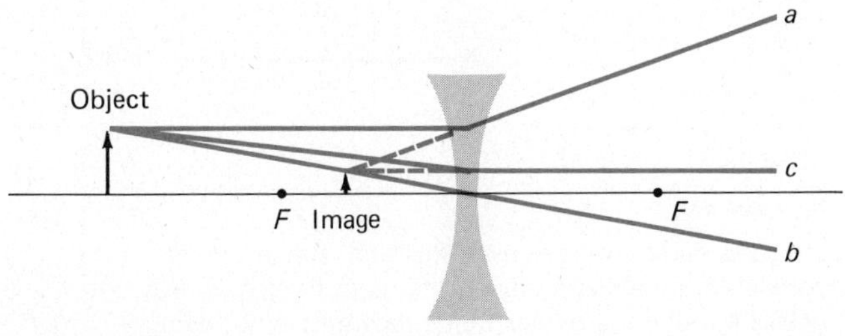

FIGURE 11.30
Image formation by a diverging lens.

with wavelength, different wavelengths emitted by the object do not focus at exactly the same point. The image, therefore, is formed at different distances for different wavelengths. This fact is known as *chromatic aberration*. In addition, a sperhical surface does not focus all the rays from a given point of an object at the same distance from the lens. As a result, blurring of the image occurs. This type of lens error is known as *spherical aberration*. Expensive optical instruments compensate for these aberrations by using a composite lens constructed of several simple lenses made from different types of glass.

A basic optical instrument that nearly everyone has used is the *camera*. Figure 11.31 depicts a simple camera. A real image of the object is focused on the film by a single converging lens. If a sharper image is wanted, the single lens is replaced by a combination of lenses in order to correct for the aberrations discussed in the preceding paragraph. With any lens, the image can be made sharper by using only the central part of the lens. The camera, therefore, has an adjustable circular diaphragm. If you close the diaphragm to a small diameter, you get a sharper picture, but less light is allowed to reach the film. The *f*-number of a lens is defined as the ratio of the focal length to the diameter of the lens opening. A large *f*-number means that the diaphragm is small, so that a better image is obtained but less light passes through the lens. However, if too large an *f*-number is used, diffraction broadening causes a less sharp image. This effect is discussed at the end of the chapter.

Another familiar optical instrument is the *astronomical telescope*. As illustrated in Fig. 11.32, two converging lenses are used. Light from a distant object passes through the objective lens and produces a real image, I_1. This image is examined and magnified by the eyepiece lens in such a way that the eye sees the virtual image, I_2. The chief disadvantage of this instrument is the inversion of the final image. This handicap is not great

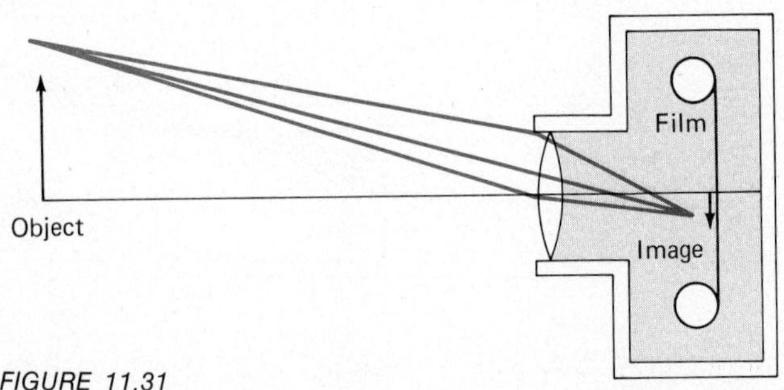

FIGURE 11.31
A camera forms an image of the subject (object) on the plane of the film.

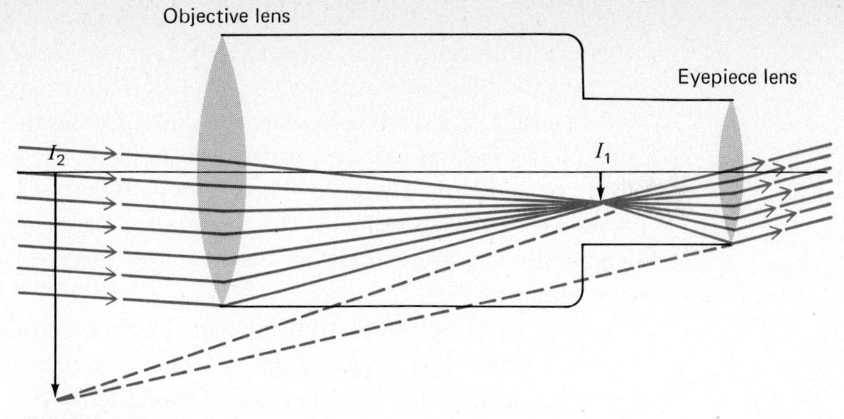

FIGURE 11.32

In this astronomical telescope light from the top of a distant object is focused at image point I_1. The second lens then forms a virtual image at I_2. The image I_2 appears magnified because it subtends a larger angle at the observer's eye than the original did.

when you are looking at stars, but it is quite annoying when you are watching a horse race, for instance.

The solution is the introduction of a third converging lens between the objective and eyepiece lenses, as shown in Fig. 11.33. The only purpose of this third lens is to invert the image a second time, so that the final image is upright. Such an instrument is called a *terrestrial telescope* or *spyglass*. With the third lens, the barrel of the telescope has to be longer than before, which is the main drawback of this device.

Another common optical instrument is the *microscope,* depicted in Fig. 11.34. Two converging lenses are used. The object is placed close to

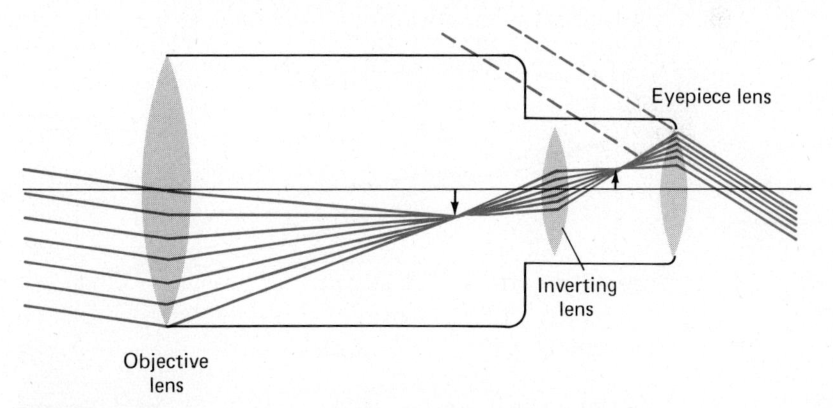

FIGURE 11.33

A terrestrial telescope. This telescope differs from the astronomical telescope of Fig. 11.32 by the extra inverting lens which produces an upright image.

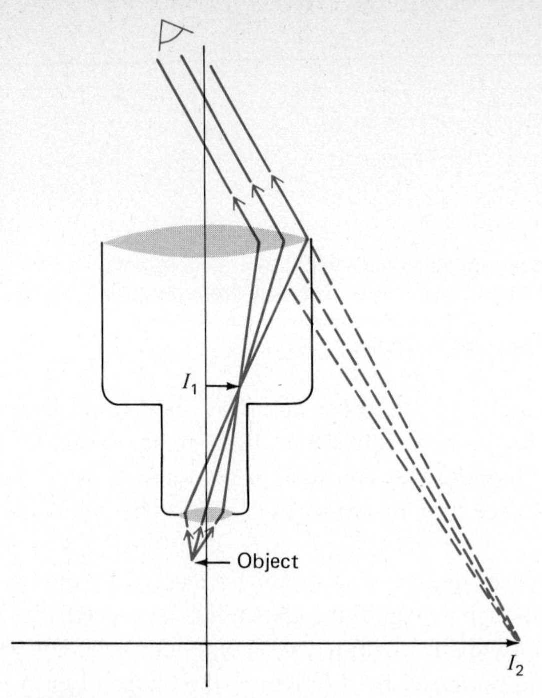

FIGURE 11.34
A compound microscope. I_2 is the final image seen by the eye.

the objective lens, which produces an enlarged, real image. This image is then examined and further magnified by the eyepiece lens. The final image is inverted. But anyone using a microscope quickly learns to compensate for this factor, so it is not really a major disadvantage.

Most large telescopes used by astronomers have a concave mirror that forms a real, inverted image. This image is then further magnified by an eyepiece lens. Although the mirrors may show spherical aberration, all wavelengths are reflected in the same way. Chromatic aberration is therefore eliminated. As another advantage, only one surface has to be carefully ground to shape. Finally, imperfections, such as air bubbles which often occur in large lenses, do not affect performance since only the front surface of a mirror is used. Consequently, all very large telescopes are built around concave mirrors. The largest such telescope now in operation is located at Mount Palomar, California, although a larger one is under construction in the Soviet Union. The mirror of the Palomar telescope has a diameter of 200 in. or $16\frac{2}{3}$ ft.

Thus far in our discussion of reflection, refraction, and geometrical optics, we have assumed that the wavelength of the light used is much smaller than the dimensions of the optical systems, so that the diffraction

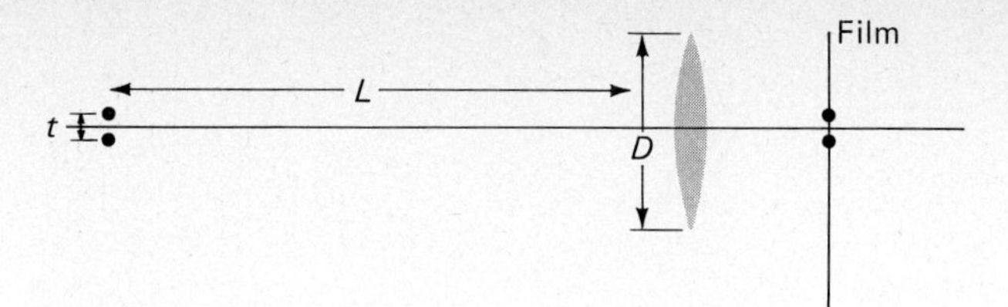

FIGURE 11.35
Two dots separated by a distance t are focused onto film by a lens of diameter D. The dots are a distance L from the lens. Diffraction broadening may cause the image of the dots to overlap.

effects caused by the wave nature of light could be ignored. Let us now briefly discuss how diffraction affects image formation by a lens.

Suppose that you wish to photograph two distant points separated by a distance t, as shown in Fig. 11.35. The points are a distance L from your lens, which has a diameter D. If diffraction were not present, the light from each of these points would be focused by the lens to a point on the film. Although a magnifying glass might be needed, they would form distinct images on the film. Diffraction changes this behavior. Because of the slight spreading produced by diffraction, the image of each point on the film is not a sharp point but a small circular region. If these regions overlap, then the images of the two points will not be distinguishable on the film. When such a situation occurs, we say that the points are too close together to be *resolved* by the lens. If we theoretically analyze this situation for a circular lens, we discover the following criteria for the resolution of two points by a lens. The smallest separation t that two points located a distance L from a lens of diameter D can have and still be resolved is given by

$$\frac{t}{L} = 1.22\,\frac{\lambda}{D}$$

where λ is the wavelength of the light used. This equation is known as Rayleigh's criterion. It places a fundamental limit on the resolving power of any lens using visible light. Diffraction broadening, the effect we are discussing here, becomes important when a camera lens is used at its largest *f*-numbers. Thus, we see that a camera will produce the sharpest pictures when used at the middle range of its available *f*-openings.

DISCUSSION QUESTIONS

1. The speed of light is very great. Discuss possible methods for measuring it.

2. Light can be reflected from a polished surface. Does this tell us whether light is a wave or a stream of particles?

3. A beam of light consists of waves of a single wavelength and frequency. Can you think of a way of changing the wavelength of this light without changing its frequency?

4. In Chapter 10 we discussed interference experiments which show that light is a wave, and yet in this chapter we find that light is transmitted in packets known as photons or quanta each of energy given by $E = hf$. Discuss whether light is a wave or particle.

5. Describe some applications of the photoelectric effect. On what properties of this effect do these applications depend?

6. If you could design the best photoelectric surface for use in the camera tube of a television transmitter, what properties would your surface have?

7. Measurements on the photoelectric effect depend greatly on the way in which the surface is prepared. Explain why this is so.

8. A number of boomerangs could be thrown so as to pass through a horizontal rail fence but not to pass through a vertical rail fence. Discuss whether or not this proves that boomerangs are transverse waves.

9. Why do various objects have definite colors, even when they are illuminated by white light which contains all colors?

10. Give an example of polarization of light used quite commonly in life.

11. If the velocity of light in a material varies with wavelength, does this mean that the index of refraction of the material is not constant?

12. If you wished to examine a small object using a lens, would you choose a converging or diverging lens?

13. A diverging lens always produces a virtual image. Think of ways using an additional lens or mirror so that a diverging lens produces a real image.

14. Discuss how you might measure the focal length of a diverging lens.

15. The *Galilean telescope* or *opera-glass* consists of a converging objective lens, and a diverging eyepiece lens. Its advantage is that it produces an upright image. Sketch a typical ray diagram for this instrument.

PROBLEMS

1. A typical frequency in the FM radio band is 10^8 cycles per second. Compute the wavelength of these waves. (*Ans.:* 3 m)

Repeat for a frequency of 10^6 cycles per second, which lies in the AM radio band.

2. A radio transmitter broadcasts at a frequency of 15,000 cycles per second. Take the velocity of radio waves to be 3×10^8 m/sec and compute the wavelength of these waves. (*Ans.:* 20 km)

3. Some radar sets used during World War II used wavelengths of 55 cm. Compute the frequency of these electromagnetic waves.

4. An electromagnetic wave is sent out by transmitter, reflected by a target, and received 10^{-4} sec later at the location of the transmitter. How far away is the target? (This is the principle upon which *radar* works.) (*Ans.:* 1.5×10^4 m)

5. Signals from two synchronized radio transmitters are received with a time interval of 10^{-3} sec between their times of arrival. Calculate how much farther you are from one transmitter than from the other. (This provides the basis for a system of navigation known as *Loran.*)

6. Compute the energy of a photon of electromagnetic radiation which has a wavelength of 3 cm, as is used in radar. (*Ans.:* 6.63×10^{-24} joules)

7. Assume (incorrectly) that sound waves of frequency 500 cycles/sec are quantized as electromagnetic waves are. Compute the energy of such a sonic photon.

8. Compute the frequency of a photon which has an energy of 10^{-18} joules. (*Ans.:* 1.51×10^{15} cps)

   Compute the wavelength of this photon.

9. A photon has a wavelength of 10^{-8} cm. Compute the energy of this photon in joules. (*Ans.:* 1.99×10^{-15} joules)

10. Light of wavelength 5×10^{-7} m strikes a surface which requires an energy of $W = 1.5 \times 10^{-19}$ joules for a photoelectron to escape. Compute the energy of the most energetic photoelectrons emitted by this material.

    Repeat the problem for radiation of wavelength 5×10^{-6} m. (*Ans.:* none emitted)

11. The longest wavelength which will cause a certain material to emit photoelectrons is 4.5×10^{-7} m. Compute the value of the photoelectric work-function of this material in joules.

12. Assume that a typical frequency of FM radio is 10^8 cycles/sec. Compute the energy of a photon of this frequency. (*Ans.:* 6.63×10^{-26} joules)

13. Compute the wavelength and frequency of a photon which has an energy of 10^{-9} joules. (*Ans.:* $\lambda = 2 \times 10^{-25}$ m)

14. A photon has a wavelength of 6×10^{-7} m. Compute the energy of such a photon in joules.

15. A photon of frequency 6×10^{14} cycles/sec hits a surface which requires an energy $W = 2 \times 10^{-19}$ joules for a photoelectron to escape. Calculate the energy of the most energetic photoelectrons emitted by this material. (*Ans.:* 1.98×10^{-19} joules)

    Repeat for the case that the incident frequency is 6×10^{13} cycles/sec.

16. An object is located 20 cm from a converging lens of focal length 15 cm. Locate the image graphically and determine its character. (Is the image erect or inverted, real or virtual, magnified or diminished, reversed or not?)

    Repeat with the object 10 cm from the lens. (*Ans.:* When the separation is 10 cm, the image is 30 cm in front of the lens, erect, virtual, enlarged, not reversed.)

17. An object is located 20 cm from a diverging lens of focal length 15 cm. Locate the image graphically and determine its character.

    Repeat with the object 10 cm from the lens. (*Ans.:* 6 cm in front of lens, erect, virtual, smaller, not reversed)

18. Light has a velocity of 2×10^8 m/sec in a certain glass. What is the index of refraction of this glass.

19. If the index of refraction of diamond is 2.42, what is the velocity of light in this material? (*Ans.:* 1.24×10^8 m/sec)

20. An object is located 5 cm from a concave, spherical mirror of focal length 10 cm. Locate the image graphically and give its character. (*Ans.:* 10 cm in back of mirror, erect, virtual, diminished, and reversed)

    Repeat with the object 25 cm from the mirror.

Repeat with the mirror convex. (*Ans.:* When the object is 5 cm from the mirror its image is 3.33 cm behind the mirror, erect, virtual, diminished and reversed.)

21. An object 2 cm tall is placed 4 cm from a converging lens of focal length 5 cm. Find graphically the size of the image. Repeat for a diverging lens of the same focal length. (*Ans.:* For diverging lens size is 1.11 cm)

22. An object is placed 6 cm from a diverging lens of focal length 5 cm. Find graphically the size of the image. (*Ans.:* 0.455 cm)

Repeat for a converging lens.

Part Four
Modern Physics

12 Relativity

During the early 20th century, physics underwent two distinct revolutions, which together spawned what is now known as modern physics. The first of these two revolutions resulted from the work of one man, Albert Einstein. His theories of special and general relativity changed our entire view of the structure of space and time. In this chapter we shall present some of the ideas of special relativity. The second revolution was the development of quantum mechanics, the subject of Chaps. 13 and 14.

12.1 THE POSTULATES OF SPECIAL RELATIVITY

Implicit in the structure of classical Newtonian physics, as formulated during the 18th and 19th centuries, are several important assumptions about the nature of physical laws and the structure of space and time. Scientists believed in an absolute space and an absolute time. The existence of an absolute time implies that time flows in precisely the same way for all observers, regardless of their motion or position. In particular, all observers would agree on the time elapsed between two particular events. Absolute space means that there is some coordinate system relative to which all motion could be defined. To a fish, for instance, the water in which he lives is a frame of reference relative to which he either is or is not in motion.

At the end of the 19th century, the existence of a fluid called the

ether was postulated. This fluid was at rest in the absolute space. The laws of nature would be in their simplest form when expressed in the coordinates and time of this absolute system. All motion was to be related to the ether. For example, light was assumed to move with its universal speed c only with respect to the ether, if an observer were moving through the ether, he would measure a different speed for light. (In fact, scientists went even further and said that the ether was the medium in which light waves propagated.) The speed of the earth in its orbit around the sun is about 18 mi/sec and the speed of light is 186,000 mi/sec; therefore, to observe the speed of the earth through the ether presumably should be relatively simple. In 1887, A. A. Michelson and E. W. Morley attempted to measure this speed by using interference methods that were sufficiently accurate to detect the effect quite easily. They failed. Regardless of the time of day or the season of the year, no change in the velocity of light because of the earth's assumed motion was observed. Motion through the ether could not be detected, and there was no way to find the absolute space that was believed to exist. Many other experiments were devised to detect the ether; all of them failed.

Einstein, however, came to the conclusion that the basic assumptions of Newtonian physics were wrong. He felt that it made no sense to define a physical quantity or concept unless it could be measured. Because the only motion that is physically measurable is motion relative to a material body, he said that we must confine ourselves to studying such motions. Thus, the vague idea of motion relative to an absolute space was given up. In its place, Einstein defined the *inertial system* or *inertial reference frame*.

An inertial reference frame is defined as one in which the Law of Conservation of Momentum holds. Thus, in an inertial reference frame, an object with no forces on it will move with constant velocity. The simplest example of an inertial system is a rocket ship with its motors turned off in deep space far from any large masses. Any other system moving at a constant velocity with respect to a given inertial system is also an inertial system, because an object moving with constant velocity in the first system will still have a constant, although different, velocity in the second system. An accelerated frame of reference is not an inertial system, because an object at rest in an inertial system will appear to be accelerated if viewed from such an accelerated frame. All inertial systems differ from one another only by a constant relative velocity.

If an absolute frame of reference existed, it would be an inertial system and all other inertial systems would be moving relative to it. However, there seemed to be no way to detect this relative motion, and so Einstein assumed that there was no preferred inertial frame. Instead, all

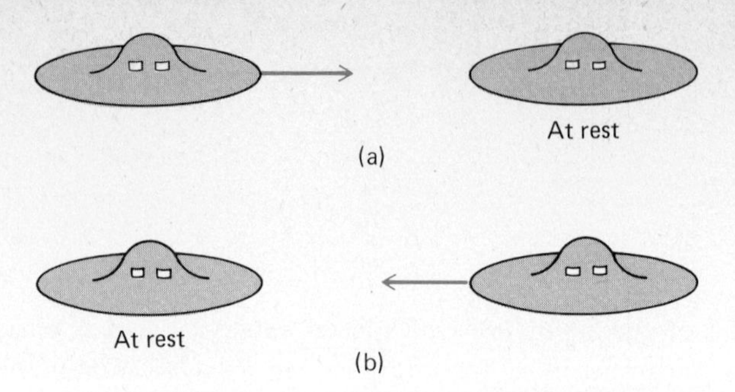

Two flying saucers moving relative to each other in deep space. It is impossible to determine which of the two pictures, (a) or (b), each showing one of the saucers at rest, is correct.

inertial frames were equivalent. To formalize what was meant by "equivalent," he postulated that

> *The laws of physics take the same form in all inertial frames.*

This *postulate of relativity* is one of the two basic assumptions of special relativity. To illustrate, suppose that a certain inertial observer (an observer in an inertial frame) measures the force on an object as $\mathbf{F}$ and finds that the momentum of the particle changes by an amount $\Delta\mathbf{p}$ after that force acts for a time Δt. He then notes that these quantities are related by the equation $\mathbf{F}\,\Delta t = \Delta\mathbf{p}$. He takes this equation to be a law of physics. (It happens to be the correct generalization of Newton's Second Law.) Another observer in an inertial frame moving with respect to the first one might look at the same particle. He would measure the force as $\mathbf{F}'$, the time interval $\Delta t'$, and the change in momentum $\Delta\mathbf{p}'$, all slightly different numbers. However, he would still find that $\mathbf{F}'\,\Delta t' = \Delta\mathbf{p}'$, and his physical law takes the same *form.*

Einstein also studied Maxwell's equations for electrodynamics and the propagation of light. He realized that Maxwell's equations *predicted* that the speed of light would be the same for all inertial observers. Feeling that this fact was fundamental to the structure of physics, Einstein formulated his second postulate:

> *The speed of light is the same in all inertial frames.*

Einstein regarded the Michelson-Morley experiments as strong experimental confirmation of this postulate. With these two simple postulates as his

starting point, Einstein, in 1905, formulated the *Theory of Special Relativity*. In the remainder of this chapter we shall explore some of the predictions of this theory.

12.2 TIME DILATION

One of the predictions of special relativity that is easiest to derive and hardest to accept is the effect known as *time dilation*. It means that, with respect to a given observer, moving clocks run slower than stationary ones. Thus, it strikes at the very heart of absolute time, a concept deeply imbedded in classical Newtonian physics. To derive this effect, we must first build a clock. Our clock must be both simple and accurate, and so we build it out of objects governed by fundamental rather than derived laws. The most fundamental law that we have is the postulate on the speed of light. Therefore, we build a *light clock*.

Imagine two parallel mirrors separated by a distance D, as shown in Fig. 12.1. This device is our "clock." A photon is bouncing back and forth between the two mirrors. The laws governing the rate at which this clock "ticks" are extremely simple. The photon always moves at exactly the speed of light c. From the definition of speed as distance divided by time, we find that the time needed for it to travel between the two mirrors is

$$\Delta t_0 = \frac{D}{c} \tag{12.1}$$

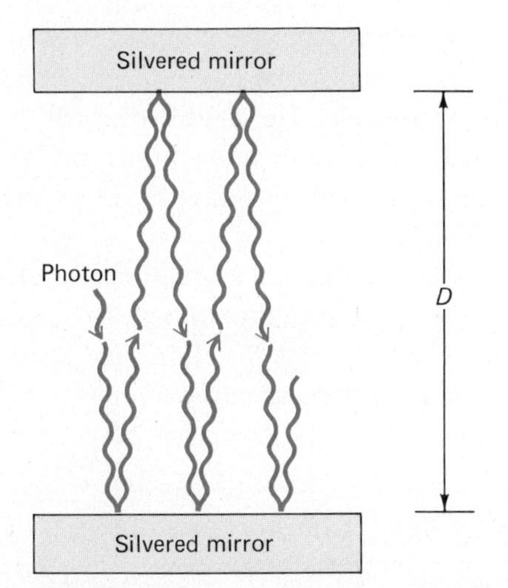

FIGURE 12.1
A light clock. The photon bounces back and forth between the two silvered mirrors taking time $\Delta t_0 = D/c$ to pass from one mirror to the other.

Because the distance between the mirrors is constant and the photon always moves at exactly the same speed, this clock will run at a precisely defined rate. We define Δt_0, the time needed for the photon to travel from one mirror to the other, as our basic unit of time. If $D = 0.30$ m, then $\Delta t_0 = 10^{-9}$ sec, and our unit of time is the *nanosecond.* Although beautiful in its simplicity, our light clock is not very practical. To use it, we would have to be able to detect the single photon each time that it made a round trip of 2 nanoseconds.

A more practical clock, using the same basic principles, can be constructed from a mirror, a strobe light, and a phototube. The mirror is mounted a distance D from the strobe light and phototube, as illustrated in Fig. 12.2. When the strobe light flashes, it sends off a pulse of light that bounces off the mirror and returns to the phototube, which again triggers the strobe light with negligible time delay. A recording device is attached to the strobe to count the number of flashes. If both the light clock and the strobe clock use the same base distance D, they will clearly tick at the same rate. However, the strobe clock is subject to inaccuracies resulting from faulty connections between the strobe light and the phototube. It also lacks the simplicity of the light clock.

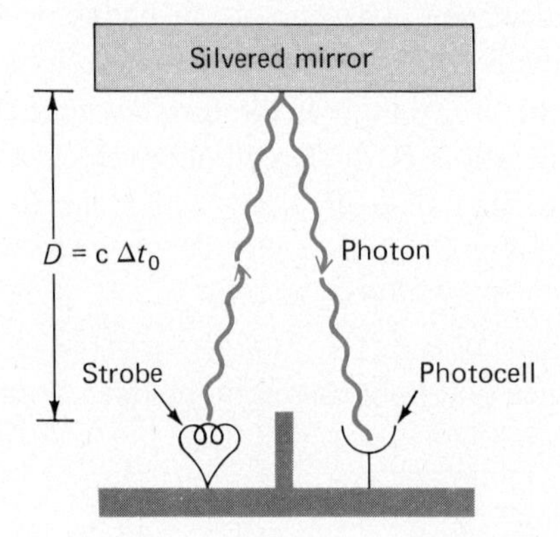

FIGURE 12.2

A more complex light clock utilizing a strobe and a photocell. If $D = 30$ cm then the strobe light would flash every 2 nanoseconds.

Suppose that we have constructed several identical light clocks and we wish to study their properties. Each one has exactly the same length D, which we take to be 30 cm. We can refer, therefore, to nanoseconds, although this reference is clearly not necessary. Operating from an inertial frame, one of the light clocks is set in motion so that it moves past us with constant velocity v. Figure 12.3 shows the clock over a time interval of

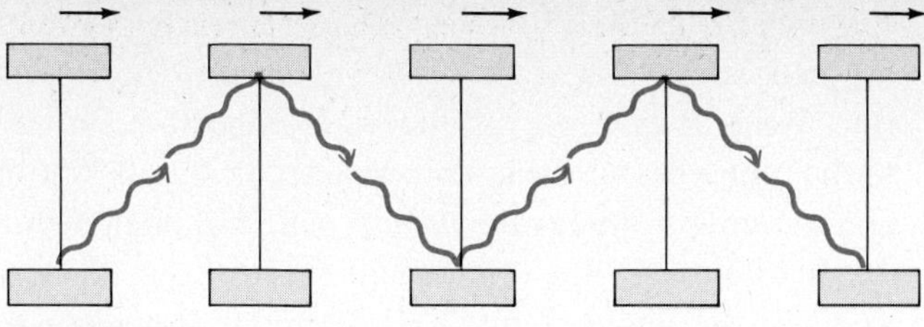

FIGURE 12.3
In this figure the light clock is moving to the right at a large velocity. The photon must travel a longer path to keep reflecting off the mirrors.

several nanoseconds. The wavy line in the figure is the path followed by the photon as it bounces between the mirrors. Notice that each time the photon is reflected, the clock has moved slightly. *Thus, the photon travels a longer path* than it would have had the clock been at rest. However, our second postulate states that the photon always moves with the same speed c; therefore, *the moving clock ticks slower.* This fact is the origin of time dilation. The time Δt between the moving clock's ticks, as measured by a stationary observer, is still defined as the time elapsed between successive reflections of the photon. Figure 12.4 depicts the position of the moving clock at two such reflections. The photon, in bouncing from the lower to the upper mirror, travels a distance s, which is the hypotenuse of a right triangle. The base d of the triangle is the distance that the clock moves in the time Δt; and the height is D, the length of the clock. Using the Pythagorean theorem, we find that $s^2 = d^2 + D^2$, which can be rewritten as

$$D^2 = s^2 - d^2 \tag{12.2}$$

Because s is the distance that light travels in the time Δt and d is the distance

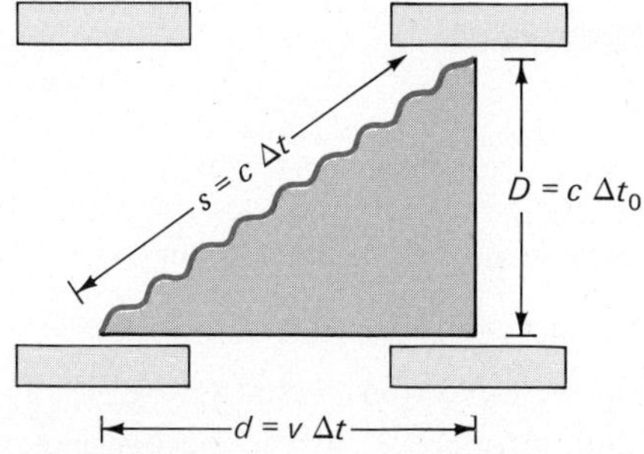

FIGURE 12.4
The moving light clock shown at successive "ticks". The time it takes to tick is found by using the Pythagorean theorem on the colored triangle.

that the clock travels in the time Δt, we can express s and d in terms of c and v as

$$\begin{aligned} s &= c\,\Delta t \\ d &= v\,\Delta t \end{aligned} \tag{12.3}$$

Furthermore, from Eq. (12.1), we have $D = c\,\Delta t_0$. Substituting this equation and Eq. (12.3) into (12.2) gives

$$(c\,\Delta t_0)^2 = (c\,\Delta t)^2 - (v\,\Delta t)^2$$

which can be written as

$$\begin{aligned} c^2(\Delta t_0)^2 &= c^2(\Delta t)^2 - v^2(\Delta t)^2 \\ &= (\Delta t)^2(c^2 - v^2) \end{aligned} \tag{12.4}$$

We want to solve Eq. (12.4) for Δt in terms of Δt_0 so that we shall know just how much the moving clock is slowed by its motion. Thus, we first divide Eq. (12.4) by c^2 on both sides, giving

$$(\Delta t_0)^2 = (\Delta t)^2\left[1 - \left(\frac{v}{c}\right)^2\right]$$

and then take the square root, so that

$$\Delta t_0 = \Delta t\sqrt{1 - \left(\frac{v^2}{c^2}\right)}$$

This equation can be solved for Δt by dividing through by the square root, and we find

$$\Delta t = \Delta t_0\left\{\frac{1}{\sqrt{1 - (v^2/c^2)}}\right\} \tag{12.5}$$

which is our final result for the time dilation.

Before we discuss the physical content of Eq. (12.5), we must introduce some notation. The quantity in the braces of Eq. (12.5) will appear in almost *all* of our relativistic formulas, so we define a special symbol $\gamma(v)$ to stand for it. Thus, we let

$$\gamma(v) = \left\{\frac{1}{\sqrt{1 - (v^2/c^2)}}\right\} \tag{12.6}$$

We can then write the time dilation equation, Eq. (12.5), as

$$\Delta t = \gamma(v)\,\Delta t_0 \tag{12.7}$$

Figure 12.5 shows the value of γ for various values of v/c, which is the ratio of the speed of the object (in this case, the clock) to the speed of light. Notice that when $v = 0$, $\gamma = 1$; when $v > 0$, $\gamma > 1$. Furthermore, as $v \rightarrow c$ and $v/c \rightarrow 1$, γ becomes infinite. Using the fact that $3^2 + 4^2 = 5^2$, we can easily show that $\gamma = 1.25$ when $v = (3/5)c$ and that $\gamma = 1.67$ when $v = (4/5)c$. In addition, although this situation is not so easy to show, when $v = 8000$ mph (about 4000 m/sec), which is faster than any of us are likely

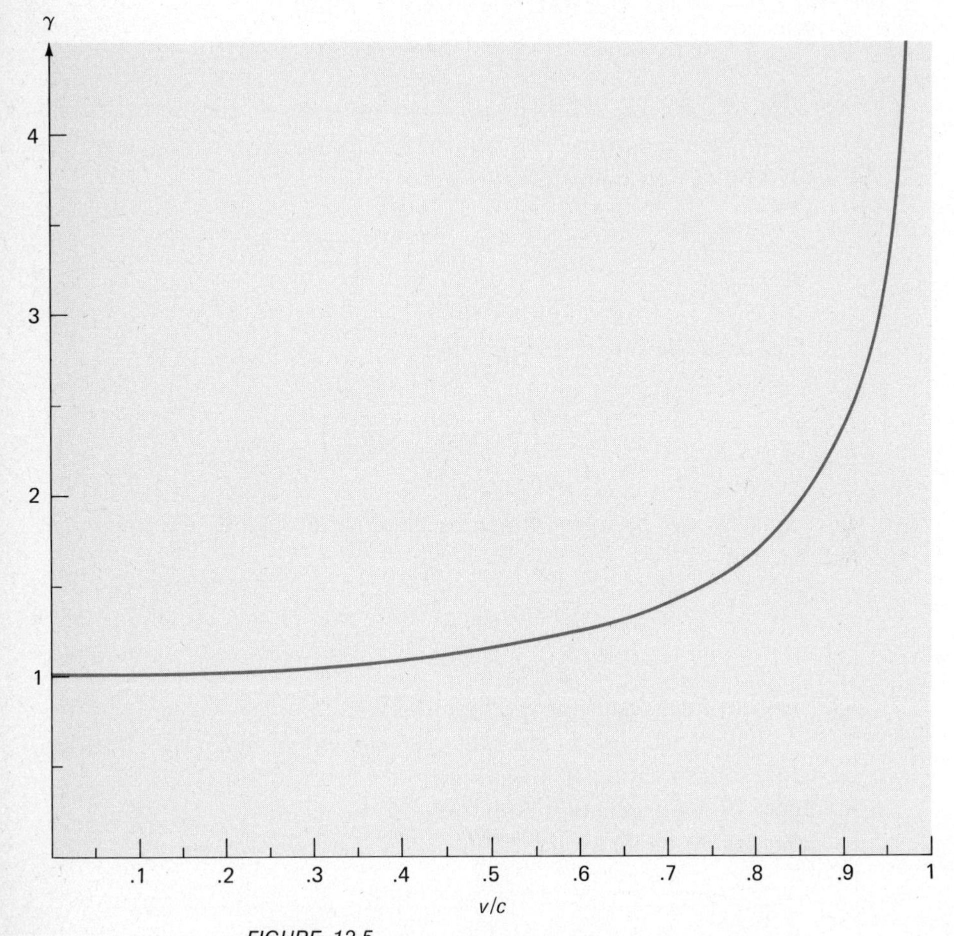

FIGURE 12.5

Values taken by the relativistic function γ for various values of v/c.

to travel, $\gamma = 1.0000000001$ or $1 + 10^{-10}$. For ordinary speeds, therefore, γ differs from unity by less than one part in 10 billion.

Returning to time dilation, we can see from Eq. (12.7) that the moving clock is running slow. The effect, however, for ordinary speeds is almost undetectable in everyday life. If the clock were moving at 8000 mph, the preceding numerical example tells us that we would have to wait 10^{10} sec, or about 300 years, for it to lose 1 second. We cannot fault the 19th-century scientists for not having noticed so small an effect.

We have derived time dilation by using a rather idealized clock. You might wonder if the effect is limited to light clocks. The answer is no. *All* clocks, whatever their construction, are subject to time dilation. This conclusion comes from the principle of relativity. Suppose that a light clock and a cesium clock are both at rest in an inertial reference frame. They are observed to be synchronized (running at the same rate). The laws of physics tell us that they will continue to run at the same rate, because both clocks are very accurate. The principle of relativity tells us that this synchronization must be possible in any inertial reference frame. If we now observe this pair of clocks from another inertial reference frame that is moving with respect to the frame of the two synchronized clocks, the light clock will appear to be running slow. But the clocks are synchronized; thus, the cesium clock (or any other accurate clock in the same inertial frame) will appear to slow down exactly as much as the light clock.

Time dilation gives rise to many interesting effects, some of which have been observed experimentally. For instance, many naturally occurring substances are unstable and decay. The best-known examples are certain nuclear isotopes, such as carbon 14. As we shall see in Chapter 14, radioactive decay rates can be characterized by the *half-life* of the substance or the time needed for 50 per cent of a given sample to decay. Half-lives range from around 10^{10} yr to 10^{-23} sec. The decay rate of a given substance provides us with a natural clock. If we observe what fraction of a sample of material has decayed, we can deduce how much time has elapsed. This method is used to estimate the age of the earth. It has also provided us with a direct test of time dilation.

One of the elementary particles produced in large quantities by the particle accelerators used in high-energy physics is the charged pion. Pions are quite unstable, having a half-life of 2.6×10^{-8} sec, and thus do not occur naturally, except in cosmic-ray showers. If time dilation applied to the internal clock of a pion, then pions accelerated to very large velocities should decay more slowly. This behavior is just what is observed. Within the limits of experimental error, the half-life of a pion, moving at an appreciable fraction of the speed of light, is increased by a factor of γ, as predicted by Eq. (12.7).

Another prediction of special relativity closely related to time dilation is the so-called "*twin paradox*." Suppose that an interstellar spaceship, with a very powerful engine, is equipped with a clock. The spaceship leaves earth, rapidly accelerates to a velocity close to the speed of light, then turns off its motor and coasts. From the point of view of an observer on earth, its clock will appear to be running quite slow. From Fig. 12.5 we see that a speed of about 87 per cent the speed of light will produce a γ of about 2; for each hour elapsed on earth, Eq. (12.7) gives the time elapsed on the ship as 30 min. After several years, the spaceship again turns on its engine, quickly reverses direction and velocity, and returns to earth at the same speed of 87 perecent the speed of light. On the return trip, it is still moving at high speed, and, from the point of view of an earthbound observer, its clock is still running slow. If 10 years elapsed on the earth between the ship's departure and return, only 5 years will have been recorded by the ship's clock, because from the point of view of an observer on earth it has been running at half the rate of earth clocks for almost the entire trip.

However, the ship's clock was not the only thing affected by the trip. Biological systems also behave somewhat like clocks. Your heartbeat is an irregular timing device; your aging process can also be used to keep time. All the life processes of the crew of the ship would be slowed along with the ship's clock. The crewmen, therefore, when they return, will have aged only 5 years to the 10 years of people on earth. Thus, a twin who took the trip will have aged less than his brother who stayed home.

The paradoxical aspects of this situation arise when an attempt is made to describe the trip from the point of view of the ship's crew. The (incorrect) argument goes as follows: From the point of view of the crew, it is the earth that accelerates away at high speed, travels into space, and then returns to the ship. While the earth is moving, its clocks will appear to run slow from the ship's point of view. Thus, when the earth "returns," it and not the ship will be younger. If the postulate of relativity is to be believed, then this description is just as valid as the one in the preceding paragraph. Each twin would be younger than the other, thus, the paradox.

The resolution of this paradox lies in the fact that the spaceship is not an inertial frame, so our simple explanations do not apply to it. When it leaves the earth, when it turns around, and when it returns, the spaceship undergoes very high acceleration, thus destroying the symmetry between it and the observer on the earth. Because the ship is not an inertial frame, a description of the trip from the point of view of the crew requires a careful argument that is more complex than we can present here. If this analysis is made however, the time elapsed on the ship does turn out to be less than the time elapsed on earth and no paradox arises.

The twin paradox clearly illustrates the way in which special relativity requires us to change our naive notions about the nature of space and time. In fact, the effect is so at variance with our intuition that the Theory of Relativity continues to produce sporadic controversy almost 70 years later, although it was correctly treated by Einstein in 1905. This controversy has not subsided because, until quite recently, there has been no experimental test of the effect sufficiently definitive to satisfy the very small number of vociferous critics.

However, in 1972, D. J. C. Hafele and R. E. Keating reported on an experiment in which they flew four cesium clocks around the world twice, once with and once opposite to the earth's rotation. The time elapsed, as measured by these clocks, was compared with the time elapsed as measured by a cesium clock that remained on the ground. Hafele and Keating used Einstein's equations to calculate the theoretical prediction for the elapsed times as measured by these three sets of clocks and compared these calculations with the actual measurements. The analysis was somewhat complicated by altitude variations, the rotation of the earth, and other practical factors. Nevertheless, the measurements agreed with the theoretical predictions to within the experimental error. The traveling clocks did lose time relative to the stationary one. This experimental verification of the twin "effect" (a better word because no true paradox exists), by using real clocks, provides a definitive test of an often misunderstood prediction.

12.3 LENGTH CONTRACTION

In the preceding section we saw how special relativity modifies our concepts of the flow of time. In this section we show how our concepts of distance must also be modified. Suppose that we wish to discuss the length of a rod. We must first carefully specify how that length is to be measured. Light clocks, whose behavior we understand, are available, so we shall devise a method of measurement that makes use of these clocks. Imagine that we have equipped one of our clocks with a spray paint attachment in such a way that every time it ticks it sprays a little paint. Let this clock move past the stationary rod at a speed v, as shown in Fig. 12.6. The length of the rod can be found from the speed of the clock times the time t that it takes to move past the rod, giving

$$L_0 = vt \tag{12.8}$$

After the clock has moved past the rod, we can find the time t by counting the number N of paint marks on the rod and multiplying this number by

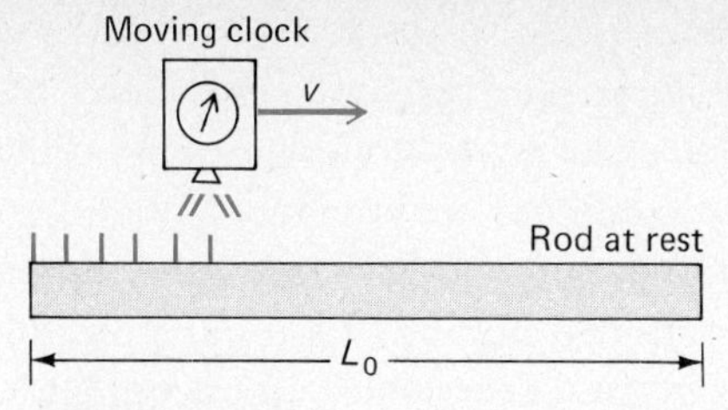

FIGURE 12.6

An experiment for measuring the length of a rod. The light clock moves past the rod at a known speed v and the length of the rod is found from the time it takes the clock to pass the rod.

the time Δt between ticks, giving $t = N\,\Delta t$. However, the clock is moving, and we must take into account the fact that it is running slow. Therefore, the time t is given by

$$t = N\,\gamma(v)\,\Delta t_0 \tag{12.9}$$

We use Eq. (12.7) to express t in terms of the time measured by a stationary clock. If we substitute Eq. (12.9) into Eq. (12.8), we find that if the rod is stationary in our inertial frame, its length will be

$$L_0 = v\,N\,\gamma(v)\,\Delta t_0 \tag{12.10}$$

We now consider this same measurement from a different point of view. Suppose that we observe the rod and clock from an inertial frame moving along with the clock. In this frame, the clock is at rest and the rod is moving with speed v, as shown in Fig. 12.7. The length of the rod is now the speed v of the rod multiplied by the time it takes to pass the clock. The speed v is unchanged from the preceding example, because the relative

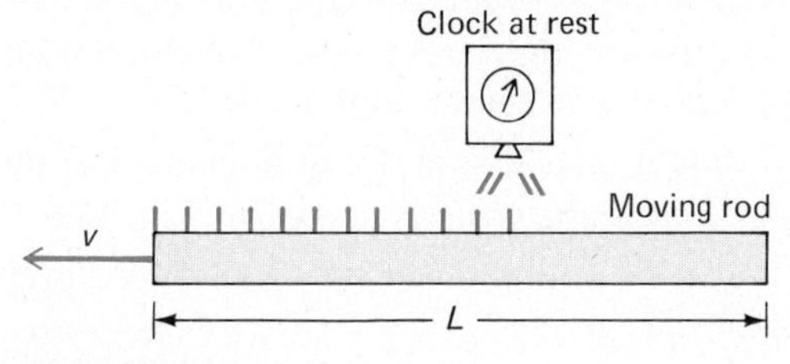

FIGURE 12.7

The same measurement as in Fig. 12.6, viewed by an observer moving with the clock so that the clock appears to be at rest and the rod moves with speed v. The length of the moving rod is found from the time it takes to pass the clock.

speed of the clock and the rod is the same from either point of view. To find the time elapsed as the rod moves past the clock, we must retrieve the rod and count the number of paint marks on it. However, we are considering the same physical event from a different point of view; therefore, the number of paint marks must be the same. Thus, the time elapsed is $N\,\Delta t_0$ because the clock was not moving. We then find the length L of the moving rod to be

$$L = v\,N\,\Delta t_0 \tag{12.11}$$

If we compare this result to Eq. (12.10) for the measured length of the stationary rod, we find that L is less than L_0 by a factor of $\gamma(v)$; that is,

$$L = \frac{L_0}{\gamma(v)} \tag{12.12}$$

We have, consequently, length contraction. Recalling that $\gamma(v)$ is always greater than 1 unless an object is at rest, we see that Eq. (12.12) tells us that if we measure a moving rod, its length L will be less than L_0, which is its length measured when at rest. This result applies to any extended object.

Although we shall not give the derivation, the result does not apply, however, to dimensions perpendicular to the motion. Only the dimensions parallel to the velocity vector are contracted. For example, consider a moving box, as shown in Fig. 12.8. The dimension of the box in the direction of motion is measured as contracted, whereas the other two dimensions are measured as unchanged. We used this fact in our discussion of the light clock when we assumed that the moving clock was the same length as the

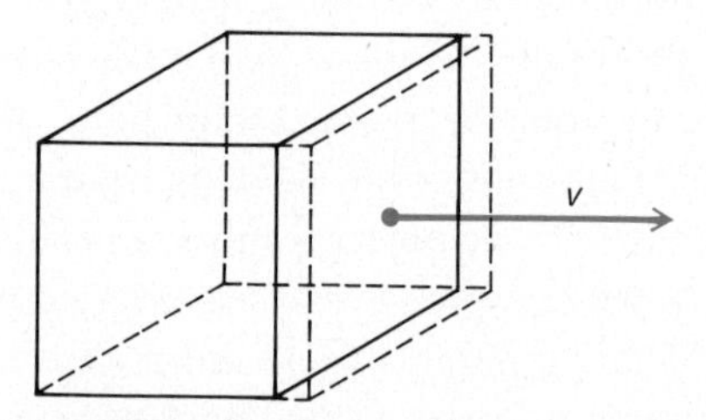

FIGURE 12.8
A block moving at high speed in the direction of the colored velocity vector. If its dimensions were measured by the methods discussed, it would appear contracted in the direction of motion and would have unaltered dimensions in the other two directions.

stationary one. At this point, you might wonder if the time dilation also depended on the orientation of the axis of the clock. The answer is no. A detailed analysis of our light clock would show that it slows down the same amount regardless of the direction of motion.

Length contraction can be derived by using other methods for measuring the length of a rod. In fact, Einstein's original derivation differs considerably from ours. However, in each case, the result is the same. Length contraction, like time dilation, is an intrinsic property of space and time.

EXAMPLE Consider a meter stick moving parallel to its length at a speed of 4000 m/sec, how much length does it lose? We mentioned in Sec. 12.2 that for this speed $\gamma = 1 + 10^{-10}$, and so

$$L = \frac{L_0}{\gamma} = 1 \times (1 - 10^{-10}) \text{ m}$$

The meter stick is about 10^{-8} cm shorter. This is roughly the diameter of a single atom!

You should note that we have neither asked nor answered the question of whether the moving rod actually *is* shorter. This question is not well formed unless a specific meaning is given for the word "is," that is, without a specific criterion for answering the question. All we have said is that if you measure the length of a moving rod in a specific manner (which happens to coincide with the standard meaning of "length"), you will obtain a number smaller than that which you would obtain by measuring the rod at rest.

As an example of how important it is to specify just what the experimental procedure for verifying a statement is, consider the question of the *appearance* of a rapidly moving object. As it stands, this statement is ambiguous and could be interpreted in different ways. However, if we specify the *visual appearance,* then the question is well put. At a given instant of time the eye forms an image by receiving light rays from an object and interpreting them in the brain. If these light rays have come from an extended object, the rays from the more distant parts of the object must have left slightly earlier than those from the closer parts in order to arrive at the eye at the same time. If the object is moving rapidly across the field of vision, the time delay will cause the eye to see different parts of the object at locations they occupied at different times, and thus its shape will appear distorted. For an object that occupies a small portion of your field of vision, this distortion takes the form of an apparent rotation of the object.* This

*If you are interested in the details of this effect, consult the excellent article by Victor Weisskopf entitled "The Visual Appearance of Rapidly Moving Objects," which appeared in the September, 1960, issue of *Physics Today* and is reprinted in *Special Relativity Theory, Selected Reprints,* published by the American Institute of Physics for the American Association of Physics Teachers.

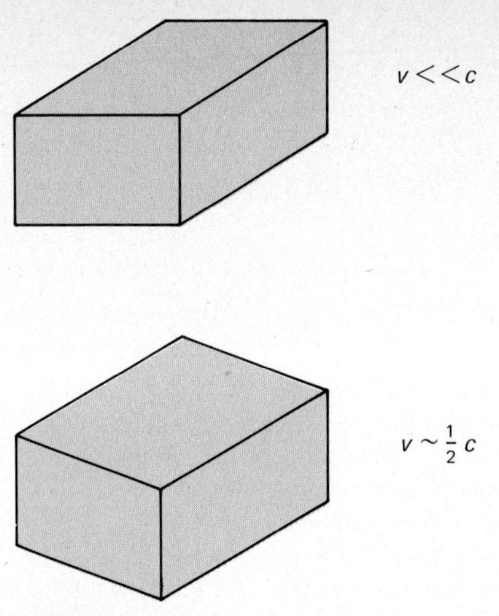

FIGURE 12.9

The shape of a moving object as perceived by the eye. In the upper figure, a rectangular object moves past at a speed much less than c. In the lower picture, it is moving at roughly one half the speed of light.

is shown in Fig. 12.9. We conclude that, because of the finite speed of light, we would not "see" the length contraction. Thus, the question of whether or not a moving object *is* contracted depends on what type of measurement is contemplated.

12.4 RELATIVISTIC DYNAMICS

Obviously the two postulates of relativity require that we make some rather far-reaching changes in our concepts about the measurement of time and distance. If we were to follow these changes through to their logical conclusion, we would have to rewrite entirely Chap. 2 on kinematics. As long as the velocities that we describe are small compared to the speed of light c, the changes would be completely negligible. However, the description of very high-speed motion must be changed to be consistent with the two relativity postulates. These changes are presented in more advanced texts on relativity, so we shall describe only one of them here.

Suppose that two spaceships are approaching us from opposite directions and that we measure their speeds as v_1 and v_2 respectively. Equation (2.10) would give the relative speed of the two ships (that is, the speed of one of them as seen by the other) as $v_1 + v_2$. This relationship is not correct relativistically however. (Note that we could produce a relative speed in excess of the speed of light by making v_1 and v_2 close to c. But

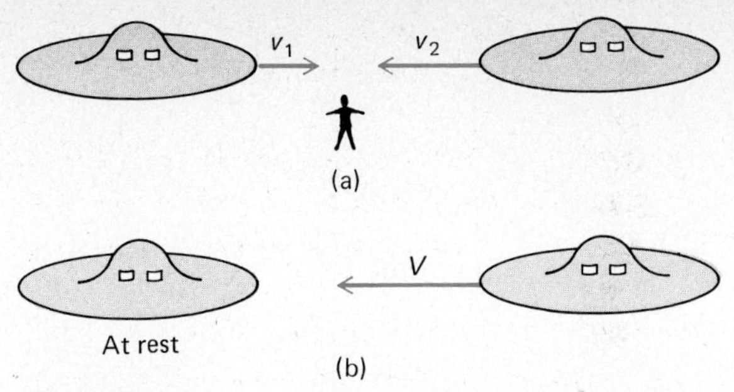

FIGURE 12.10

In the upper picture an observer measures the speed of two spaceships approaching him to be v_1 and v_2. In the lower picture the left spaceship measures the relative speed of the other to be V. The relation between these measurements is given by Eq. (12.13).

the following equation will show that such speed is impossible.) The correct relative speed is given by the relativistic law for the addition of velocities, which states that the relative speed V is

$$V = \frac{v_1 + v_2}{1 + \left(\frac{v_1 v_2}{c^2}\right)} \tag{12.13}$$

When v_1 and v_2 are much smaller than c, the term $(v_1 v_2)/(c^2)$ is negligible, and Eq. (12.13) reduces to Eq. (2.10). On the other hand, a little algebra will convince you that if either v_1 or v_2 is equal to c, then Eq. (12.13) gives $V = c$. In other words, an object moving at the speed of light will move at that speed for all observers.

The dynamical laws of motion that we studied in Chap. 3 must also be modified if they are to be consistent with the two postulates of relativity. We shall not discuss these changes in detail but only describe how they affect the three main dynamical quantities: mass, energy, and momentum.

The experimental method that we described in Sec. 3.1 for defining mass remains valid relativistically (which is one reason that we chose it). We call the mass found in this manner the *rest mass* of an object. As with length and time measurements, measurements of mass on high-speed objects yield different values than we might expect. If we observe that a given particle has rest mass m_0, then the mass m of an identical particle moving past us with speed v is found to be

$$\boxed{m = m_0 \gamma(v)} \tag{12.14}$$

where $\gamma(v)$ is the function that appeared in both the time dilation and length contraction effects presented in the first part of this chapter. It is defined by Eq. (12.6) and shown in Fig. 12.5. The mass defined by Eq. (12.14) is called the *dynamical* or *relativistic mass* of the particle. As the speed of a particle increases, so does its mass. As was the case for time dilation and length contraction, this mass increase amounts to only a few parts per billion for ordinary speeds. However, as the speed of a particle approaches the speed of light, γ becomes very large, and at the speed of light it will be infinite.

In terms of the relativistic mass, the definition of momentum, Eq. (3.1), is unchanged. We have

$$\mathbf{p} = m\mathbf{v} = m_0\gamma(v)\mathbf{v} \tag{12.15}$$

Thus, the momentum of a particle, as well as its mass, approaches infinitely as the particle approaches the speed of light. With this definition, the Law of Conservation of Momentum remains valid. As we mentioned in Sec. 12.1, the correct generalization of Newton's Second Law is

$$\Delta\mathbf{p} = \mathbf{F}\,\Delta t \tag{12.16}$$

which is the same as Eq. (3.2). In the relativistic case, the steps used in Sec. 3.2 to derive the simpler form, $\mathbf{F} = m\mathbf{a}$, break down, because the mass is no longer constant but changes as the speed changes. However, the content of the law is the same: A force is required to change the momentum and thus the velocity of a particle.

Imagine that you are exerting a constant force on a massive object initially at rest. From our discussions in Chap. 3, we know that at first the object will move off in the direction of the force with constant acceleration. This behavior follows from Eq. (12.16), which says that the object will gain equal amounts of momentum in equal time intervals. As long as the speed of the object is much less than c, the speed of light, the momentum is directly proportional to the velocity. Equal momentum increases will produce equal velocity increases. This motion will continue until the speed of the object begins to approach the speed of light at which point, relativistic effects begin to become important, γ starts to grow, and the mass of the object starts to increase. Your constant force will still produce a constant rate of increase in the object's momentum, but this increase is no longer accompanied by a constant acceleration because of the mass increase. The momentum of the object is its *relativistic* mass times its velocity. As the mass gets larger, smaller and smaller increases in the velocity are needed to produce the constant rate of increase in the momentum. The constant force is pushing

on an object that is getting more and more massive, and so the resulting acceleration is less and less. The object will never reach the speed of light. As its speed approaches c, its mass approaches infinity, and its acceleration (the rate at which it gains speed) becomes vanishingly small. We thus conclude that *it is impossible for any massive object to move at the speed of light.* The speed of light represents an absolute upper limit on the speed of any material object in our universe. An object can approach arbitrarily close to this speed, but it can never reach it and can certainly never exceed it.

Let us consider this example from the point of view of energy conservation. The constant force that you exerted on the object certainly did work on the object. Even as the object approached the speed of light, work was being done. Where was this work going? If we believe in conservation of energy, this work was not lost but was stored in the motion of the object. It clearly was not going into increased speed. The only other possibility was that it was being stored in the increased mass of the object. Although we shall not do so, it can be shown that the energy of a mass m_0 moving at speed v is given by

$$E = mc^2 = m_0\gamma(v)c^2 \qquad (12.17)$$

This equation is Einstein's famous mass-energy relation. It tells us that energy can be stored in the form of mass. Equation (12.17) does not look very much like $\frac{1}{2}m_0v^2$, which is the nonrelativistic kinetic energy of a mass. However, if we use mathematical approximation techniques for $\gamma(v)$, it is possible to show that, for a v that is very much smaller than c, Eq. (12.17) becomes

$$E = m_0c^2 + \tfrac{1}{2}m_0v^2 \qquad (12.18)$$

The first term, m_0c^2, in Eq. (12.18) is called the *rest energy* of the object. It is the energy that the object possesses because of its rest mass. The second term is our familiar *kinetic energy*. Thus, when we say that the energy of a moving particle equals mc^2, we are counting the rest energy that the particle would have, even when stopped, along with the kinetic energy that it would have from its motion. If work is done on a particle by forces that accelerate it, then its kinetic energy K is given by

$$W = mc^2 - m_0c^2 \qquad (12.19)$$

The work is stored in the increase in the particle's energy over its rest energy.

If the speed of the particle is small compared with the speed of light, then Eq. (12.18) can be used to find the kinetic energy. If, however, the speed is relativistic, then Eq. (12.17), with the exact $\gamma(v)$, must be used.

You could now ask: If energy can be stored as increased mass, can the rest mass of a particle be used to produce energy? The answer is yes—some of it. In certain nuclear reactions that we shall discuss in Chap. 14, the mass of the reacting system decreases and the lost mass is converted into energy. If Δm is the net mass decrease, then the energy produced is found by the equation

$$E = \Delta m\, c^2 \qquad (12.20)$$

In Eq. (12.20), if mass is measured in kilograms and the velocity of light is expressed in meters per second ($c = 3 \times 10^8$ m/sec), then the energy comes out in joules. One gram (10^{-3} kg) of mass is equivalent to almost 10^{14} joules of energy. We see, therefore, that mass and energy, which once were considered as entirely different, are really closely related. Mass can be changed into energy, and vice versa.

We now must drop the Law of Conservation of Mass that we briefly discussed in Sec. 3.1 and generalize the Law of Conservation of Energy to include the energy stored in the rest mass of all matter. The equivalence between mass and energy has been thoroughly confirmed by numerous experiments in nuclear physics. The most notable example is the atomic bomb. In the case of a nuclear explosion, releasing energy equivalent to 20,000 tons of TNT, approximately 3 grams of mass are converted into energy. But even aside from this spectacular example, most nuclear phenomena would be inexplicable if mass could not be made equivalent to energy. The source of the sun's enormous amount of energy, for instance, is the conversion of mass into energy at the rate of about 4.4 million tons per second. In addition, the relativistic mass increase is observed daily in the large elementary-particle laboratories, such as Brookhaven, where accelerators produce protons and electrons moving at more than 99 per cent of the speed of light and having relativistic masses many times their rest masses.

The Theory of Relativity has had abundant experimental verification, so that it forms one of the cornerstones of modern physics. Its various predictions may seem intuitively unreasonable, simply because intuition is based on everyday experience, whereas relativity becomes important only in situations far removed from everyday life. Nevertheless, intuitive judgments are no basis for deciding that the theory is not correct.

DISCUSSION QUESTIONS

1. Discuss changes in our life if the speed of light had the value $c =$ 100 km/hr.

2. Two spaceships pass each other in interstellar space with a relative speed of $0.9c$. Discuss what each set of spacemen will observe if they measure the meter sticks and clocks of the other set. Is there a contradiction here?

3. Is it possible to increase your lifespan by travelling at relativistic speeds? What can be achieved?

4. A spaceship is moving away from the earth at a speed of $0.8c$. Will the crew of the ship observe any length contraction or time dilation in the ship's equipment?

5. Modify the statement "matter can neither be created or destroyed" so that it is correct.

6. Suppose that another civilization is detected 100 light years from the earth. What kind of a dialogue could we establish with that civilization and would an exchange of ambassadors be practical?

7. How many grams of mass would need to be converted into energy to supply the United States with energy for a year?

8. Discuss the scientific content of the statement "Einstein said everything is relative."

PROBLEMS

1. Calculate the contraction in length of a train 1 mi long moving at a speed of 90 mph. Would it be possible to measure this contraction in length, and, if so, how?

2. Compute the speed at which a passing car would have an observed length half of the length which it would when it was at rest relative to the observer.

3. Compute the amount of mass which has to be converted into energy for a total amount of energy equal to 10^{20} J to be released. If this energy is released, in a time of 2×10^{-4} sec, what is the average power.
(*Ans.:* 1111 kg)

4. What is the mass of an electron moving at a speed of 2.9×10^{8} m/sec?
(*Ans.:* 3.6×10^{-30} kg)

5. Compute the speed that an electron must have in order for its mass to be equal to the rest mass of a proton which is 1.67×10^{-27} kg.

6. Assume that the speed of the earth in its orbit around the sun is 10^{-4} times the speed of light and that the diameter of the earth is 8000 mi. Compute the contraction of the size of the earth as measured by someone on the sun.
(*Ans.:* 2.5 in.)

7. Find the energy released when 3 gm of mass disappear.

8. Calculate the speed of an electron if its mass is to be twice its rest mass.
(*Ans.:* 2.6×10^{8} m/sec)

9. Suppose that an elementary particle has a lifetime of 10^{-9} sec. One of these particles is observed to travel a distance of 30 meters before decaying. Find the ratio of its mass while moving to its rest mass.

10. The best atomic clocks are accurate to 1 part in 10^{10}. How fast must one of these clocks move past the other for time dilation to be observable?

11. Recall that it requires 80 kcal of heat to melt 1 kg of ice at 0°C. Find how much the mass of 1 kg of ice increases when it melts.

12. A free neutron (one not in a nucleus) decays after about 1000 sec. The radius of Uranus' orbit is about 3 billion km. How much energy must a neutron emitted by the sun have in order to reach Uranus?

13 Quantum Mechanics and the Atom

During the first 25 years of this century, two new physical theories were developed that modified and extended the physics of the 18th and 19th centuries. The first of these theories, special relativity, we studied in the last chapter. We found that it modified the description of motion at very high speed and required us to change some of our basic ideas about the flow of time. The second theory, quantum mechanics, modified the laws of physics in a different area, that of the nature of matter.

In probing the structure of various natural chemical compounds 19th-century scientists discovered that all these substances could be formed by using only 92 basic building blocks called *elements*. These elements were arranged into a periodic table by Dmitri Mendeleev about 1870. He organized the elements in such a way that their properties, similarities, and differences could be predicted from their position on the tables. The elements were made up of individual units called *atoms* (from the Greek word for "indivisible"). Atoms could then be combined in various ways to form the individual *molecules* of the known chemical compounds. For instance, scientists understood that water (H_2O) is formed from two hydrogen (H) atoms and one oxygen (O) atom; carbon dioxide (CO_2), from one carbon (C) atom and two oxygen atoms. Although empirical rules for combining atoms into molecules were known, the internal structure of the individual atoms, and consequently the theoretical basis for these rules, were completely unknown. During the 50 years following Mendelcev's periodic

tables, a wealth of new experimental data became available about the structure of the atom. The interpretation of this data required the revision of classical physics and the development of quantum mechanics, which is the subject of this chapter. The first experimental development that we shall present involved the discovery of the electron.

13.1 THE ELECTRON

We have already discussed the electron as the carrier of electric current in metals. However, in the 19th century when the basic laws of electromagnetism were discovered, electrons were not known. Although the early applications of electricity to communications, power generation, and power transmission began to transform society, they did not depend on knowing whether electricity was carried by individual particles or by a continuous fluid. Now let us look at the evidence leading to the discovery that electric current (in its most common form) is formed from discrete negatively charged particles: *electrons.*

An important field of research during the second half of the 19th century was the study of electrical discharges through gases at low pressure. A typical piece of equipment would consist of two metal plates (electrodes) inside a glass tube in which the gas pressure and nature of the gas could be varied. An arrangement of this sort is shown in Fig. 13.1. The experimenter could vary the gas pressure and the applied voltage, among other variables. If the gas pressure were low and the voltage were high, a current would flow and the gas would emit colored light. (This colored light has many applications in modern society. The very intense yellow sodium vapor and blue mercury vapor lights used in highway illumination and the multicolored "neon" signs are examples.) If the pressure were reduced further, the color mainly disappeared and the glass of the tube glowed green.

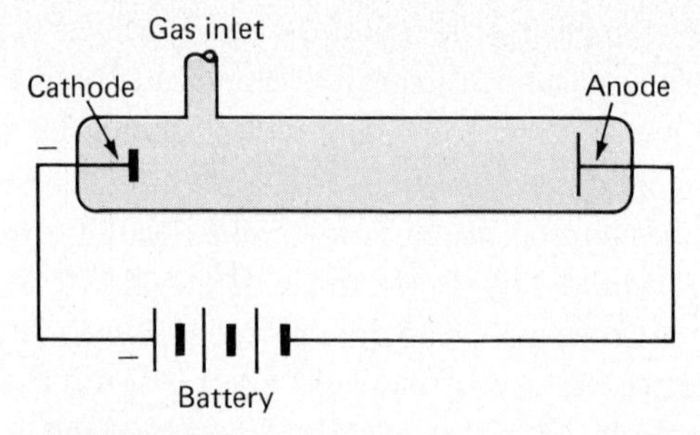

FIGURE 13.1

A cathode ray discharge tube used for studying electrical discharges.

Because current would still be flowing through the tube, this "discharge" would cause certain materials, such as zinc sulfide and various phosphors, to glow or "fluoresce." (This principle is used today in television tubes, as we shall discuss later.) This electrical discharge appeared to begin at the negative electrode (cathode) of the tube and was more or less independent of the location of the positive electrode (anode); therefore, it became known as a *cathode ray*. The tube using it was called a *cathode-ray tube*. This interesting effect was the subject of many investigations.

As early as 1858, Julius Plücker, a German mathematician and physicist, observed that cathode rays could be deflected by a magnetic field. From this behavior, it was concluded that the rays were similar to an electric current or moving charges. Furthermore, the direction in which the beam of cathode rays was deflected in either a magnetic or an electric field indicated that the rays were negatively charged. Other experiments demonstrated that the cathode-ray beam carried momentum and could exert a force, thus showing that cathode rays consisted of a stream of negatively charged particles.

In 1897, an English physicist, Sir J. J. Thomson, performed quantitative experiments in which he subjected beams of cathode rays to electric and magnetic fields. From his data he could calculate the ratio of the charge to the mass of these cathode-ray particles. He found the same value for this ratio, regardless of the gas used, the material of the electrodes, or the values of pressure and voltage used. The conclusion from Thomson's experiments is that cathode rays consist of negatively charged particles, all with the same ratio of charge to mass. These particles are our electrons (from the Greek word for "amber"). Here we should point out that Thomson's work did not show that these particles are identical. His experiments only showed that all of them have the same ratio of charge to mass. The proof that they are identical had to wait until the actual charge on one of them was measured.

In 1911, Robert A. Millikan, an American physicist, began a series of experiments to measure the charge of an electron independently of its mass. Millikan's apparatus, illustrated schematically in Fig. 13.2, consisted of two parallel metal plates to which a voltage could be applied, producing a known electric field between the plates. Into the region between the plates he sprayed small charged drops of oil. When the electric field was turned off, these drops would fall slowly through the air. If Millikan watched a drop through a microscope, he could determine its weight W. He then would turn on the electric field, which would exert a force $F = qE$ (q is the net charge) on the drop. By adjusting the strength of the E field so the drop was suspended with the weight of the drop just balanced by the electrical force, he was able to measure the charge on the drop. With luck, the charge

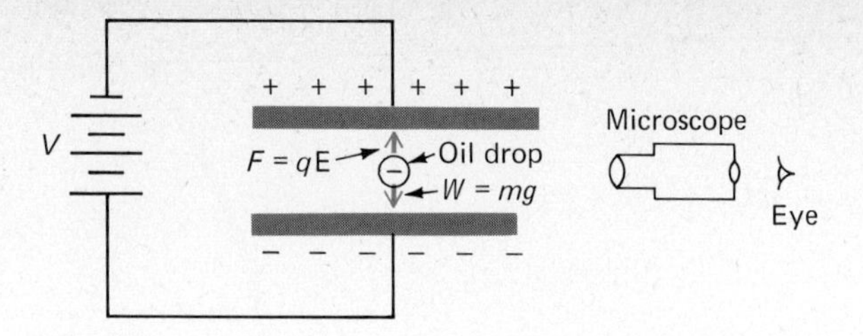

FIGURE 13.2

Millikan's oil drop apparatus. The drop shown is in equilibrium between the downward pull of gravity and the upward electrical force.

on the drop could be changed a number of times, and each charge could be measured. In addition, charges on many drops could be measured. Millikan did this procedure many thousands of times and found a fascinating result. To within the limits of experimental error (and occasional mistakes), each charge measured was in integral multiples of 1.6×10^{-19} coulombs. For instance, no charge was measured with 2.4×10^{-19} coulombs. Millikan concluded that the charge on an electron was exactly 1.6×10^{-19} coulombs and that the charges on his oil drops depended on how many extra electrons they had. Since Millikan's original experiment, his observations have been repeated many times and in many different ways with the same result.

Thomson's work showed that all cathode-ray particles (electrons) have the same ratio of charge to mass; therefore, they must be identical. In fact, physicists now believe that electrons are completely indistinguishable. They all have precisely the same mass (9×10^{-31} kg) and precisely the same charge. Later we shall discuss other charged particles, such as the proton, which is positively charged and 2000 times heavier than the electron. Each of these particles will always have a charge that is some multiple of e, the charge on the electron. Thus, nature has provided us with one basic unit of electrical charge from which all other observed charges are constructed. As a result, we say that charge is *quantized*, because it can only occur in definite, discrete amounts. The question of why nature has allowed only one basic unit is unsolved; we only know that it is so.

Cathode rays, or *electron beams* as we now call them, have many diverse applications in modern electronic technology. Perhaps the best-known and most ubiquitous application is the television tube, shown in Fig. 13.3. In a television tube, electrons, produced by a hot cathode, are focused into a very thin beam that is aimed by magnets. This beam "paints" the picture on the phosphors coating the face of the tube. When four separate beams and three colors of phosphor are used, color pictures are produced.

Another similar device is the oscilloscope, a common laboratory instrument illustrated in Fig. 13.4. The oscilloscope works on the same

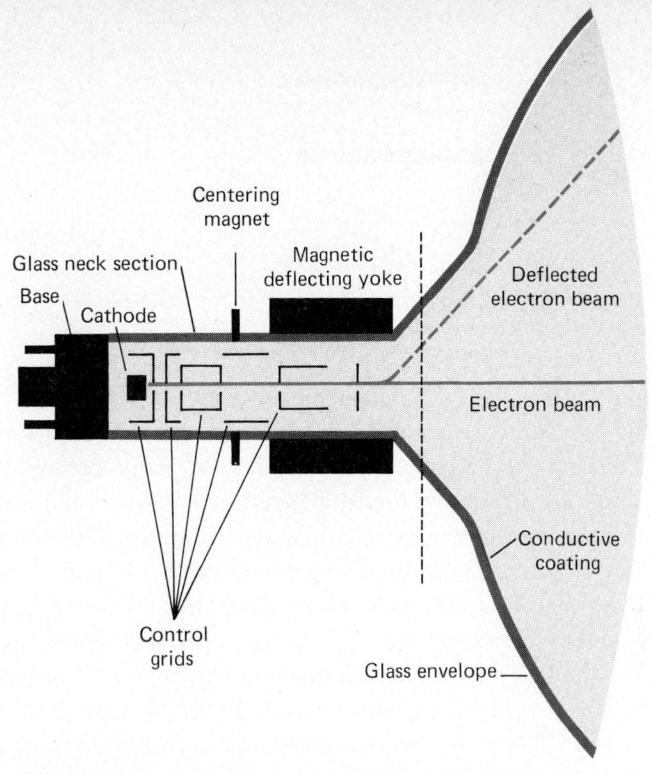

FIGURE 13.3

Cross section of a television tube. The electron beam is produced by the cathode. The control grids accelerate and focus the beam which is then aimed by the magnetic deflecting yoke.

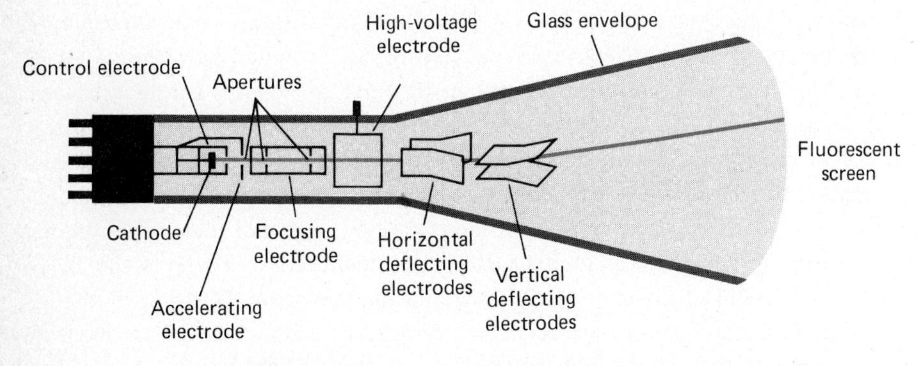

FIGURE 13.4

Cross section of an oscilloscope tube. The operation is similar to the television tube except that the beam is deflected by electric fields produced by the two sets of deflecting electrodes, one controlling the horizontal and the other the vertical motion.

principle as the television tube except that the beam is deflected by electrostatic forces rather than magnetic.

A last example is the vacuum tube. Shown in Fig. 13.5, a simple vacuum tube (a triode in this illustration) consists of a hot cathode that emits electrons, which flow to the cold positive anode, thus producing a current. Between the cathode and the anode is a grid of fine wire. When a relatively small negative voltage is applied to the grid, it repels the electrons and stops the current. Thus, a small voltage can control a large current and, when properly wired into a circuit, produce amplification of a signal. Although vacuum tubes are now frequently replaced by solid-state transistors, which serve the same function, they played an indispensable role in the development of our modern electronic technology.

In many electronic devices, electrons or other charged particles are accelerated, and thus given energy, by passing them through a potential difference. Because the charge on any particle is always an integral multiple of e (the charge of an electron), it is convenient to define a new unit of energy, the *electron volt*. One electron volt (eV) is defined as the energy given to an electron (or any other particle with the same charge) when it passes through one volt of potential difference. Using Eq. (8.6), we see that

$$\begin{aligned} 1 \text{ eV} &= (1.6 \times 10^{-19} \text{ coulombs})(1 \text{ V}) \\ &= 1.6 \times 10^{-19} \text{ joules} \end{aligned}$$

Obviously, the electron volt is a very small unit of energy. Nevertheless, it is very useful for discussing the energies of atomic and nuclear systems.

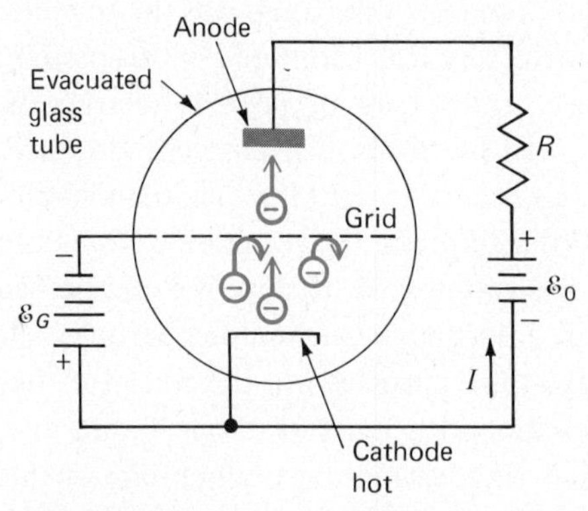

FIGURE 13.5

A vacuum tube. Electrons are boiled off the hot cathode and are attracted to the anode producing the current I. If the grid (a fine mesh of wire between the anode and cathode) is made negative by $\mathcal{E}_G$ then it repells the electrons as shown and decreases or stops the current.

Two common multiples of the electron volt are the MeV, equal to a million (10^6) eV, and the GeV, equal to a billion (10^9) electron volts. Note that the conversion from eV into joules is numerically the same as the charge on an electron in coulombs. To find the energy in electron volts of a charged particle that has been accelerated through a known potential difference, multiply the charge measured in units of e (the magnitude of the electron's charge) by the potential difference (in volts).

13.2 ATOMIC MODELS

Having discovered that electrons were one of the fundamental constituents of matter and thus of the individual atoms forming it, physicists began to make models for atoms that incorporated electrons into the structure. The earliest models for the atom, which were constructed during the early 19th century, visualized atoms as small, hard, indivisible spheres that could be put together to form molecules. However, by the time the electron was identified, enough experimental data had been accumulated to show that this model was too simple. From various experiments and theories scientists knew that atoms had a diameter of roughly 10^{-8} centimeters and masses ranging from 1 amu for hydrogen to 235 amu for uranium. (One atomic mass unit, amu, equals 1.66×10^{-27} kg.) Because the mass of an electron is only $\frac{1}{2000}$ amu, the mass of an atom is clearly not attributable to electrons. Atoms are electrically neutral; therefore, they must contain a positive charge, and the mass associated with this positive charge must account for most of the mass of the atom.

One of the early models of the atom was the so-called "plum pudding" model, proposed in 1897 by J. J. Thomson. In this model, the electrons were embedded in a uniform sphere of positive charge. However, by 1911, experiments performed by the British physicist Ernest Rutherford and his coworkers, Hans Geiger and Ernest Marsden, demonstrated the inadequacy of this model. In order to probe the structure of the atom, they took a very thin piece of gold foil (using gold because it could be beaten into very thin leaves) and bombared it with a beam of alpha particles. (In the next chapter, we shall learn that alpha particles are the nuclei of helium atoms. They have a charge of $+2e$ and a mass of about 4 amu and are produced by radioactive materials such as radium.) When one of these particles struck a fluorescent screen of zinc sulfide, for instance, a flash of light was emitted. Thus, by observing the screen, the experimenters could determine exactly where the particle had hit the screen. The experimental arrangement is illustrated schematically in Fig. 13.6. When the gold foil was not in place, all of the alpha particles struck a small spot at the center of the fluorescent

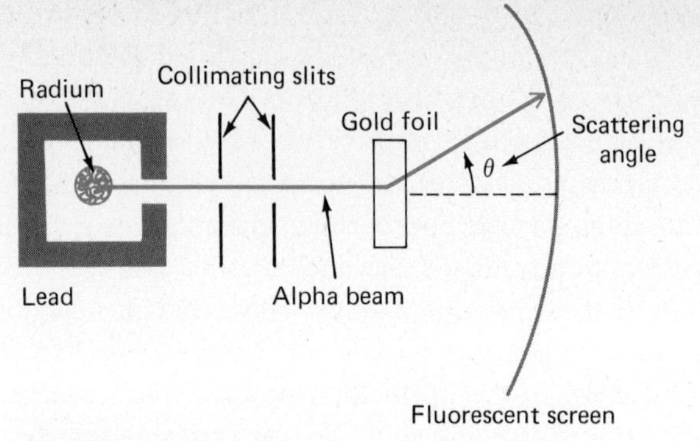

FIGURE 13.6
Rutherford scattering experiment. Alpha particles emitted by the radioactive radium are collimated into a beam, strike the gold foil, and are scattered to the fluorescent screen where they are observed.

screen. However, when the metal foil was put in the path of the beam, quite a few flashes occurred well away from the center of the screen. This observation indicated that these particles had been deflected or scattered by the material in the metal foil.

Because he knew the properties of the alpha particles and the nature of electrical forces, Rutherford was able to calculate the amount of deflection expected from various models of the atom. For the Thomson model, he predicted that the alpha particles would be scattered by no more than 1 or 2 deg. He based his prediction on the fact that the fluid of positive charge largely neutralized the negative electrons; consequently, very little net force would be exerted on the alpha particles. This situation is shown in Fig. 13.7. On the other hand, deflections through large angles were observed fairly frequently, with the alpha particles sometimes being scattered back towards the source. The plum pudding model was wrong.

Because the Thomson model of the atom failed to agree with the experimental data, Rutherford proposed a different model that would predict

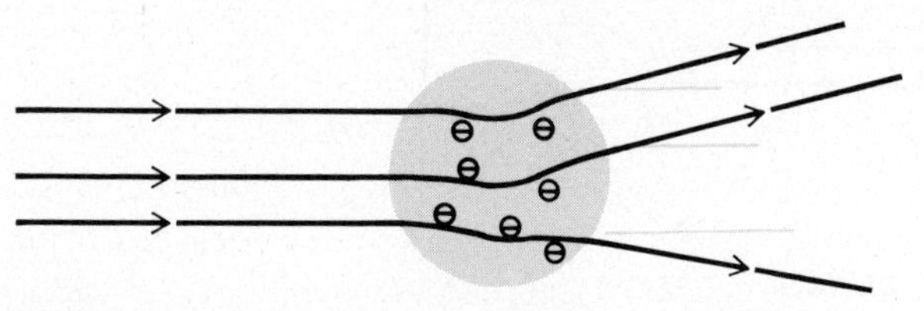

FIGURE 13.7
Alpha particle scattering in the Thomson model. The electrons in the atom are embedded in a sea of positive charge and none of the alpha's are deflected more than a few degrees.

the occasional large-angle scattering observed. He suggested that an atom might consist of a small, massive core, called the *nucleus,* which would be positively charged and comprise most of the mass of the atom. Moving around the nucleus in various "orbits" would be a cloud of light, negatively charged electrons. This model correctly predicted the observed scattering. When an alpha particle approached such an atom, it usually passed through the electron cloud, missed the nucleus, and was scattered through a small angle, as in the Thomson model. However, an occasional alpha particle would pass very near the nucleus, as shown in Fig. 13.8. Inside the electron cloud and close to the unshielded positive nucleus, it would experience a strong repulsive force and could be scattered through very large angles, thus explaining the observations. By careful analysis of his data, Rutherford was able to calculate the size of the nuclei of various atoms. He found that the average diameter of a nucleus was about 10^{-12} centimeters, or 1/10,000 the size of the atom. This result is remarkable because the nucleus contains virtually the entire mass of the atom. (Recall that the electron mass is $\frac{1}{2000}$ amu.) Atoms are almost completely empty. The nucleus of an atom, therefore, bears the same relation to the size of the atom as the sun does to the entire solar system measured out to the orbit of Pluto.

At this point, we must explain why we used quotation marks when referring to the "orbit" of an electron around the nucleus. It is very tempting to picture the atom as we do the solar system, with the electrons "orbiting" the nucleus like planets, held by the attractive coulomb force between them and the nucleus. (Recall that the coulomb force has the same $1/r^2$ dependence that gravity has.) In fact, Rutherford and many others originally had a very similar view. However, such a picture *cannot be correct.* According to classical electrodynamics, an electron in a circular orbit would continually emit electromagnetic radiation, thus losing energy and spiraling into the nucleus. According to classical physics, atoms cannot be stable. But because

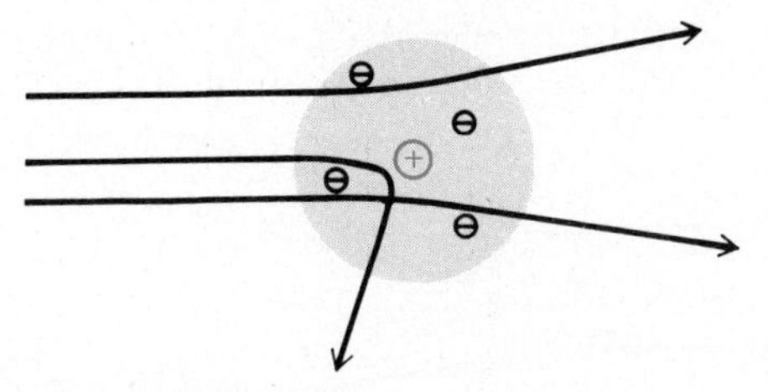

FIGURE 13.8

Alpha particle scattering in the Rutherford model. Most of the alpha particles miss the very small nucleus and are scattered only a few degrees. However an occasional particle passes near the nucleus and is strongly deflected as shown.

atoms demonstrably are stable, we must conclude that classical physics cannot give an adequate description of the structure of an atom. Within 15 years after the acceptance of the Rutherford model, Erwin Schrödinger, Werner Heisenberg, Niels Bohr, and others had developed quantum mechanics, which then supplanted classical physics in the realm of atomic particles. In order to understand the other developments that led to this revolution, we must return to the electron and its sometimes strange properties.

13.3 DE BROGLIE WAVES

In our discussion of light, the presence of diffraction and interference effects was viewed as almost conclusive evidence that light was a wave. Only after Einstein explained the photoelectric effect was it discovered that light could also behave the way particles do. Until about 1920, electrons were regarded unquestionably as particles. They had a definite mass (although the smallest known) and a definite charge. They could carry momentum and kinetic energy. Furthermore, they were observed to move along well-defined trajectories. (All of these properties are associated with particles.) On the other hand, the particle theory was unable to explain the role of electrons in even the simplest atoms. By 1923, physicists were ready to admit that light, usually viewed as a wave, had properties that could only be explained in terms of particles (photons). In 1924, reasoning by analogy with photons, Louis de Broglie, a French physicist, made the bold suggestion that perhaps electrons (and other particles as well) had wave properties. Because photons carry momentum, then particles with momentum $p = mv$ should have associated with them a wavelength λ given by

$$\lambda = \frac{h}{p} = \frac{h}{mv} \tag{13.1}$$

where h is Planck's constant from the photoelectric effect. This brilliant guess was verified experimentally by C. J. Davisson and L. H. Germer of the United States and by G. P. Thomson of England. Their work demonstrated that under proper conditions a beam of electrons showed destructive interference, which is a property of waves. Furthermore, the wavelengths observed agreed with those computed from Eq. (13.1). Later it was shown that heavier particles and even atoms possessed wavelengths in agreement with de Broglie's hypothesis.

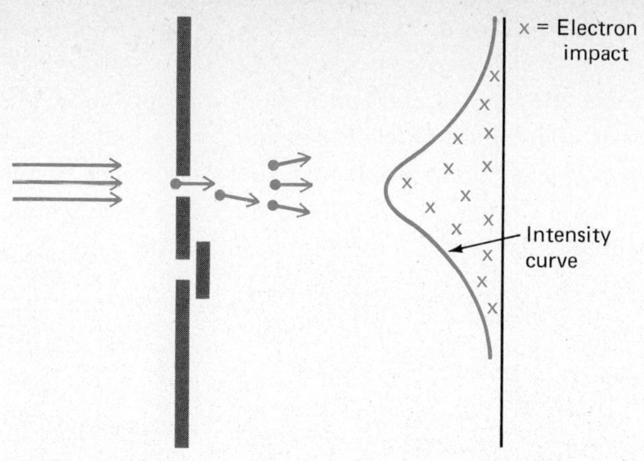

FIGURE 13.9

An electron beam is incident on a slit. After passing through the slit, the beam spreads out. Each colored cross represents the impact of a single electron on the detecting screen and the curve represents the average number.

Figures 13.9 and 13.10 depict a simplified electron interference experiment. A beam of electrons is incident on a pair of small slits. (Compare this setup with that in Fig. 10.16.) In Fig. 13.9 one of the slits is blocked, and the beam must pass through the single slit. On the far side of the slits is a phosphor screen. The transmitted electrons are observed striking this screen. Each electron strikes a given place, causing the phosphor to emit a flash of light. If the slit is small compared with the de Broglie wavelength of the electrons, the transmitted beam will spread. Then the electrons, rather than all striking opposite the slit, will form the pattern shown on the screen. (This phenomenon is known as *diffraction broadening* and is itself evidence of wave nature. Compare with Fig. 10.15.) When the second slit is opened, we would classically expect the pattern to be just the sum of the patterns from the two slits, because particles do not interfere with each other. Instead, we see the interference pattern shown in Fig. 13.10. At point *1*, which is one-half a de Broglie wavelength farther from slit *A* than from slit *B*, the electrons are interfering destructively. With only one slit open, electrons strike this point. With both slits open, they do not.

The interference pattern shown in Fig. 13.10 is clearly a wave-type effect, so it is not due to the presence of many electrons in the beam. If we were to send the electrons through the slits one by one, each would strike the screen at a particular point. After we had sent many electrons through, the intensity pattern shown would start to appear. More electrons would have hit the screen at the points where the pattern is maximum, and none would have struck at the points where the pattern is minimum. However, we must be careful here. Based on our experience with golf balls, rain drops,

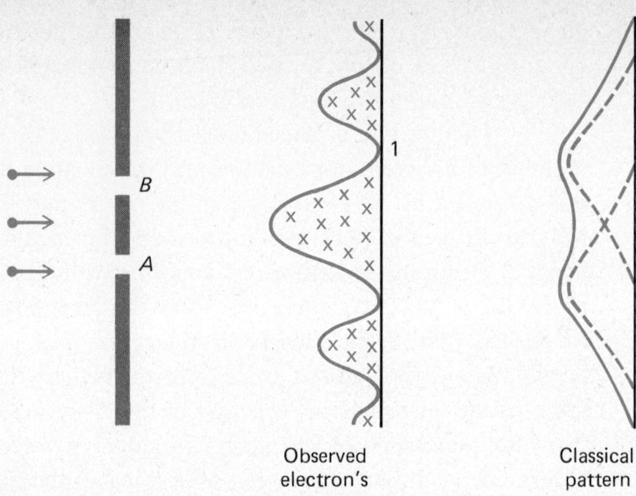

FIGURE 13.10

Twin-slit interference of electrons. Rather than the classical pattern (a sum of two curves like Fig. 10.9) a clear interference pattern results. Point 1 is one-half de Broglie wavelength further from slit *A* than from slit *B* and destructive interference results.

marbles, and other small macroscopic objects, we might argue intuitively that the single electron *must have gone through one or the other of the two slits* and thus could not have known whether the other slit was open or closed. However, we can only make the italicized statement if we have some method of experimentally verifying it. It can be shown that any experiment, no matter how clever, that is designed to detect which slit the electron passed through will destroy the interference pattern. Any experiment to detect the electron as it passes a given slit must interact with the electron. In other words, the electron must affect the detection apparatus sufficiently to produce an observable reading, and the apparatus must, in turn, by our conservation laws affect the motion of the electron. This change in the electron's motion is always just enough to upset the phase relations that must hold if interference is to occur.

Why has our intuition failed? There are two reasons. The first reason is that h, Planck's constant, is a very small number. Just as the energy of individual photons was far too small to be measured until recently, the wavelength of everyday objects is so small that no interference or diffraction effects are ever observed in ordinary experience. The second reason involves the measurement process. We are accustomed to thinking that we can measure some property of an object without disturbing it. For objects as small as an atom or an electron, such measurement is not possible. Any measurement of a system must disturb the system in some way. For large

objects, this disturbance is negligible; but for electrons, it is not. In attempting to measure which slit the electron passed through, we disturbed the interference pattern enough to erase it.

The limitation placed on our knowledge by the fact that any measurement of a system must disturb and thus change the system in some way was described by W. Heisenberg in his *Uncertainty Principle.* Suppose that we wish to localize electrons into a very thin beam. We can pass a large beam of electrons through a very small slit, thus producing a narrow beam. If the slit width is Δx, then just after the electrons emerge from the slit, we can say that they are located at the center of the slit with an uncertainty in their lateral position of Δx. However, if the width of the slit is of the order of the de Broglie wavelength of the electron, as shown in Fig. 13.11, then the same type of diffraction broadening discussed in Sec. 10.3 will occur. The beam will spread out after it leaves the slit. Just after they leave the slit, therefore, the electrons will have a momentum vector (which points in the direction of their motion) that may point either up or down. Consequently, we are uncertain about the precise value of the part of the momentum transverse to the direction of the beam. If we make Δx smaller in an attempt to further localize the beam, this diffraction spreading will increase and the uncertainty Δp in the beam's transverse momentum will increase. A mathematical analysis of the diffraction broadening gives the very simple result that the uncertainties in position and momentum are related by

$$\Delta x \, \Delta p \geq h \tag{13.2}$$

where again h is Planck's constant.

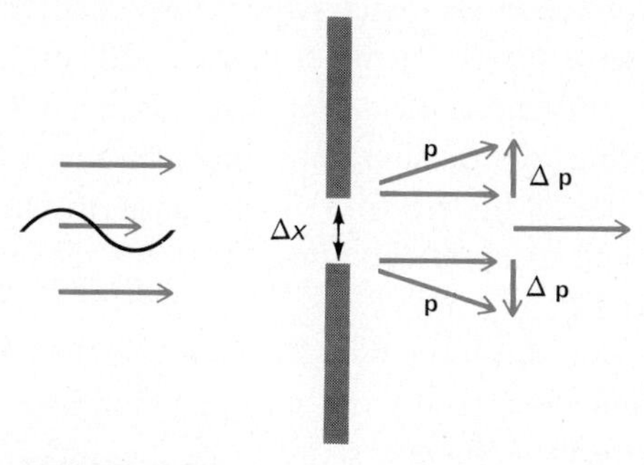

FIGURE 13.11

Electron diffraction. A beam of electrons is incident on a slit whose width is comparable to the de Broglie wavelength of the electrons. After the beam emerges from the slit, its transverse position has been localized to within Δx, the width of the slit. However, the beam is spreading and thus has some transverse momentum, but of an uncertain amount Δp.

Notice that as the beam moves away from the slit and broadens, the uncertainty in position increases; that is, the product $\Delta x \, \Delta p$ gets larger. If we try to make Δx infinitely small, the electrons will leave the slit in all directions and with unknown momenta. Heisenberg's Uncertainty Principle states that we can never know the position and momentum of a particle simultaneously with a precision greater than that given in Eq. (13.2). This result can be derived rigorously from quantum mechanics and represents a fundamental fact of nature.

As a second example of the Uncertainty Principle, consider a particle moving with a known momentum. We shall try to measure its position by using photons. If we wish to observe the position of the particle accurately, we must use light with a very short wavelength and high frequency. The individual photons with this frequency have great energy and momenta. When they bounce off the particle and return to our microscope, they will change the momentum of the particle, leaving us uncertain as to its precise value. If we use less energetic photons to decrease the effect of the collision on the particle, their wavelength will be longer. Consequently, the position of the particle is still not accurately known. If this experiment is carefully analyzed, it is again found that the uncertainty in position is related to the uncertainty in momentum by Eq. (13.2).

Although the Heisenberg Uncertainty Principle severely restricts the type of atomic models that we may construct out of electrons, it is essentially irrelevant for ordinary objects. (In particular, we shall see shortly that the planetary model of an atom is meaningless.) Consider an object with a mass of 1 kg. Its momentum is $p = mv$; therefore, the uncertainty in its momentum is related to the uncertainty in its velocity by $\Delta p = m \, \Delta v = \Delta v$. We find the relation between the uncertainties in position and velocity as follows:

$$\Delta x \, \Delta p = \Delta x \, \Delta v = h = 6.6 \times 10^{-34} \tag{13.3}$$

Thus, if the position of this 1-kg object is known to an accuracy of 10^{-16} m, its speed can only be known to an accuracy of 10^{-17} m/sec. These numbers are so small, however, that they place no real limitation on our knowledge. We cannot measure position and velocity to 16 decimal places. For an electron, the situation is quite different. Because $\Delta p = m\Delta v = 9 \times 10^{-31} \, \Delta v$, Eq. (13.2) gives us

$$\Delta x \, \Delta v = \frac{h}{m} = \frac{6.6 \times 10^{-34}}{9 \times 10^{-31}} \simeq 10^{-3} \tag{13.4}$$

If we have measured the position of an electron to an accuracy of 1 mm (10^{-3} m), then we have introduced an uncertainty in its velocity of 1 m/sec.

Contrasting Eq. (13.3) with Eq. (13.4), we see that the uncertainty relations restrict only our description of an electron, not an ordinary object.

13.4 BLACKBODY RADIATION

All substances, when heated or otherwise excited, emit visible light. (You may recall that the visible light corresponds to wavelengths ranging from about 4×10^{-7} meters for violet light to about 7×10^{-7} meters for red.) The collection of wavelengths emitted by a particular substance is called its *spectrum* (plural: *spectra*). The study of the spectra is called *spectroscopy*. Almost every solid, when heated, emits a continuous spectrum containing all the visible wavelengths in varying amounts. The character of these spectra depends largely on the temperature, rather than the particular substance involved. For example, all solids start to glow red at roughly the same temperature. This continuous thermal radiation given off by a hot solid is called *blackbody radiation*. It was intensively studied at the end of the 19th century.

A body that completely absorbs all of the radiation hitting it is called a *blackbody*. The mechanism by which a body is a perfect absorber, or blackbody, is clearly unimportant, because all such bodies have the same properties. One way to construct a blackbody experimentally is to line a cavity with lampblack and drill a small hole in the wall of the cavity. Any radiation from the outside that enters the hole has a very small chance of getting out again after it hits the interior of the cavity because of the high absorption power of lampblack. The small amount of radiation reflected from the lampblack has a very slight chance of being aimed at the hole and emerging. Thus, the hole in the cavity is almost a perfect absorber of radiation incident on it.

It is easy to show that all blackbodies at the same temperature have the same properties for emitting radiation. Consider two blackbodies at the same temperature enclosed by a perfectly reflecting cavity. Let us assume that body *1* transfers more energy to body *2* than body *1* receives from *2*. This assumption violates the Second Law of Thermodynamics, because there is a transfer of energy between the bodies without a difference in their temperatures. Thus, we conclude that each body has the same emission characteristics, so that neither body gains energy from the other. In other words, all blackbodies have both the same absorption properties as well as the same emission properties when their temperatures are the same. This fact should not be surprising. A perfect absorber is an ideal body; we could hardly expect one ideal body to have better emission characteristics than another.

Figure 13.12 shows a plot of intensity versus wavelength for the

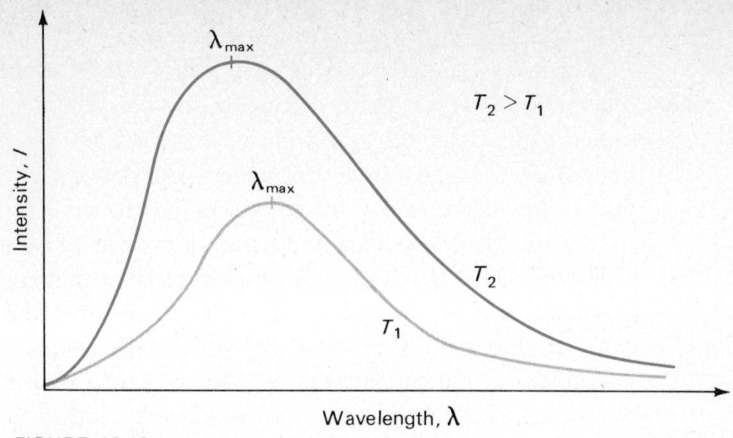

FIGURE 13.12
Intensity versus wavelength for a blackbody at two temperatures T_1 and T_2.

blackbody radiation given off by a solid at two different temperatures. The graph clearly illustrates two important characteristics of blackbody radiation. First, there is a characteristic wavelength, λ_{max}, at which the radiation is most intense. By comparing the two graphs, we see that λ_{max} decreases as the temperature increases. The decrease is described by Wilhelm Wien's *Displacement Law,* which states that at temperature T

$$\lambda_{max} T = 2.90 \times 10^{-3} \text{ m-°K}$$

That is, λ_{max} is inversely proportional to the absolute temperature of the object. This law explains the common experience in which an object starts to glow a deep red after it is heated; then, with higher and higher temperatures, it becomes orange and finally "white hot." The second characteristic of blackbody radiation is that, as the temperature increases, the area under the plot [in Fig. 13.12] increases. This area is proportional to the energy that the object radiates per second at all wavelengths. The *Stefan-Boltzmann Law* (1884), states that the power P (in watts or joules per second) emitted by each square meter of the surface of a blackbody heated to a temperature T is given by

$$P = \sigma T^4$$

where $\sigma = 5.67 \times 10^{-8}$ W-m^{-2}-°K^{-4}. From this equation, we see that doubling the temperature of a blackbody produces a 16-fold increase in the power it radiates.

The form of the Wien Displacement Law and the Stefan-Boltzmann Law can be predicted by using classical electrodynamics and thermodynamics. However, these derivations do not yield a numerical value for either of the constants appearing in the two laws. These constants had to be evaluated experimentally. Furthermore, classical physics was unable to predict the shape of the blackbody radiation curve shown in Fig. 13.12.

In 1900, Max Planck discovered the correct formula for blackbody radiation. He first found an empirical formula that could be made to fit the data; then he derived his formula. In his derivation, Planck made a complete break with classical physics. Planck assumed, as had previous investigators, that the surface of a hot radiating object was composed of electric oscillators vibrating at all frequencies and in thermal equilibrium with each other. This assumption was reasonable because the blackbody radiation was largely independent of the particular substance involved. These thermally excited oscillators then radiated electromagnetic energy at a frequency equal to their natural frequency f. To compute the amount of radiation at a given wavelength, an average was taken over the possible states of excitation of the oscillator at the given temperature. It was in taking this average that Planck discarded classical physics. Classically, an oscillator could have any energy, depending on the amplitude of its vibration. Planck knew that this assumption would give the wrong answer, so he instead assumed that the oscillators could have only discrete *quantized* energies given by

$$E = nhf \qquad (n = 1, 2, 3, \ldots)$$

where n is an integer, f is the frequency of the oscillator, and h is a constant whose value is the same for all the oscillators. He then averaged the discrete quantized energy levels of the oscillators and obtained his formula.

Planck's formula for the spectrum of radiation from a blackbody at temperature T contained the arbitrary constant h, now known as *Planck's constant.* If h equaled 6.625×10^{-34} joule-sec, then Planck's calculation agreed precisely with the measured blackbody radiation. In addition, this formula predicted both the Wien Displacement Law and the Stefan-Boltzmann Law. The two constants that appear in these laws could be evaluated in terms of Planck's constant. Again the theory agreed with experiment. Planck's treatment of blackbody radiation was completely successful.

Planck's derivation of the correct formula for blackbody radiation led to the subsequent development of quantum physics. Although the details of his argument are too difficult for us to follow here, he was the first to resolve an apparent conflict between theory and experiment by using non-

classical arguments. Over 25 years later, when the problem of the oscillator had been solved by using the newly developed theory of quantum mechanics, it was shown that his assumption about the quantization of energy levels was completely correct. During these 25 years, many other examples of quantization were discovered. In each case, Planck's constant h played a similar role in setting the scale of the quantum effects. The most important example was Einstein's explanation, in 1905, of the photoelectric effect. As we saw in Chap. 11, the energy delivered by electromagnetic radiation is quantized, each photon carrying energy $E = hf$. We shall treat other examples later in this chapter.

13.5 ATOMIC SPECTRA AND THE HYDROGEN ATOM

When a low-density gas of individual atoms or simple molecules is heated or excited electrically (as in a gas discharge tube), the spectra of emitted light is totally different from the blackbody radiation that we studied in the last section. Rather than observing a continuous range of wavelengths, we see many discrete lines when the light is spread out in a spectrometer. These *line spectra* are different for each of the elements and provide a unique way of identifying the elements, much as fingerprints identify individuals. Thus, we can analyze the various elements present in a material by observing the various wavelengths of light that the material emits. For instance, the yellow color that you see in a gas flame when a pot on the stove boils over is caused by the sodium in the salt used to flavor the food. When sodium is used in a gas discharge tube, yellow light results, as mentioned in Sec. 13.1. Similarly, when you burn driftwood (or add commercial "coloring" to a fire), you observe various pretty colors that are attributable to the minerals absorbed by the wood from seawater (or that you added). "Neon" signs are another example. A tube is filled with neon gas at low pressure and excited electrically. It then gives off the reddish color of the element neon. In general, whenever we add energy to a substance in large enough quantities, we vaporize the material and excite its elements so that they emit their characteristic line spectra.

To determine experimentally the line spectra of the elements, spectroscopists worked for many years around the turn of the century. Numerous important discoveries resulted as by-products of this work. For example, in 1895, Wilhelm Roentgen, working with high-voltage cathode-ray tubes, discovered a strange penetrating radiation that, for lack of a better name, he called X rays. These rays could fog photographic plates and penetrate many substances. The medical applications of these rays were immediately obvious; within three months X rays were being used by physicians to aid in setting fractures. By 1912, experimenters had shown that X rays were

uncharged and could interfere destructively, producing diffraction effects. X rays were then identified as electromagnetic radiation with very short wavelengths, from 10^{-10} to 10^{-12} m, and photon energies of 1 to 100 keV (1 keV = 1000 eV).

Spectroscopists carefully measured the wavelengths present in the various spectra and tried to find patterns in their numbers. The first pattern discovered was in the spectrum of hydrogen, the simplest element. In 1884, J. J. Balmer noticed that frequencies of all the lines in the visible spectrum of hydrogen, shown in Fig. 13.13, could be found from the remarkably simple formula

$$f = R\left(\frac{1}{2^2} - \frac{1}{n^2}\right) \qquad (n = 3, 4, 5, \ldots) \tag{13.5}$$

where $R = 3.289 \times 10^{15}$ cps. If we let $n = 3$ in Eq. (13.5), we obtain the frequency of the H_α line in Fig. 13.13. The wavelength of this line could then be calculated, using $c = f\lambda$ [Eq. (10.3)]. Similarly, if $n = 4$, we obtain the H_β line, and so on. The quantity n, which can take on only integer values, is known as a *quantum number.* In general, whenever we have a quantity that can take on only certain discrete values and none in between, we refer to it as a *quantized variable.* The quantized frequencies given by Eq. (13.5) are known as the Balmer series of hydrogen.

Somewhat later, two other series, named after their discoverers, Lyman and Paschen, were found in the hydrogen spectrum. The Lyman series is in the ultraviolet and the Paschen in the infrared. Both of these series satisfy formulas similar to the Balmer series. However, the factor $1/2^2$ in Eq. (13.5) is replaced by $1/1^2$ (Lyman) and $1/3^2$ (Paschen). Balmer also suggested that *all* of the spectral lines of hydrogen could be represented by the single formula

$$f = R\left(\frac{1}{m^2} - \frac{1}{n^2}\right) \tag{13.6}$$

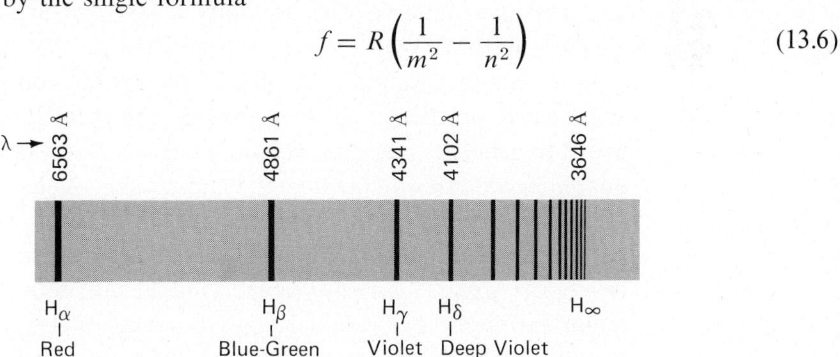

FIGURE 13.13

The Balmer series for hydrogen. The H_α, H_β, and H_γ lines are easily seen through a spectrometer. The H_δ is almost at the limit of the visible and hard to see while the rest are all in the ultraviolet. H_∞ is known as the *series limit.*

where n and m are integers and n is greater than m to give positive frequencies. Observations on the spectral series of other elements showed that many of these spectra could be accounted for by equations that were modifications of Eq. (13.6). The various wavelengths could be calculated by allowing some quantum number to take on successive integral values, such as 2, 3, 4, and so on. The simplicity and generality of these equations suggest that the same mechanism must be responsible for all the wavelengths emitted by all elements. The job of the theorist has been to construct a single theory that would predict the spectra of the elements. Classical physics is totally inadequate; it cannot even explain stable atoms, much less atomic spectra. We must, therefore, return to quantum physics for our explanation.

If we consider the spectra of hydrogen from the photon point of view, we find that the excited hydrogen atoms are emitting photons whose *energy* is quantized. The energy of a photon is related to its frequency by Eq. (11.1), $E = hf$. Using Eq. (13.6), we find

$$E = \frac{hR}{n^2} - \frac{hR}{m^2} \qquad (m, n = 1, 2, 3, \ldots, m \text{ greater than } n) \qquad (13.7)$$

for the energies of the photons in the hydrogen spectrum. It is natural to assume that each of the photons in the spectrum is associated with a single hydrogen atom. Using the conservation of energy law, we see that when a hydrogen atom emits a photon, its energy changes by a discrete quantized amount.

In 1913, Niels Bohr put forth a hydrogen atom model that explained these observations. From Rutherford's scattering experiments, Bohr knew that the hydrogen atom was composed of a single electron orbiting a very small nucleus 2000 times heavier than itself. According to classical electrodynamics, these orbits could not be stable, and the electron would have to spiral into the nucleus. But hydrogen was known to be stable. Thus, Bohr assumed that there existed certain discrete "orbits" or *states* in which the electron had a definite energy and was stable. In the first form of his model, Bohr justified this assumption by postulating that the stable orbits were those whose *angular momentum* was an integral multiple of $h/2\pi$. Ten years later, when de Broglie suggested that electrons had wave properties, Bohr realized that there was an equivalent postulate: The electrons formed circular standing waves with an integral number of wavelengths equal to the circumference of the circular orbit, as shown in Fig. 13.14. Using his model, Bohr was able to calculate that energy levels associated with the quantized states of hydrogen were given by

$$E_n = -\frac{E_0}{n^2} \qquad (n = 1, 2, 3, 4, \ldots) \qquad (13.8)$$

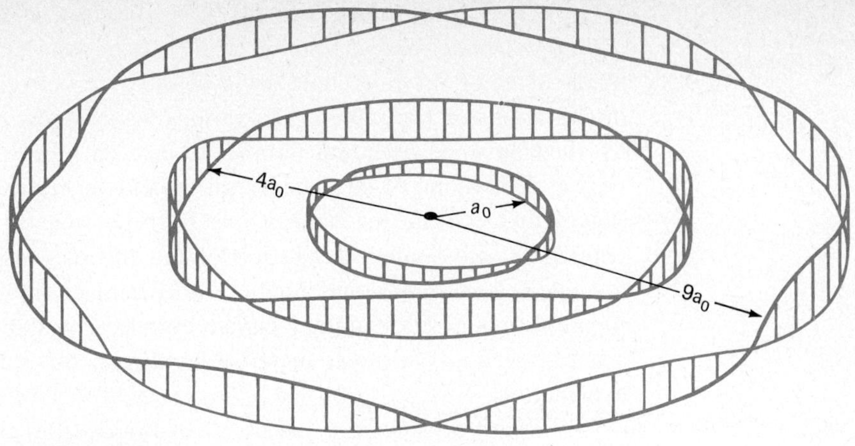

FIGURE 13.14
The first three Bohr orbits for hydrogen. The inner orbit is a standing wave whose wavelength is equal to the circumference of the circle. In the next two orbits the circumference is equal to two and three wavelengths respectively.

where E_0, the ground state energy, could be calculated from the masses of the electron and the nucleus, their charges, and Planck's constant. In addition, by assuming that the quantized states of the electron were in fact neat circular orbits, he showed that the radius of the nth orbit could be found from

$$r_n = n^2 a_0 \qquad (n = 1, 2, 3, \ldots) \tag{13.9}$$

where $a_0 = 5.3 \times 10^{-11}$ m. If we take $n = \infty$ in Eqs. (13.8) and (13.9), we find that E_∞, the energy of a free electron (one with an infinitely large orbit), is zero. For finite n, the energies are all negative, because the electron states are *bound;* that is, the electron is held near the nucleus by the attractive coulomb force. If we wish to remove it, we must do work *on* the electron by *adding* energy to the system. Normally, the electron in hydrogen is in its ground state ($n = 1$), with energy $-E_0$. To remove this electron completely, and produce a hydrogen *ion,* E_0 joules of energy must be added to the atom. (This is known to chemists as the *ionization potential* of hydrogen and is equal to 2.18×10^{-19} joules or 13.6 eV.)

The greatest success of the Bohr model was its prediction of the emission spectra of hydrogen. When hydrogen is excited by adding energy to the gas (which could be heated or excited electrically), the electrons in the atoms jump into higher energy states. Then they spontaneously drop back into lower energy states, emitting a photon whose energy is just equal to the energy jump. Suppose that an electron has been excited into the mth energy state where its energy is given by Eq. (13.8) as $E_m = -E_0/m^2$. It

then drops to the nth state (n less than m), where its energy is $E_n = -E_0/n^2$. The emitted photon will have an energy E given by

$$
\begin{aligned}
E = E_m - E_n &= \frac{-E_0}{m^2} - \frac{-E_0}{n^2} \\
&= \frac{E_0}{n^2} - \frac{E_0}{m^2} \qquad (m, n = 1, 2, 3, \ldots, m \text{ greater than } n)
\end{aligned}
\tag{13.10}
$$

If E_0 is equal to the constant hR in Eq. (13.7), then Eq. (13.10) predicts exactly the spectrum of frequencies observed for hydrogen. In particular, the Balmer series corresponds to electrons dropping in to the first excited state ($n = 2$) from various higher states ($m = 3, 4, \ldots$), as shown in Figs. 13.15 and 13.16. The Lyman series corresponds to electrons dropping directly into the ground state ($n = 1$) from higher states ($m = 2, 3, 4, \ldots$). The Paschen series corresponds to *transitions* ending in the second excited state ($n = 3$). Experimentally, the constant hR in Eq. (13.7) is equal to E_0, and so the Bohr model correctly predicts the spectrum of hydrogen with great precision.

The success of the Bohr model for hydrogen immediately prompted attempts to use it to explain the spectrum of more complex atoms. With one notable class of exceptions, this attempt failed completely. The spectra of singly ionized helium (with E_0 replaced by $4E_0$) and doubly ionized

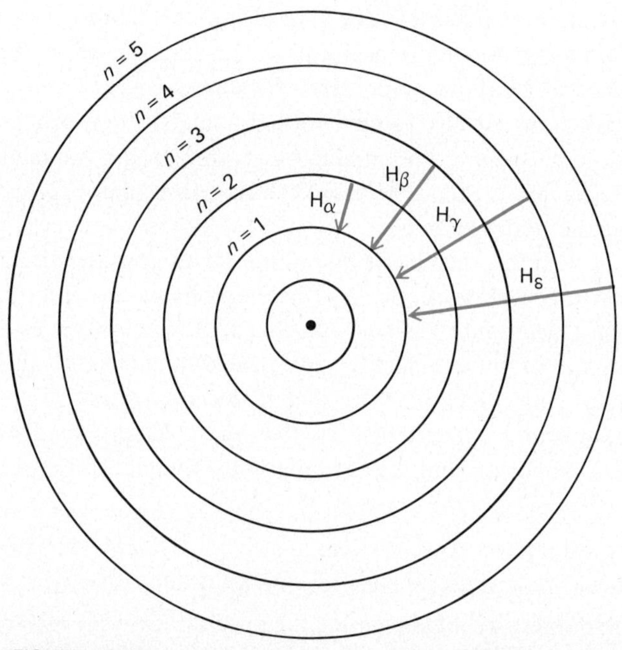

FIGURE 13.15

The orbital transitions that produce the Balmer series.

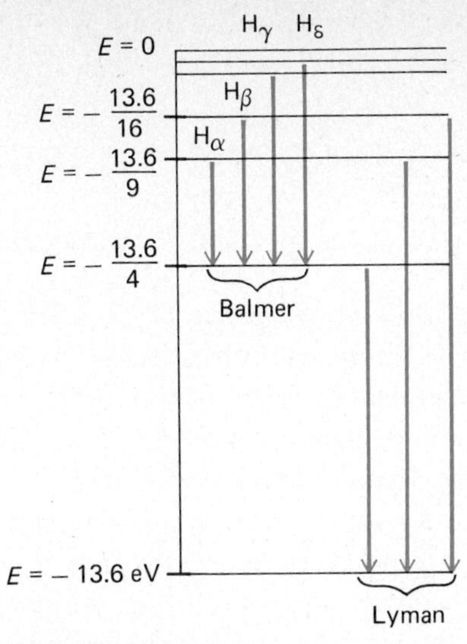

FIGURE 13.16
The energy level transitions producing the Balmer and Lyman series in hydrogen.

lithium (with E_0 replaced by $9E_0$) are given correctly by Eq. (13.10). Helium normally has two electrons orbiting a nucleus of charge $+2e$, and lithium has three electrons orbiting a nucleus of charge $+3e$. Singly ionized helium is helium that has had one of its electrons removed, leaving only one behind. Similarly, doubly ionized lithium has had two of its electrons removed. In general, the Bohr model gives the correct energy levels, and thus the correct spectrum, for any "atom" with a single electron orbiting the nucleus. But as soon as more than one electron are present, the model fails completely. Therefore it cannot be correct, although it does provide a heuristic picture of the hydrogen atom.

To understand its failure, let us consider the ground state of the hydrogen atom, as pictured in the Bohr model, from the point of view of the Uncertainty Principle. In Sec. 13.3, we saw that as a result of the wave nature of matter, electrons must satisfy the Heisenberg Uncertainty Principle. This principle states that if the momentum of an electron is known to within an uncertainty Δp, then the position must be uncertain to within Δx, where Δx and Δp are related by Eq. (13.2); that is,

$$\Delta x \, \Delta p \geq h \tag{13.11}$$

where h is Planck's constant. To apply this relation to the ground state of hydrogen, we must estimate the uncertainty in the momentum. Figure 13.14 shows that in the ground state of hydrogen, the de Broglie wavelength of the electron is equal to $2\pi a_0$, the circumference of its circular orbit. The

de Broglie wavelength is related to the momentum by Eq. (13.1), giving $h/p = \lambda = 2\pi a_0$. Thus,

$$p = \frac{h}{2\pi a_0}$$

for the magnitude of the electron's momentum in its ground state. However, we do not know the direction in which the momentum points, because we can never know just where on the circle the electron is. Therefore, the uncertainty in its value in some particular direction is $\Delta p = 2p = h/\pi a_0$, and we have

$$\Delta x \geq \frac{h}{\Delta p} = \pi a_0 \tag{13.12}$$

The uncertainty in position is as large as the orbit itself! It makes no sense to speak of the orbit as a circle. According to the Uncertainty Principle, we cannot know the position of the electron accurately enough to make such a statement. For this reason, we refer to electron "states" rather than "orbits."

The Bohr model, nevertheless, is correct in the following respects: The electrons may exist only in certain quantized states; each of these states has a definite quantized energy; and when electrons make transitions between different states, they emit or absorb a photon whose energy is just equal to the energy difference between the states. However, Bohr's idea of the electrons moving in circular orbits, and thus the physical basis for his calculations of the energy levels, is incorrect. It is just a fortunate historical coincidence that it gives the correct answer for hydrogen. The Bohr model, even though it cannot predict the energy levels in atoms with more than one electron, has still been useful in visualizing the structure of complex atoms.

13.6 QUANTUM MECHANICS

During the first quarter of the 20th century, a wealth of new experimental data and phenomenological theories showed conclusively that classical Newtonian physics did not apply to nature on the atomic level. The state of atomic physics was as undeveloped as mechanics was when Kepler was working. At that time, Galileo had observed and described the motion of objects and Kepler himself had set forth the laws of planetary motion, but Newton had not yet formulated the fundamental laws that would bring all these phenomena together. About 1926, Erwin Schrödinger and Werner Heisenberg independently published theories that were interpreted by Niels Bohr, Max Born, and others. These theories provided the formulas needed

to explain atomic physics. Now known as *quantum mechanics,* this theory is the theoretical framework for modern physics and the other sciences that are based on physics. Modified in the 1930's, 1940's, and 1950's to include special relativity, quantum mechanics seems at present to be such a sufficiently rich and flexible theory that it will continue to provide the theoretical framework for physics in years to come.

Although the conceptual basis of quantum mechanics is simple and elegant, the mathematics in which it is clothed is formidable. We shall content ourselves, therefore, with discussing qualitatively the relation of quantum mechanics to those topics that we have already discussed in this chapter. You must realize that Newtonian mechanics did not suddenly cease being a good description of the macroscopic world, nor did classical electricity and magnetism fail. Quantum mechanics, if it was to be a truly general basis for all physics, had to reproduce these theories in the areas where they are valid. This principle, called the *Correspondence Principle* by Bohr, has been satisfied. When quantum mechanics is applied properly to macroscopic objects, the laws that we studied in the first three parts of this book emerge in their familiar forms. The correspondence occurs when the separation between the quantized energy levels of a system is very small compared with the energy of the system. For example, in an oscillator with $f = 1$ cps and an energy of 1 joule, the separation between energy levels is 6×10^{-34} joules, which is completely undetectable. We have seen similar effects twice before. In the first instance, when the speed of an object is small compared to the speed of light, relativistic effects are negligible and the equations of special relativity reduce to those of Newton. In the second case, when the wavelength of light is much smaller than the size of the object with which it interacts, interference and diffraction effects disappear and the laws of geometrical optics apply. Thus, quantum mechanics does not change the laws that we have studied governing macroscopic objects but it does require far-reaching changes in the laws governing the microscopic world.

The central equation in quantum mechanics is the Schrödinger equation, named after its discoverer. The Schrödinger equation describes the time evolution and energies of the waves of matter suggested by de Broglie. In the interpretation of these waves, the square of their amplitude at a given point (the *intensity* of the wave) gives the *probability* of finding the particle at that point. The information that the waves contain is essentially statistical. The Schrödinger equation makes it possible to derive rigorously the Heisenberg Uncertainty Principle, Eq. (13.2), and to explain the other interference effects discussed in Sec. 13.3.* It is also possible to

* As the name implies, the uncertainty relations were originally derived by Heisenberg from his matrix form of quantum mechanics, which was later proved equivalent to Schrödinger's theory.

explain the structure of the hydrogen atom and other more complex atoms with this equation.

De Broglie was originally motivated to suggest that particles might have wave properties by reasoning in analogy with photons. In Sec. 11.2, we saw that the energy of a photon was related to its frequency by $E = hf$ and that this same relation holds for particles. The energy of a particle is related to the frequency of its wave by

$$E = hf \tag{13.13}$$

If the electron in the hydrogen atom could exist only in states vibrating with certain discrete quantized frequencies, then the energy levels would be explained. In Sec. 10.2 we saw that one of the characteristics of all waves was the ability to form standing waves under the proper conditions. For systems such as a stretched string or a drumhead, these "stationary" vibrational modes had a basic mode with a low "fundamental" frequency and many higher modes, the "overtones." In the hydrogen atom, the attractive coulomb force holds the electron near the nucleus, and the stable "stationary" states are electron standing waves in three dimensions. Figure 13.17 shows the "fundamental" or *ground state* of hydrogen and three "overtones" or excited states. The ground state has the lowest energy level. The shaded regions represent the places that the electron is most likely to be found. Notice that these standing-wave states bear little or no resemblance to Bohr orbits. Each of the states shown has its own frequency and thus, by Eq. (13.13), its own energy. When the exact calculation is done, the energy levels predicted by the Schrödinger equation are identical to those of the Bohr model, even though the picture of the electron's "orbits" is completely different.

The energy levels, E_n, of hydrogen are given by Eq. (13.8) where n, the *principal quantum number,* can be any positive integer. These energy levels have an interesting property called *degeneracy.* With the exception of the ground state ($n = 1$), each of the higher energy states, E_n, has more than one standing wave associated with it. The second level, E_2, has four distinct standing-wave patterns, each having the same vibrational frequency and hence the same energy. The third state, E_3, has nine patterns. When a given energy level has more than one standing-wave state associated with it, the level is said to be degenerate. For hydrogen, the nth state has n^2 distinct standing waves, each with the same energy. We now use this degeneracy to help explain the structure of more complex atoms.

Consider an atom whose nucleus has charge Ze. Z, called the *atomic number,* is the number of positive charges in the nucleus. Because atoms are electrically neutral, this atom will normally have Z electrons surrounding it. If we assume that these electrons interact only with the nucleus and not

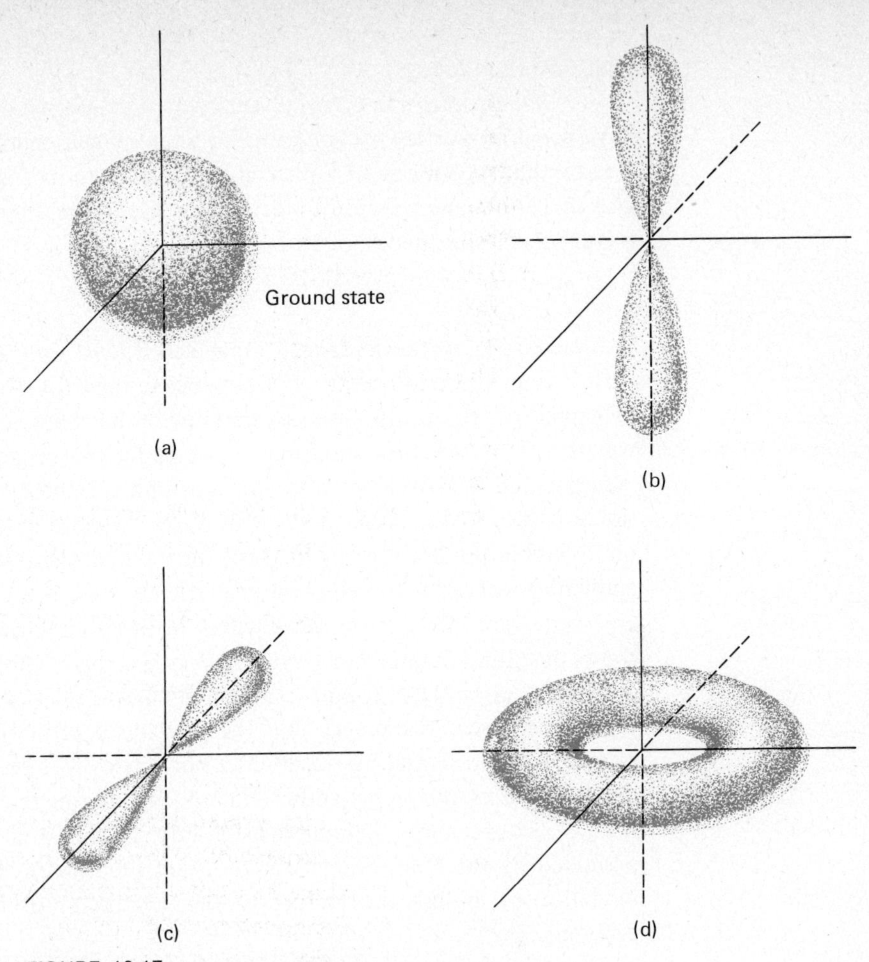

FIGURE 13.17
The electron configuration for the ground state and three possible excited states of hydrogen. The colored regions represent the places where the electron is most likely to be found.

with each other, then each electron can exist in standing-wave patterns similar to those in hydrogen. This assumption is clearly an approximation; the electrons will in fact repel each other because of their negative charge. However, this approximation will give us a correct qualitative picture. In Sec. 13.5 we saw that electrons tend to drop down into the lowest energy level available to them. They emit photons as they do so. Thus, we might expect that in a complex atom all the electrons would exist in the $n = 1$ ground state. However, this is not the case. In helium, with $Z = 2$, both of the electrons are found in the lowest $n = 1$ state. In lithium, with $Z = 3$,

two of the electrons are found in the $n = 1$ ground state, but the third is found in the $n = 2$ state. In fact, for each of the eight elements from lithium to neon ($Z = 10$), two of the electrons are found in the $n = 1$ state and the rest are found in the $n = 2$ state. As a final example, in sodium ($Z = 11$), two of the electrons are in the $n = 1$ state, eight are in the $n = 2$ state, and the last one is in the $n = 3$ state. Figure 13.18 shows the electronic configurations for these 11 elements. Why are the electrons in these atoms not dropping down into the lowest possible energy state, the state with $n = 1$? The answer to this question was given in 1925 by Wolfgang Pauli when he formulated what is now known as the *Pauli Exclusion Principle.* Before we discuss this principle, we must introduce one last property of the electron: spin.

In 1925, G. Uhlenbeck and S. Goudsmit pointed out that certain unexplained features in atomic spectra could be understood if it was assumed that the electron carried a small amount of intrinsic angular momentum in addition to its mass and charge. In classical terms, angular momentum is usually associated with rotation; therefore, this property was called *spin.* We need not be concerned with the details of their suggestion and the data that it explained; it is enough to say that their suggestion was correct. The Bohr theory indicated that the natural quantum of angular momentum is $h/2\pi$; however, the electron always carries angular momentum equal to $h/4\pi$. Consequently, the electron is said to have spin-$\frac{1}{2}$. Just as classical rotation has an axis and thus a direction associated with it, so does spin. Unlike classical rotation, where the various directions that this axis can point to are independent, there are only two independent directions to which the electron's spin can point. These directions are usually taken to be "up" and "down." The other possible directions can be formed as a combination of these two. When an electron is in a particular standing-wave state, its spin can take on either of these two directions. Thus, each standing-wave state is doubly degenerate. In the case of the hydrogen atom, where there are n^2 standing waves for each energy level, the total degeneracy of the nth level is then $2n^2$.

To return to the observation that not all the electrons in a complex atom end in the ground state, the Pauli Exclusion Principle postulates that *no more than one electron can occupy exactly the same state.* Thus, the $n = 1$ ground state of an atom with only one standing-wave state can be occupied by no more than two electrons, one with spin up and the other with spin down. In a similar manner, the $n = 2$ state, which has four independent standing-wave states, can be occupied by eight electrons, two for each standing wave. When electrons are added to a sodium nucleus, for example, the first two will occupy the $n = 1$ ground state; the next eight must then go into the $n = 2$ state; and the last one will be in the $n = 3$ state. Thus,

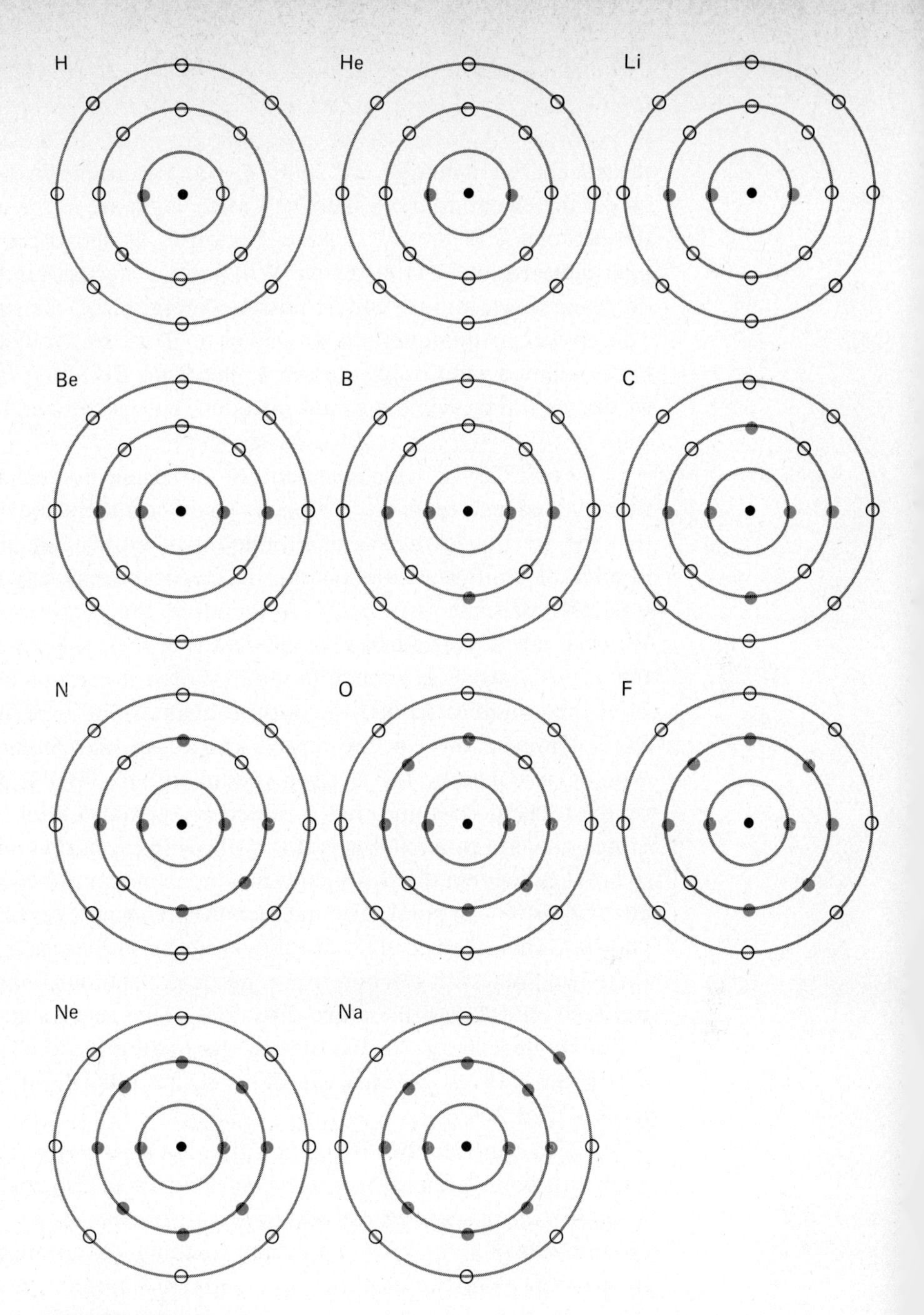

FIGURE 13.18

The ground state electron configuration for the first 11 elements.

in a complex atom, electrons will first fill the lowest energy levels (often called *shells*) and then fill the higher ones.

In our highly simplified discussion, we have been assuming that the electrons in an atom exist in hydrogen-like standing waves. Such an assumption would indicate that all atomic spectra are similar to that of hydrogen. However, this is clearly not the case. The standing waves of a real atom are modified mainly by two effects. The first is the effect of the electrons on each other. In an atom with more than one electron, each electron feels the electrical force of the other electrons as well as the electrical force of the nucleus. The second effect comes from its spin. As a result, the electron has a small magnetic field and thus also changes the standing-wave states. When these effects are taken into account, the $2n^2$ degeneracy of the nth energy level is removed, and each of the standing waves has its own energy. Splitting the energies of the nth level produces *fine structure* in the spectrum of the atom. This fine structure can be observed with a high-resolution spectroscope. For atoms with Z over about 18, the splitting is large enough for the original energy levels to overlap. Because the electrons always go into the lowest empty level, the shells just mentioned are not always filled in order.

Although the effects discussed in the preceding paragraphs cannot be calculated exactly, by using computers and the Schrödinger equation scientists can calculate them approximately. Then the observed structure of the periodic table emerges, along with the spectra of complex atoms. Unlike the Bohr theory, quantum mechanics can be used on complex atoms. By examining the electronic structure of atoms, as predicted by the Schrödinger equation, we can discover the chemical properties of atoms. Thus, the laws of chemistry can be understood and given a firm theoretical basis. Quantum mechanics can also be used to interpret the structure and properties of solids and to explain the existence of, for instance, conductors and insulators and such strongly magnetic materials as iron and cobalt.

An interesting effect caused by the spin of the electron occurs in hydrogen. We shall see in the next chapter that the proton (the nucleus of hydrogen) is also a spin-$\frac{1}{2}$ particle and, like the electron, has two possible independent orientations of its spin axis, up or down. When hydrogen is in its ground state ($n = 1$), the electron spin can be either parallel or antiparallel to the proton spin. These two states have slightly different energies, and transitions from one state to the other give rise to photons with a wavelength of 21 cm. This transition produces an *emission line*. The 21-cm emission line of hydrogen has played an important role in astronomy. At this wavelength the interstellar hydrogen "glows." Furthermore, interstellar dust, which is opaque at visible wavelengths, is transparent to 21-cm

radiation. As a result, radio astronomers have been able to map the spiral arms of our galaxy.

Another application of quantum mechanics and energy levels is the *laser*, an acronym for *l*ight *a*mplification by *s*timulated *e*mission of *r*adiation, invented by C. H. Townes. When a gas is excited, it emits characteristic spectral lines. Its electrons, excited into high energy states, constantly make transitions back into lower states and emit photons whose energy equals the energy difference between energy levels for each transition. The timing of a given transition occurs randomly, and there is no relation between the many photons emitted. Consequently, even if we use only one of the spectral lines, such as the H_α line of hydrogen, and thus have monochromatic light, the various photons have randomly related phases. (Recall that being "in phase" means that crests match crests.) Light of this sort is called *incoherent*. Until the invention of the laser, all the known sources of visible light were incoherent. To produce coherent radiation (in which all the photons are in phase), it is necessary to control the vibrations or transitions producing the radiation; such control could be established only over those frequencies that could be generated electronically (i.e., microwave frequencies or less).

The heart of one type of laser is a ruby crystal. The electrons in the ruby have certain energy levels in which they can exist, just as the electron in hydrogen did. We can simplify the situation somewhat by assuming that there are only two levels.* Under normal circumstances, most of the electrons are in the ground state and only a few are in the excited state. If we passed light whose frequency f matched the frequency $(E_1 - E_0)/h$ that is characteristic of the energy gap, then some of this light would be absorbed as electrons made transitions into the excited state. However, if light of this frequency hits an electron that is already in an excited state, the electron is stimulated to make the transition into the ground state and emit another photon. This behavior is stimulated emission. Normally this process is relatively unimportant because most of the electrons are in the ground state. However, if it can be arranged that most of the electrons are in the excited state, the situation changes. In a ruby laser, the ruby crystal is energized electrically in such a way that most of the electrons are in the excited state, as shown in Fig. 13.19. Then a small amount of light of the proper frequency is introduced at one end. As this light passes electrons, the electrons are stimulated to drop into the ground state, emitting a photon *exactly in phase with the stimulating light*. As the light beam passes down the crystal, it is strongly amplified by the addition of many additional photons, all of which are in phase with each other. In a typical laser, the

*This approximation is reasonable because only two levels are important for laser action in the real crystal.

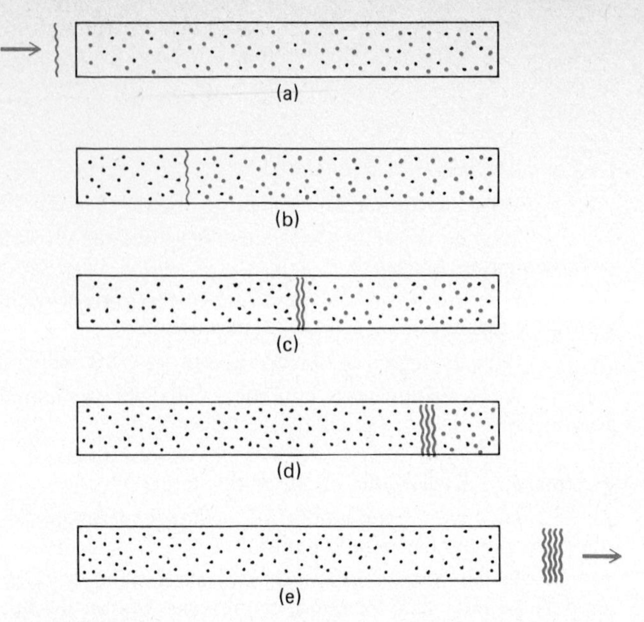

FIGURE 13.19

A ruby laser. In (a) the ruby has been excited so that most of its atoms are in their upper energy state. In (b) a light pulse has entered and is moving across the crystal gaining photons by stimulating their emission. In (e) it exits from the crystal amplified and coherent.

two ends of the ruby are partially silvered, so that the light is reflected many times before emerging from the end as an absolutely parallel coherent beam of pure red light. (It is red not because ruby is red but because the transition that produces laser action happens to give a frequency in the red part of the spectrum.)

In the short time since its discovery, the laser has had many diversified applications, ranging from eye surgery to measuring the distance from the earth to the moon within an accuracy of a few inches. Lasers are used to keep tunnels straight and to cut fabric for clothing manufacturers. In addition, lasers form the light source for making holograms, which are three-dimensional pictures formed without lenses.

DISCUSSION QUESTIONS

1. "The color of an electron is green." Discuss whether or not this statement is physically meaningful.

2. "On Monday, Wednesday, and Friday we treat the electron as a particle, and on Tuesday, Thursday, and Saturday we treat it as a wave." Does this statement have an accurate physical meaning?

3. Discuss the great discoveries which came about because various men studied the apparently trivial topic of the conduction of electricity through gases. Describe a number of important ways in which these discoveries have had an impact on our lives today.

4. What are the experimental difficulties in using the Stefan-Boltzmann Law as a definition of temperature?

5. Why are the walls of a thermos bottle silvered.

6. Why do you think that Planck proposed the revolutionary ideas expressed in the equation $E = nhf$?

7. If sound waves are quantized in the same way that light waves are, do you think this could be detected experimentally?

8. Give examples of objects in your own life which are quantized.

9. An electric heater or hotplate usually glows a dull red. At what wavelengths do you think most of the energy is transmitted?

10. The study of the masses of the various atoms of elements is called mass spectroscopy. Explain the origin of this term.

11. How would you undertake to show experimentally that atoms and molecules are usually electrically neutral?

12. In the various spectra of a given element, a particular energy level will serve as the final state for one spectral series and for the initial states of a number of wavelengths in other spectral series. Discuss how this method of internal consistency is used in analyzing spectra.

13. Discuss the appearance of a light bulb if e had the value 1 coulomb.

14. X rays are absorbed most effectively by elements of high atomic number, and the amount of absorption depends on the amount of the material used. For these reasons, lead, which has atomic number 82 and high density, is most commonly used to absorb x rays. Do you think it would be worthwhile to look for some other element which would be as good an absorber as lead, yet would make a lighter shield?

15. Discuss changes in our life if Planck's constant had the value $h = 10^{-20}$ joule-sec.

PROBLEMS

1. Compute the electrical force on an oil droplet carrying a charge of 3.2×10^{-19} coulombs when it is located in an electric field of 2000 V/m. (*Ans.:* 6.4×10^{-16} N)

Compute the mass and weight of such a droplet which would be in equilibrium under the action of such an upward electrical force and the force of gravity.

2. Compute the wavelength at which electromagnetic radiation is most intense from a human body at a temperature of 40°C. (*Ans.:* 9.27×10^{-6} m)

If the surface of a human body is 2 m^2, compute the total radiation from the human body, treating it as an ideal radiator.

3. The most intense radiation from our sun occurs at a wavelength of about 5×10^{-7} m. Compute the surface temperature of the sun. (*Ans.:* 6000°K)

4. An electric heater is rated at 100 watts. If its surface area is 300 cm^2, compute its temperature.

5. Calculate the wavelengths of the three longest wavelengths of the Balmer series of hydrogen.
(*Ans.:* 6.56×10^{-7} m; 4.86×10^{-7} m; 4.34×10^{-7} m)

6. Calculate the value of the shortest wavelength in the Balmer series.

7. If the longest wavelength in the Balmer series of hydrogen is 6.56×10^{-7} m, calculate the energy change which produces this wavelength.
(*Ans.:* 3.01×10^{-19} joules)

8. An oil droplet has a mass of 10^{-17} kg and carries an electrical charge of 4.8×10^{-19} coulombs. Compute the value of the electric field which will hold this droplet in equilibrium against the force of gravity.

9. A furnace operates at a temperature of 1500°C. Compute the value of the wavelength which is most intensely emitted by this furnace.
(*Ans.:* 1.64×10^{-6} m)

   If the door of this furnace measures 30 × 60 cm, compute the amount of the radiation emerging (in watts) when the door is fully open.

10. At the earth's surface the radiation received by sunlight is 1.3 kilowatts/m^2 on a surface at right angles to the sunlight. Treat the human body as a black body of area 0.75 m^2. Compute the energy received in 0.5 hour at noon by a sunbather.

11. Calculate the value of the shortest wavelength emitted by hydrogen.

12. Calculate the value of the energy change when mercury emits its resonance line at a wavelength of 2.536×10^{-7} m. (*Ans.:* 7.87×10^{-19} joules)

13. Calculate the least energy which must be given to an unexcited hydrogen atom for the longest wavelength of the Lyman series to be emitted.

14. Calculate the minimum energy which must be added to an unexcited hydrogen atom for the longest wavelength of the Balmer series to be emitted.

15. Calculate the least energy which must be given to an unexcited hydrogen atom for the electron to be removed from the atom entirely.
(*Ans.:* 2.17×10^{-18} joules)

16. Calculate the frequency of an x-ray of wavelength 10^{-11} m.

   Calculate the energy of a photon of such a wavelength.
(*Ans.:* 6.63×10^{-23} joules)

17. Calculate the energy change so that a photon of wavelength 10^{-13} m will be produced.

18. Calculate the momentum associated with a photon of wavelength 10^{-13} m.
(*Ans.:* 6.63×10^{-21} kg-m/sec)

19. Calculate the wavelength associated with a particle of mass 1 kg moving at a speed of 10 m/sec.

20. If a particle moves with a speed of 10^7 m/sec, what must its mass be if its associated wavelength is to be 1 cm?

21. A particle of mass 5 kg has an associated wavelength of 10^{-30} m. What is the velocity of the particle? (*Ans.:* 1.32×10^{-4} m/sec)

22. Some cosmic rays have energies of 10^{-10} joules. Calculate the frequency of such a photon. (*Ans.:* 1.51×10^{23} cps)

Calculate the wavelength of such a photon. Calculate its momentum.
(*Ans.:* 3.32×10^{-19} kg-m/sec)

23. How many electrons are needed to form 1 coulomb of charge?

24. An oscillator constructed with a mass and a spring has a frequency of 5 cycles/sec. It is oscillating with an energy of 3 joules. If this oscillator were quantized, what value would n have. Would the separation between adjacent energy levels be observable?

25. The speed of an electron is measured with an uncertainty of 0.1 m/sec. Find the uncertainty in its position. Repeat for a proton.

14 The Nucleus

The Rutherford model correctly interpreted the structure of the atom by picturing it as a cloud of electrons surrounding a very small, heavy, positively charged core: the *nucleus*. With the development of the Bohr model and quantum mechanics, the structure of the electron cloud was understood and the chemistry of atoms explained. However, for the next 20 years the structure of the nucleus itself remained a mystery. In this chapter we shall discuss the properties of nuclei and some of the applications that nuclear physics has had in our society.

14.1 BASIC NUCLEAR PROPERTIES

The most important characteristic of a nucleus is its charge. As we discussed in Sec. 13.1, all charges come in multiples of e, the charge on the electron. The nucleus is no exception. We can thus write the charge of any nucleus as Ze, where Z is an integer called the *atomic number* of the nucleus. The values for Z range from $Z = 1$ for hydrogen to $Z = 92$ for uranium, the most highly charged of the naturally occurring nuclei. The atomic number is important because it determines how many electrons surround the nucleus in a given atom. A nucleus with charge $+Ze$ will have Z electrons orbiting around it; these electrons have total charge $-Ze$, making the atom electrically neutral. The atomic number determines the place of the element in the periodic table, Fig. 14.1.

								1 H 1.0080									2 He 4.003
3 Li 6.939	4 Be 9.012											5 B 10.81	6 C 12.011	7 N 14.007	8 O 15.999	9 F 18.998	10 Ne 20.183
11 Na 22.990	12 Mg 24.31											13 Al 26.98	14 Si 28.09	15 P 31.974	16 S 32.06	17 Cl 35.453	18 Ar 39.948
19 K 39.102	20 Ca 40.08	21 Sc 44.96	22 Ti 47.90	23 V 50.94	24 Cr 52.00	25 Mn 54.94	26 Fe 55.85	27 Co 58.93	28 Ni 58.71	29 Cu 63.54	30 Zn 65.37	31 Ga 69.72	32 Ge 72.59	33 As 74.92	34 Se 78.96	35 Br 79.909	36 Kr 83.80
37 Rb 85.47	38 Sr 87.62	39 Y 88.91	40 Zr 91.22	41 Nb 92.91	42 Mo 95.94	43 Tc (99)	44 Ru 101.1	45 Rh 102.91	46 Pd 106.4	47 Ag 107.870	48 Cd 112.40	49 In 114.82	50 Sn 118.69	51 Sb 121.75	52 Te 127.60	53 I 126.90	54 Xe 131.30
55 Cs 132.91	56 Ba 137.34	57 *La 138.91	72 Hf 178.49	73 Ta 180.95	74 W 183.85	75 Re 186.2	76 Os 190.2	77 Ir 192.2	78 Pt 195.09	79 Au 196.97	80 Hg 200.59	81 Tl 204.37	82 Pb 207.19	83 Bi 208.98	84 Po (210)	85 At (210)	86 Rn (222)
87 Fr (223)	88 Ra (226)	89 †Ac (227)															

	58	59	60	61	62	63	64	65	66	67	68	69	70	71
*** Lanthanide series**	58 Ce 140.12	59 Pr 140.91	60 Nd 144.24	61 Pm (147)	62 Sm 150.35	63 Eu 152.0	64 Gd 157.25	65 Tb 158.92	66 Dy 162.50	67 Ho 164.93	68 Er 167.26	69 Tm 168.93	70 Yb 173.04	71 Lu 174.97
† Actinide series	90 Th 232.04	91 Pa (231)	92 U 238.03	93 Np (237)	94 Pu (242)	95 Am (243)	96 Cm (247)	97 Bk (245)	98 Cf (251)	99 Es (254)	100 Fm (253)	101 Md (256)	102 No (254)	103 Lw (257)

FIGURE 14.1

Periodic table of the elements. The number above the chemical symbol is the atomic number Z; the number below is the atomic weight (different isotopes are averaged in according to their natural abundance).

The mass of a given atom is made up almost entirely of the mass of its nucleus. The electrons contribute less than one part in 2000 of the total. Because the charge of any nucleus is an integral multiple of e (the charge on an electron), you might wonder whether atomic and, thus, nuclear masses have a similar regularity. If we measure the masses of individual atoms in atomic mass units (1 amu $= 1.660 \times 10^{-27}$ kg), we find that all atomic masses are within $\frac{1}{10}$ amu of being integral. Thus, we can assign an *atomic mass number* A to each atom where A is the integer closest to the atomic mass in amu.

As early as 1903, it was known that atoms with the same chemical properties (i.e., the same Z) could have different masses. Atoms with the same Z but different A are called *isotopes*. Ordinary hydrogen has $A = 1$, but careful investigation shows that about 0.015 per cent of naturally occurring hydrogen has $A = 2$. This heavy isotope of hydrogen is named deuterium and is sometimes given the special symbol D. Figure 14.2 shows all the known isotopes. Notice that A is slightly more than twice Z for most stable atoms. In order to distinguish different isotopes, we use a superscript on their chemical symbol and denote a particular isotope by X^A, where X is the chemical symbol and A the atomic mass number. Thus, ordinary hydrogen is H^1 and deuterium is H^2 or D^2. The most abundant isotope of helium is He^4; its other lighter isotope is He^3. When we wish to include the atomic number Z, we write $_ZX^A$ (i.e., $_1H^1$, $_2He^4$, $_6C^{12}$, etc.). The most abundant isotope of carbon is C^{12}, and the atomic mass unit is defined in such a way that the mass of C^{12} is exactly 12 amu. Similarly, the mass of H^1 would be 1.007825. Although different isotopes of the same element may have very different *nuclear* properties (such as radioactivity), they all have the same *chemical* properties, because they all have the same Z and, thus, the same number of electrons around their nucleus.

The extremely small size of the nucleus makes it difficult to give a precise meaning to the concept of a "nuclear radius." However, most methods of measurement agree that the radius of a hydrogen nucleus is around 1.2×10^{-15} meters, whereas heavy nuclei have radii around 6×10^{-15} meters. The volume of a nucleus is roughly proportional to its mass. Therefore, the *density* (mass/volume) is approximately constant at 5×10^{14} gm/cm^3. When compared to the density of water (1 gm/cm^3), this figure is rather large.

The early models for the structure of the nucleus used the only elementary particles that were known at the time: the *electron*, which we have discussed, and the *proton*. The proton is the nucleus of the lightest isotope of hydrogen, $_1H^1$, and its symbol is p. All the heavier nuclei were correctly believed to be composite structures, that is, made up of more elementary objects. If a given nucleus were made up of protons and elec-

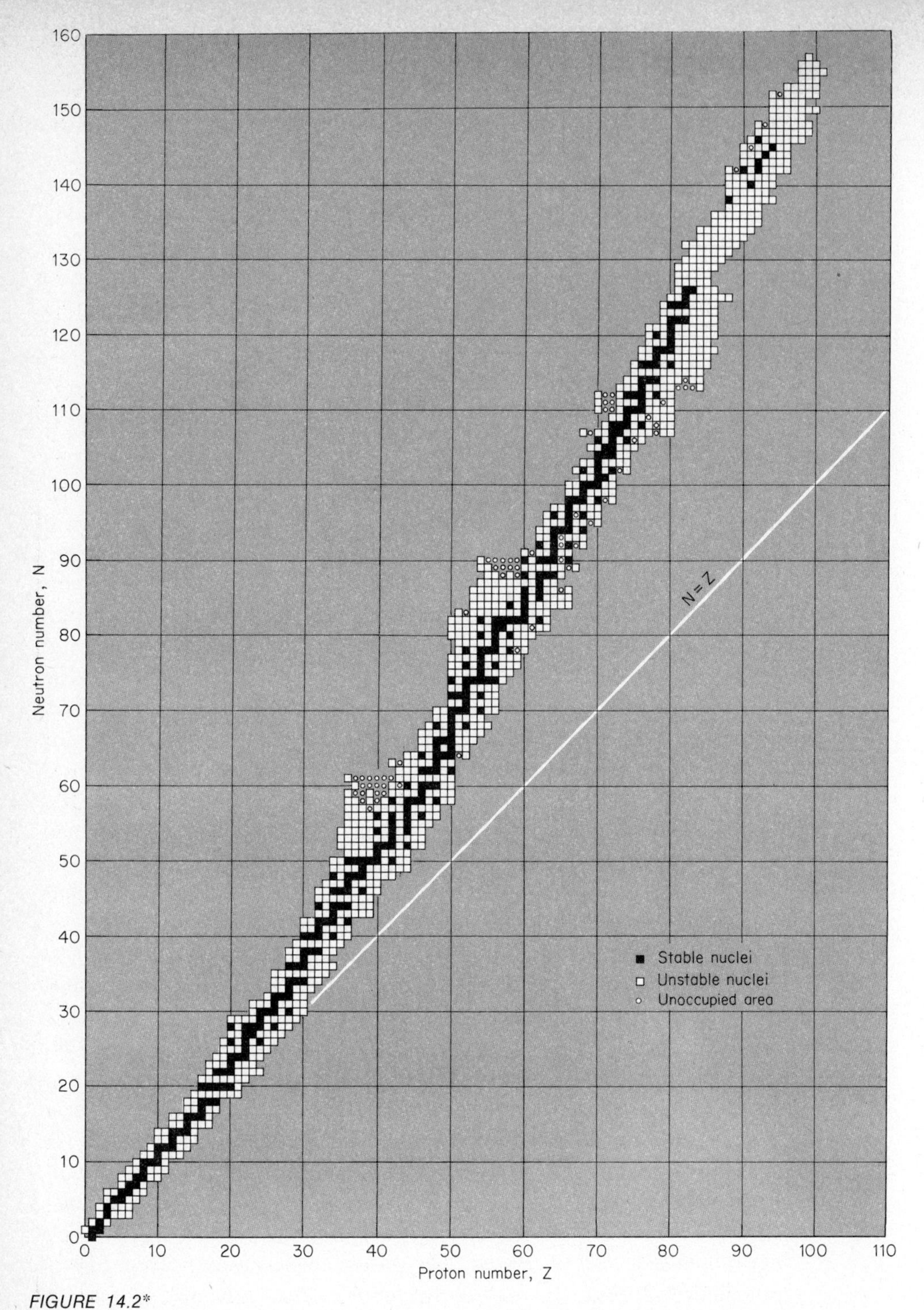

*FIGURE 14.2**

The number of neutrons, N, versus the number of protons Z for the known isotopes. The mass number A is $N + Z$.

*From Blanchard, Burnett, Stoner & Weber, *Introduction to Modern Physics,* 2nd edition © 1969. Reproduced by permission of Prentice-Hall, Inc., Englewood Cliffs, New Jersey.

trons, it would have to have A protons to account for its mass and $A-Z$ electrons to give it a net charge Z. Most nuclei have an atomic mass number A that is slightly more than twice Z, the atomic number; therefore, they would be composed of A protons and roughly $A/2$ electrons, held together in some unspecified manner. This model accounts for the mass and charge of the nucleus. It has one fatal problem however. According to the Uncertainty Principle, when an electron is confined to a region of 10^{-15} m, it will have a tremendous uncertainty in its momentum and thus in its energy. This calculation, although relatively simple, must be done relativistically. Consequently, an electron cannot be bound in a nucleus. If it were, it would have so much energy (by the Uncertainty Principle) that it would escape immediately. Thus, the simple model of a nucleus made up of electrons and protons cannot work.

The composition of the nucleus remained a mystery until 1932 when Sir James Chadwick discovered a new elementary particle. He observed that a neutral particle of about the mass of the proton is emitted in certain nuclear reactions. This particle is now called the *neutron,* and its symbol is n. With the discovery of the neutron, the way was open to understanding the constituents of the nucleus. A nucleus with atomic number Z and mass number A has Z protons, which account for its charge, and $N = A-Z$ neutrons, which make up the remainder of its mass. Thus, the nucleus of deuterium, ${}_1D^2$, differs from that of ordinary hydrogen by having one neutron in addition to one proton. The nucleus of ${}_2He^4$ has two protons and two neutrons, whereas the nucleus of ${}_{92}U^{238}$ (uranium) has 92 protons and $238 - 92 = 146$ neutrons. Clearly, different isotopes of the same element have the same number of protons and different numbers of neutrons.

Having learned that nuclei are made up of protons and neutrons, known collectively as *nucleons,* we now turn to the question of how they are held together in the nucleus. Although there is a small gravitational attraction between nucleons, this attraction is weaker by a factor of 10^{36} than the electrical repulsion between the protons. Some new attractive force, therefore, must be present to hold together a nucleus. This new force is called the nuclear force or the *strong interaction.* Physicists still do not understand its precise nature, but several of its general characteristics can be stated. First, the nuclear force has a very *short range.* The electrical force between charged particles correctly describes their behavior down to distances of around 10^{-14} meters, so the nuclear force must act only when particles are closer than this distance. However, when they are closer, the strong interaction produces an attractive force 100 times stronger than electrical forces and thus holds the nucleus together. Second, the strong interaction acts almost identically on protons and neutrons but does not affect electrons at all. Particles that are affected by the strong interaction are called *hadrons,*

and we shall study some of them at the end of this chapter. Finally, the nuclear force becomes strongly repulsive when nucleons approach much closer than 10^{-15} m; thus, nuclear matter is almost incompressible.

In order to see how strongly the nuclear force binds together the protons and neutrons in a nucleus, let us consider $_2He^4$. The mass of this most common isotope of helium is measured as 4.0026 amu. It is made up of two protons ($m_p = 1.0078$ amu) and two neutrons ($m_n = 1.0087$ amu). If we calculate the mass of its constituents taken separately, we find that they add up to 4.0330. Thus we see that helium has a *mass defect* of 0.03 amu; that is, its mass is 0.03 amu less than the mass of the protons and neutrons composing it. To understand this fact, we must recall Einstein's relation between mass and energy. Einstein found that mass and energy were related by Eq. (12.17):

$$E = mc^2 \tag{14.1}$$

To separate the protons and neutrons in the He^4 nucleus, we would have to supply enough energy to equal 0.03 amu of mass. Using the conversion

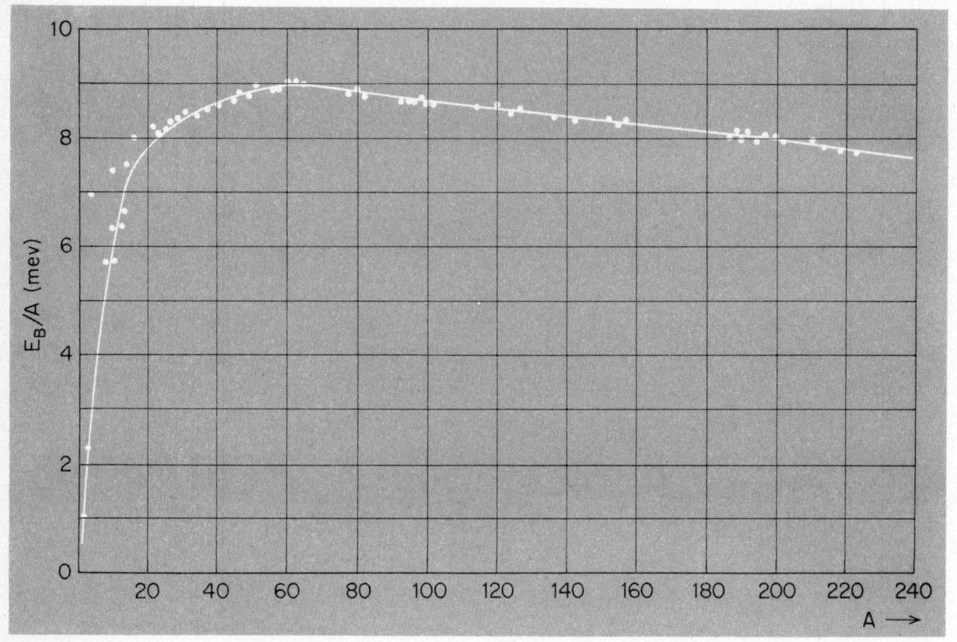

*FIGURE 14.3**

The binding energy per nucleon as a function of mass number.

*From Blanchard, Burnett, Stoner & Weber, *Introduction to Modern Physics,* 2nd edition © 1969. Reproduced by permission of Prentice-Hall, Inc., Englewood Cliffs, New Jersey.

that 1 amu equals 931.5 MeV of energy (1 MeV $= 10^6$ eV $= 1.6 \times 10^{-13}$ joules), we find that roughly 28 MeV is needed. This energy is known as the *binding energy* of the nucleus. It should be compared to the binding energy of atomic hydrogen (the energy needed to completely remove hydrogen's single electron), which is 13.6 eV. Alternatively, if we were to combine two protons and two neutrons to form ${}_2He^4$, 28 MeV of energy would be released. As we shall see later in this chapter, this *fusion* reaction is probably the energy source for the sun.

In general, the mass of a nucleus with Z protons and $N = A - Z$ neutrons is always less than the mass of the individual nucleons that compose it. We can use this mass defect to calculate from Eq. (14.1) the binding energy of the nucleus. Figure 14.3 shows the binding energy per nucleon (i.e., the binding energy divided by A, the atomic mass number) of the various nuclei. The relation between the mass defect of a nucleus and its binding energy is one of the many experimental verifications of Einstein's mass-energy relation. Note that the binding energies of nuclei are roughly a million times larger than the binding energies of atoms, and so we would expect nuclear reactions to involve far larger releases of energy than chemical reactions between atoms.

14.2 RADIOACTIVITY

Historically, the first discovery in nuclear physics was that of radioactivity. In 1896, the French physicist Antoine Becquerel found that uranium was able to fog photographic plates even when they were wrapped in black paper. He named this strange effect *radioactivity.* Two years later, in 1898, Pierre and Marie Curie (who discovered radium) showed that a number of other elements also possessed the ability to affect photographic plates and that some of these elements were much more active than uranium. Becquerel's discovery was made by accident, but it opened up an entire new field of physics. The exploration of this field is still continuing.

Numerous experiments have uncovered other properties of radioactivity. In addition to affecting photographic plates, radiation from radioactive elements can ionize gases and cause fluorescence. Both of these effects are used in the detection of radiation. The ability of radiation to ionize materials means that it can damage the complex organic molecules in living cells. Thus, *ionizing radiation,* as it is sometimes called, can be extremely dangerous to work with. Scientists have found that the intensity of radioactivity depends not on the state of chemical combination of an element, nor on the temperature of the material, nor on the application of an electric or magnetic field. They have concluded, therefore, that radioactivity must

be a property of the part of an atom that is best shielded from external influences; the nucleus.

The nature of the radiation emitted by a radioactive source is more complex than the electromagnetic radiation that we have studied earlier. Suppose that we put a small piece of radioactive material, such as radium, at the bottom of a hole drilled in a lead block, as pictured in Fig. 14.4. Because lead is one of the best absorbers of radiation, only a narrow beam of radiation will emerge from the hole. If we now apply an electric field from left to right, as shown in the diagram, we find that the radiation is split into three distinct beams. These three types of radiation were called alpha, beta, and gamma rays by the early investigators. From the directions of the deflections of the beams, we see that alpha rays are positively charged particles (called *alpha particles*), beta rays are negatively charged particles, and gamma rays are uncharged. The three types of radiation can also be distinguished by their penetrating powers in various materials. Although penetrating power does depend on the energy of the beam and the material, alpha particles are stopped rather easily, whereas gamma rays are the most penetrating. Moreover, these three types of radiation differ in the ionization that they produce in materials. Alpha particles produce the greatest amount of ionization, gamma rays produce very little ionization, and beta particles are intermediate in their ionization.

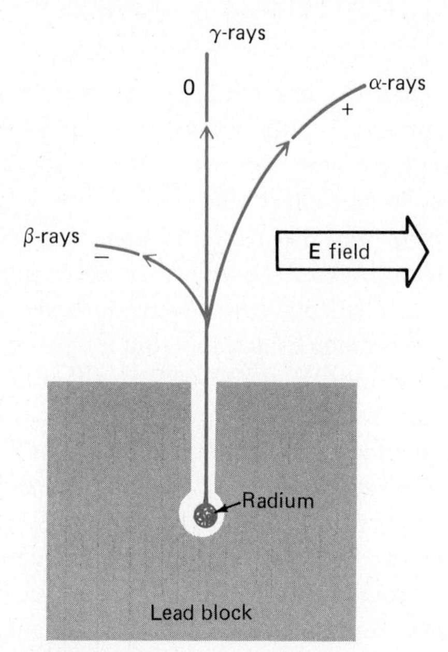

FIGURE 14.4

The three types of radiation emitted by a radioactive source.

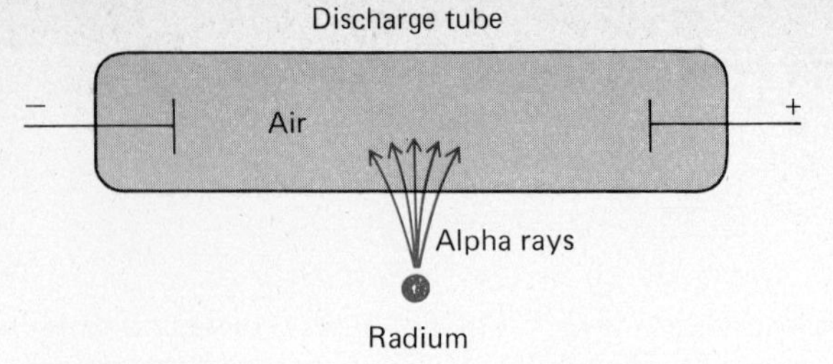

FIGURE 14.5

Rutherford's experiment for determining the nature of alpha-rays. Initially only the spectrum of air was observed. After some time the spectrum of helium began to appear.

Many years of experimentation were necessary to discover the exact nature of these three types of radiation. In 1908, Rutherford allowed alpha particles to pass into a thin-walled tube containing air, as shown in Fig. 14.5. A spark was passed through the tube, and the emission spectrum was observed. At first, only the spectrum of air was seen. As time passed, however, the spectrum of helium appeared more and more strongly. Rutherford also found that alpha particles have a charge of $+2e$ and a mass of 4.0026 amu. Evidently, therefore, they are the nuclei of ${}_2He^4$. The other types of radiation were also analyzed. It was found that beta particles are ordinary electrons emitted from the nucleus, whereas gamma rays are very short wavelength photons.

A basic law in chemical reactions is the conservation of the number of atoms of each element present. Thus, if 50 atoms of oxygen and 100 atoms of hydrogen are combined, the result is 50 molecules of H_2O (water). Until the discovery of radioactivity, no reaction had ever broken this law. Atoms of one element could not be changed into atoms of a different element. However, consider the following experiment. If a stream of air flows past pure ${}_{88}Ra^{226}$ (radium), which emits alpha particles; the air shows radioactivity. Study of this air reveals that it contains a radioactive gas called *radon*. When the radon is analyzed, it has atomic number $Z = 86$ and mass number $A = 222$. Thus, it is a different element that was not present in the original sample of radium or in the air. Clearly, in some way, one element, radon, is produced when another element, radium, emits an alpha particle. The problem is to find the law governing this type of *radioactive decay*.

A partial answer to this problem came in 1912 when Ernest Rutherford and Frederick Soddy suggested that a new nucleus with two less units of charge and four less units of mass is formed when a nucleus emits an alpha particle. Thus, when ${}_{88}Ra^{226}$ emits an alpha particle (${}_2He^4$), the resulting nucleus must have $Z = 88 - 2 = 86$ and $A = 226 - 4 = 222$,

which is exactly the description of radon. In the form of an equation, we write

$$_{88}Ra^{226} \longrightarrow {}_{2}He^{4} + {}_{86}Rn^{222} \tag{14.2}$$

Today we know that such *nuclear decay processes* are governed by three conservation laws.* The first law is *conservation of charge,* which states that the charge of the decay products must equal the charge of the original nucleus. The charge of a nucleus is given (in units of e) by Z, the atomic number; therefore, according to this conservation law, the sum of the atomic numbers of the decay products must equal the atomic number of the original nucleus. The second conservation law is known as *conservation of baryon number.* This rather imposing title means, in the present context, that the number of nucleons is conserved in a nuclear decay. Recalling that nucleons are protons and neutrons, we see that conservation of baryon number means that the number of protons plus the number of neutrons is constant. Because the number of protons plus the number of neutrons is the mass number A, the sum of the mass numbers of the decay products must equal the mass number of the original nucleus. The last conservation law is a familiar one, *conservation of energy,* which says that the total energy before a decay must equal the total energy after decay. It is absolutely essential when using this law to include the energy stored in the masses of the particles by virtue of Eq. (14.1).

Let us see how these laws apply in the decay of radium, Eq. (14.2). The atomic number of radium is 88, and the atomic number of helium is 2. Therefore, the other decay product have to have atomic number 86 to conserve charge. In a similar fashion, radium is composed of 226 nucleons, whereas helium has 4 nucleons. Thus, 222 nucleons remain for radon. The mass of Ra^{226} is 226.0254 amu. It decays into He^{4} with a mass of 4.0026 amu and Rn^{222} with a mass of 222.0175 amu. The net mass of the decay products is then $222.0175 + 4.0026 = 226.0201$, which is 0.0053 amu less than the original nucleus. Using the fact that $(1 \text{ amu}) \times c^2 = 931$ MeV, we find that 4.9 MeV of energy is left over. The alpha particles emitted by Ra^{226} are found to have a kinetic energy of 4.8 MeV, and thus energy is also conserved. (The remaining 0.1 MeV is the recoil kinetic energy of the radon nucleus.) In any decay process, the net mass of the decay products must be less than that of the original nucleus; otherwise, conservation of energy forbids the decay. The missing mass always shows up as kinetic energy. Another example of alpha decay is U^{238}, which decays into Th^{234} according to

$$_{92}U^{238} \longrightarrow {}_{2}He^{4} + {}_{90}Th^{234}$$

*There are actually others that we shall not discuss.

You can easily verify that this calculation satisfies our three conservation laws if the kinetic energy of the alpha particle is 4.2 MeV.

The same three conservation laws also govern beta decay. Because the electron emitted in this decay is not a nucleon, the net mass number of the nucleus is unchanged. On the other hand, its charge and thus its atomic number Z must increase by one unit to conserve charge. An example is the decay of Pa^{234}, which proceeds as

$$_{91}Pa^{234} \longrightarrow e^- + {}_{92}U^{234} + ? \tag{14.3}$$

where we have denoted the electron by e^- as is conventional rather than by $_{-1}e^0$, which would be consistent with our notation for nuclei.

You are probably curious about the ? in Eq. (14.3). To see why it is there, let us consider the masses in this decay. The mass of Pa^{234} is 234.043 amu, whereas the mass of U^{234} is 234.041 amu, leaving 0.002 amu or 1.86 MeV of energy for the electron. The mass of an electron is equivalent to 0.511 MeV. Conservation of energy then requires that the electron emitted in this decay have a kinetic energy of 1.35 MeV. In alpha decay, the emitted alpha particle always has just the energy predicted by the mass difference and the equation $E = mc^2$. However, in the decay of Pa^{234}, the electrons are observed to have a spread of energies, as shown in Fig. 14.6. According to conservation of energy, *all* the emitted electrons should have an energy of 1.35 MeV. Experimentally, however, this figure represents the maximum energy observed; the majority of electrons have energies that are considerably less. Energy is not conserved by this reaction!

In 1931, Wolfgang Pauli proposed a solution to this problem. He suggested that a third particle, which he called a *neutrino* (meaning little neutral one), was emitted in beta decay. This particle, denoted by ν, carried away the missing energy. This bold postulate of a new elementary particle

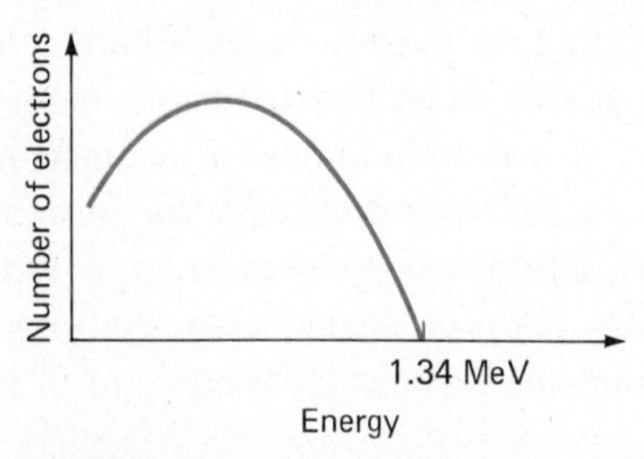

FIGURE 14.6
The number of electrons emitted in the beta decay of Pa^{234} plotted against their energy. According to conservation of energy all the electrons should have energy 1.35 MeV.

saved the Laws of Conservation of Energy and Momentum, both of which were violated by the existing beta decay data. The neutrino is perhaps the most elusive of the elementary particles. It has no charge, no mass, and always moves at the speed of light. It is not affected by electromagnetic or strong forces, and a new force, the *weak interaction,* must be introduced to explain its interactions. Its only characteristic is its very small amount of intrinsic angular momentum (spin). Whereas most particles are absorbed by a few centimeters of lead, it would require a block of lead *one light-year thick* to stop a neutrino. Given these strange properties, it is not surprising that the neutrino resisted attempts to detect it directly until 1956; even today its detection is an elaborate procedure.

The decay of Pa^{234} is now written as

$$_{91}Pa^{234} \longrightarrow {}_{92}U^{234} + e^- + \bar{\nu}$$

where we have written a bar over the neutrino because it is actually an antineutrino that occurs in this reaction. We shall discuss such antiparticles later. The 1.35 MeV of energy left over from the mass of Pa^{234} is shared by the electron and the neutrino. We now know that *all* beta decays involve the emission of a neutrino as well as an electron. The simplest beta decay is that of the neutron, which decays into a proton, an electron, and a neutrino:

$$n \longrightarrow p + e^- + \bar{\nu}$$

In fact, it is this process that occurs within the nucleus in all beta decays.

The third type of radiation, the emission of a gamma ray, or photon, does not change either the charge or the number of nucleons in a nucleus. Thus, N and Z are unchanged, although the energy of the nucleus decreases. The neutrons and protons in a nucleus exist in specific energy states, just as the electrons surrounding the nucleus do. Normally, all the nucleons are in their lowest possible energy states, but when a nucleus is created, perhaps by alpha or beta decay, it may be that one of its nucleons is in an excited state. As a result, the nucleon will drop in to its ground state, emitting a photon whose energy equals the change in the energy of the nucleon. Thus, gamma decay is analogous to an excited atom emitting its characteristic spectral lines, except that the photons have energies of a few MeV rather than a few eV.

The number of particles emitted per second by a radioactive substance is called its *activity*. The activity of all radioactive substances decreases with time in a very simple manner, as illustrated in Fig. 14.7. This

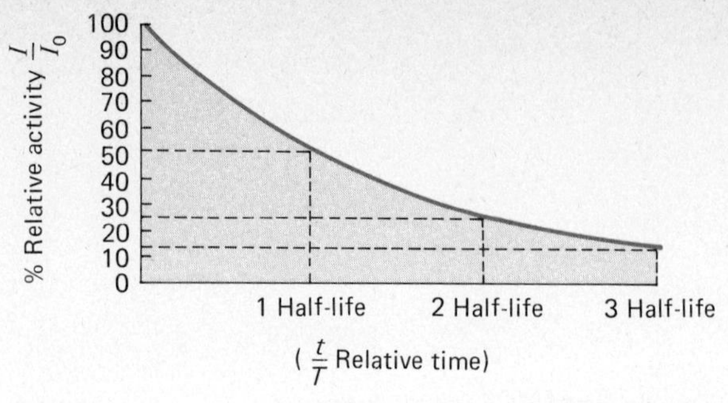

FIGURE 14.7
The percent activity of a radioactive substance as a function of time.

decrease in activity is characterized by the *half-life, T,* of the substance. The half-life is the time required for the activity to decrease to one-half of its initial value. Figure 14.7 shows that, after $2T$, the activity is one-fourth of its original value and, after $3T$, it is one-eighth. For example, consider a hypothetical alpha-particle emitter with a half-life of 1 hr. If a sample of the substance were initially emitting 320 alpha particles/sec, its activity would have decreased to 160 particles/sec after 1 hr. In 2 hrs, it would be emitting 80 particles/sec; in 3 hrs, 40; in 4 hrs, 20; and so on. Thus, the activity of a given sample decreases by 50 per cent in a time T of one half-life. The half-lives of the various radioactive isotopes range from a few millionths of a second to several billion years.

Although the radioactive decay law precisely describes how the activity of a large sample of radioactive material decreases, it does not tell us when a given nucleus will decay. In fact, such a prediction is impossible. The timing of a nuclear decay is governed by statistical laws. We can only say that there is always a certain probability that the decay will occur in a given time. After one half-life has elapsed, there is a 50 per cent chance that a nucleus will have decayed. The isotopes Ra^{226}, U^{238}, and Pa^{234}, whose decays we have already discussed, have half-lives of 1622 yr, 4.5 billion yr, and 6 hrs 40 min, respectively. The decay of U^{238} is the first step in a long chain of radioactive decays that ends in a stable isotope of lead, Pb^{206}. This series of decays, shown in Fig. 14.8, has been used by scientists to estimate the age of the earth. You might have noticed that the half-life given for Pa^{234} in Fig. 14.8 is much shorter than the one just given. This discrepancy is caused when Pa^{234} is formed from the decay of Th^{234}. It is then in an excited state and decays more rapidly. As a last example, a free neutron has a half-life of 12 min. However, when bound in a stable nucleus, the neutron is stable.

$_{92}U^{238}$	4.51×10^9 yr
$\alpha \downarrow$ 4.18 MeV	
$_{90}TH^{234}$	24.1 days
$\beta \downarrow$ 0.193 MeV	
$_{91}PA^{234}$	1.175 min
$\beta \downarrow$ 2.31 MeV	
$_{92}U^{234}$	2.48×10^5 yr
$\alpha \downarrow$ 4.76 MeV	
$_{90}TH^{230}$	8.00×10^4 yr
$\alpha \downarrow$ 4.68 MeV	
$_{88}Ra^{226}$	1622 yr
$\alpha \downarrow$ 4.78 MeV	
$_{86}Rn^{222}$	3.825 days
$\alpha \downarrow$ 5.49 MeV	
$_{84}Po^{218}$	3.05 min
$\alpha \downarrow$ 6.00 MeV	
$_{82}Pb^{214}$	26.8 min
$\beta \downarrow$ 0.65 MeV	
$_{83}Bi^{214}$	19.7 min
$\beta \downarrow$ 1.65 MeV	
$_{84}Po^{214}$	1.64×10^{-4} sec
$\alpha \downarrow$ 7.68 MeV	
$_{82}Pb^{210}$	19.4 yr
$\beta \downarrow$ 0.017 MeV	
$_{83}Bi^{210}$	5.00 days
$\beta \downarrow$ 1.17 MeV	
$_{84}Po^{210}$	138.4 days
$\alpha \downarrow$ 5.30 MeV	
$_{82}Pb^{206}$	stable

FIGURE 14.8

The chain of decays leading from $_{92}U^{238}$ to $_{82}Pb^{206}$. The half-life, decay mode, and energy of the emitted particle are shown. Some of the isotopes decay by more than one mode, introducing loops in the diagram. These are not shown since they represent less than a fraction of a percent of the main chain.

14.3 NUCLEAR REACTIONS

The discovery of radioactivity opened the door to many new experiments. In 1911, E. Marsden and in 1913 E. R. Rutherford bombarded different elements with alpha particles from various radioactive sources. The reactions observed always obeyed the same laws as the decays that we discussed in

the last section. For example, when nitrogen was bombarded with alpha particles, a proton and an isotope of oxygen were observed. Rutherford explained this reaction as

$$_2He^4 + _7N^{14} \longrightarrow _8O^{17} + p$$

Recalling that the proton is $_1H^1$, we see that both charge and baryon number (the number of nucleons) is conserved in this reaction. Another important example was

$$_2He^4 + _4Be^9 \longrightarrow _6C^{12} + n$$

because this reaction led to James Chadwick's discovery of the neutron in 1932.

In early experiments protons and neutrons were also used as projectiles. Neutrons are particularly useful in such work. Unlike protons and alpha particles, they are uncharged and therefore are not repelled by the electrical charge of the target nucleus. Many new isotopes not observed in nature were produced in these experiments. All of the new isotopes were unstable, so this *induced radioactivity* was studied intensely. Today, using accelerators to produce beams of protons, electrons, alpha particles, and other projectiles, scientists have extended the periodic table to $Z = 103$. They even think that a nucleus with $Z = 114$ may be possible. The number of different isotopes is close to 1000. These artificial *radioisotopes* have many useful properties.

Artificial radioactivity obeys the same rules as natural radioactivity, except that a new decay mode is observed. Before we discuss this new type of decay, we must mention another discovery. In 1932, Carl D. Anderson discovered a new elementary particle. In studying cosmic rays (radiations coming from outside the earth), he found a particle with properties identical to the electron but with opposite charge. This particle, the antiparticle of the electron, was called the *positron* and given the symbol e^+. Like the electron, the positron is a stable particle. It is not normally observed in nature, however, because it reacts rapidly with electrons. In this reaction, called *pair annihilation,* two (or three) photons are produced according to

$$e^+ + e^- \longrightarrow \gamma + \gamma \qquad (\text{or } \gamma + \gamma + \gamma)$$

Many of the new artificial isotopes are unstable against positron emission. When boron, for instance, is bombarded by alpha particles, an unstable isotope of nitrogen is produced:

$$_2He^4 + _5B^{10} \longrightarrow _7N^{13} + n$$

The N^{13} then decays with a half-life of 10 min to C^{13} by the emission of a positron and a neutrino:

$$_{7}N^{13} \longrightarrow {_6}C^{13} + e^{+} + \nu$$

Almost all of the elements have at least one stable isotope. For a given Z, the isotopes with more neutrons than the stable isotopes have are generally unstable and decay by beta-particle emission, whereas those with less neutrons decay by positron emission. Positron emission is not observed in naturally radioactive elements, because any nucleus that can decay by positron emission can also decay by a reaction called *electron capture,* in which it absorbs one of its innermost electrons. An example is Be^7, which decays as

$$_{4}Be^{7} + e^{-} \longrightarrow {_3}Li^{7} + \nu$$

These two reactions occur with such sufficiently short half-lives that all the naturally occurring e^+ emitters have long since decayed. Alpha decay tends to produce nuclei that have too many neutrons and thus are unstable against beta decay. The very long-lived alpha-particle emitters, like U^{238}, provide a source of decay products that replenish the supply of beta-particle emitters, and therefore beta-decay does occur naturally.

If the net mass of the particles in a nuclear reaction decreases, according to the conservation of energy, the mass decrease must be converted into energy. Reactions of this sort are called *exothermal.* On the other hand, when the net mass increases, the reaction is called *endothermal,* and energy present in the initial state of the system is converted into the extra mass present in the final state. Exothermal nuclear reactions are important because the energy released can be very large, and probably men can put it to use. We shall discuss endothermal reactions at the end of this chapter when we discuss the production of elementary particles.

One class of exothermal nuclear reactions are the *fusion* reactions. These reactions depend on the fact that, if it occurred, the reaction

$$2p + 2n \longrightarrow {_2}He^{4}$$

would release 28 MeV of energy. This reaction would almost never occur in nature, because the probability of getting four particles whose motion cannot be controlled together at the same time and place is exceedingly small. However, nature has provided us with these particles in pairs; 0.015 per cent of the hydrogen occurring in nature is the isotope deuterium (H^2 or D^2). The oceans supply an almost inexhaustible supply of deuterium

in the form of heavy water, D_2O. Two deuterium nuclei can react together to form He^3 or H^3, according to the reactions

$$_1D^2 + {}_1D^2 \longrightarrow {}_2He^3 + n$$

$$\text{or} \quad {}_1D^2 + {}_1D^2 \longrightarrow {}_1H^3 + p$$

with a release of roughly 2.5 MeV in either case. (The H^3 formed would then decay to He^3 by emitting beta particles.) If 2 gm of deuterium (which contains 6×10^{23} atoms) were combined in this way, approximately 10^{10} joules of energy would be released. However, it is difficult to produce these reactions because both of the deuterium nuclei are positively charged. Until they are close enough for the strong nuclear force to attract them, they repel each other and do not react. To force the reaction and extract energy from it, the deuterium must be enclosed at a temperature of several million degrees. This procedure cannot be done in an ordinary container; any solid would vaporize at such temperatures. In many countries, research is presently underway to devise magnetic bottles to contain the hot deuterium and produce controlled fusion energy. When this complex problem is solved, virtually unlimited energy will be available from the deuterium stored in the earth's oceans. At present, the only way to extract large amounts of energy from deuterium is to produce the high temperatures by using a thermonuclear fission bomb as the trigger. (We shall discuss fission at the end of this section.) This process suddenly releases the stored energy, and the resulting device is a hydrogen bomb.

Another set of fusion reactions is the proton-proton chain, which proceeds as

$$\begin{aligned} p + p &\longrightarrow H^2 + e^+ + \nu \\ H^2 + p &\longrightarrow He^3 \\ He^3 + He^3 &\longrightarrow He^4 + p + p \end{aligned}$$

This reaction, however, requires even higher temperatures and densities than the deuterium reactions just discussed. As its net result, four protons are combined into an alpha particle (He^4 nucleus) and two positrons. This releases the full binding energy of helium and produces about 28 MeV of energy for each resulting He^4. Scientists believe now that this proton-proton chain is the energy source of our sun, which burns hydrogen into helium to provide its energy. The sun expends energy at a rate of about 10^{34} joules per year, which is equivalent to converting 10^{17} kg of mass into energy. You might wonder how long this process can continue. In a billion (10^9) years, the sun's mass will decrease by 10^{26} kg, or about 50 earth masses. This figure, however, is only 0.01 per cent of the total mass of the sun, which is

2×10^{30} kg. Astronomers estimate that the sun has at least 5 billion years left before it begins to die.

As a last example of exothermal nuclear reactions, let us consider fission. When the rare isotope of uranium, ${}_{92}U^{235}$, absorbs a neutron, it becomes unstable and breaks into two parts, a process known as *fission.* The daughter nuclei are not uniquely determined, so many different combinations result. For example, we might have

$$n + {}_{92}U^{235} \longrightarrow {}_{92}U^{236} \longrightarrow {}_{54}Xe^{137} + {}_{38}Sr^{90} + 9n$$

with a release of around 200 MeV. Note that the charge is distributed between the two daughter nuclei, and extra neutrons are released. These extra neutrons offer the possibility of a chain reaction. In principle, each of them could be absorbed by another U^{235} nucleus, producing 9 fissions, each of which in turn could produce 9 more fissions, and so on. This process would increase geometrically with a vast release of energy. In practice, many difficulties occur. First, U^{235} absorbs only slow neutrons easily. The neutrons produced in the fission are quite energetic and must be slowed down before they can be absorbed and produce further fission. Second, U^{235} is a very rare isotope of uranium and is quite difficult to separate from its much more abundant isotope U^{238}. The first problem is solved by using moderators to slow down the fast neutrons so that they can be absorbed. The second problem can be solved by either slowly separating U^{235} from U^{238}, producing *enriched* uranium, or using breeder reactors, which we discuss below.

When sufficient U^{235} (or other fissionable material, such as U^{233} or Pu^{239}) is assembled, it produces enough neutrons to cause an increasing number of fission events. It is said at this point to have gone *critical.* At the critical stage, the chain reaction can be controlled by inserting *control rods* to absorb neutrons and keep the reaction at a predetermined level so that heat can be produced. A nuclear reactor works in this way. Alternatively, the reaction can proceed unchecked, in which case an explosion results. Generally this explosion disperses the fissionable material, so the reaction stops producing minor explosions and perhaps wrecking the power plant. Preventing dispersion of the fissionable material is very difficult. Therefore, a major part of the engineering necessary to produce a *thermonuclear* (or, by misuse of language, *atomic*) bomb is devoted to keeping the material together long enough for all of it to react.

The fuel for a reactor must be the rare U^{235}, or either U^{233} or Pu^{239}, which do not occur naturally at all. These last two substances decay by alpha-particle emission with half-lives of 1.6×10^5 yr and 2.4×10^4 yr, respectively, short enough to ensure that none remains. However, both Pu^{239} and U^{233} can be produced artifically in *breeder reactors.* If the common

isotope of uranium, U^{238}, is bombarded by neutrons (as exist in a reactor), it is transformed into ${}_{92}U^{239}$, which then decays by two successive beta-particle emissions to ${}_{94}Pu^{239}$. In a similar manner, the naturally occurring isotope ${}_{91}Th^{232}$ becomes ${}_{91}Th^{233}$ and the beta particles decay to ${}_{92}U^{233}$. A properly designed breeder reactor can produce more fissionable material than it uses. The raw material would be the nonfissionable but more abundant isotopes Th^{232} and U^{238}.

The output of any nuclear reactor is heat. This heat can be used to produce steam and generate electricity. As the world supply of fossil fuel becomes depleted and the environmental effects of its combustion become more noxious, nuclear power plants will probably produce an increasing amount of the world's electrical power. Fission reactors will have to be used until the problems of producing controlled fusion are solved. Then men could tap the virtually unlimited source of power that fusion reactions can release.

14.4 APPLICATIONS

By exploring the nature of the nucleus, scientists have opened new areas of research and found new ways to apply their discoveries in physics and in other fields. Some of these applications, such as nuclear power, we have already discussed. Let us now see what other applications have been developed.

About 1945, W. F. Libby worked out a method of dating organic compounds by using carbon 14. *Radiocarbon dating* is based on the observation that cosmic rays produce a small amount of C^{14}, a radioactive isotope of carbon, by the reaction

$$n + N^{14} \longrightarrow C^{14} + p$$

Carbon 14 decays by beta-particle emission back to N^{14} with a half-life of 5770 years. This reaction occurs in the high atmosphere. The C^{14} produced mixes down and oxidizes to carbon dioxide, which is then absorbed by plants and thus makes its way into all living matter. If we assume that the rate of production of C^{14} is *constant,* then, as the C^{14} decays, it will be continually replenished. All living organisms, therefore, will have a constant ratio of C^{14} to ordinary carbon. This ratio is found to be about $1:10^{12}$. When the organism dies, it ceases to absorb new C^{14}, and the ratio begins to drop. After 5770 years, it will have half the C^{14} that it had when it died. By measuring the amount of C^{14} present in a sample of organic material, we can estimate its age. This technique has been used to measure ages as old

as 25,000 years. Recently, by comparing C^{14} dates to dates from tree rings, scientists have shown the incorrectness of Libby's original assumption that the rate of production of C^{14} was constant. The rate of production of C^{14} has actually varied, possibly because changes in the earth's magnetic field have produced changes in the cosmic-ray flux. These variations have been estimated, and today radiocarbon dates are corrected for this effect. In archeological work radiocarbon dating has been an invaluable tool, enabling workers to date prehistoric sites and determine their proper place in history.

Artificial radioisotopes have also played an important role in modern science. Because the chemical properties of a radioactive isotope are the same as those of the stable isotope, a radioactive isotope behaves the same way in chemical reactions. However, due to their radioactivity, these isotopes can be detected in minute quantities and followed through a sequence of reactions. For instance, suppose that you wish to measure how much phosphate plants absorb from fertilizer. If you mix a small amount of radioactive phosphorus with the stable phosphorus, you can then follow the progress of these *tagged atoms* by observing their radioactivity. It happens that there is an isotope of phosphorus with a half-life of about two weeks, which is just about right in terms of the growing season of a plant.

In a similar manner, organic compounds can be synthesized with radioactive atoms at particular places in their structure. These tagged molecules, and even particular pieces of the molecules, can then be observed as they are metabolized by a plant or animal. In this manner, scientists can unravel complex chains of chemical reactions. Radioisotopes have the inherent disadvantage that their radiation may damage tissue or destroy organic molecules. Consequently, stable isotopes are also utilized in modern research. Deuterium, carbon 13, and nitrogen 15, all rare but stable, are used to build organic molecules for biological and chemical applications.

There are numerous medical applications of radioisotopes. Many of these applications utilize the fact that cancer cells, like ordinary cells, can be destroyed by ionizing radiation. Thus, if the radiation can be concentrated in a particular region, a cancer there may be arrested. In addition, certain organs of the body tend to concentrate certain elements. For instance, iodine is absorbed mainly by the thyroid gland. If a patient takes a dose of radioactive iodine, it will concentrate in his thyroid and may destroy a cancer there while doing less damage to surrounding tissue.

14.5 ELEMENTARY PARTICLES

The discovery in 1932 of the neutron (by Chadwick) and the positron (by Anderson) marked the beginning of a real theoretical understanding of

nuclear structure. Moreover, a new branch of physics, *elementary particles*, took form. We define an elementary particle as an object that has definite mass, charge, and spin and is not made up of other particles. By this definition, nuclei, with the exception of H^1, are not elementary, because all of them are made up of protons and neutrons. This definition is open to debate. Some physicists would drop the condition that an elementary particle not be a composite, thus including deuterons and alpha particles as well as nuclei. On the other hand, there is some evidence that hadrons themselves may be composite. (Obviously, the question of what is elementary is not closed.) Note that elementary particles need not be stable, and, as we shall see, most of them are not.

The discovery of the positron was particularly interesting because its existence had, in a sense, been predicted. In 1928, P. A. M. Dirac, a brilliant theoretician, published an equation that described the properties of the electron. The Dirac equation not only predicted the electron's spin and its magnetic properties but also, if interpreted correctly, predicted the existence of another particle having the same mass and spin but opposite charge. At first, because no such particle had been observed this prediction was regarded as a defect in the equation. However, when the positron was discovered with just the predicted properties, it became a great triumph. The positron is an example of an *antiparticle.* In relativistic quantum mechanics (of which the Dirac equation is an example), every particle has associated with it an antiparticle. This antiparticle has the same mass and spin but opposite charge. In some cases, like the photon, the antiparticle is identical with the particle itself, but generally it is distinct. Antiparticles have been observed for almost all of the "stable" particles that we shall discuss in this section. We shall denote an antiparticle by placing a bar over the symbol for the particle (e.g., $\bar{p}$, $\bar{n}$).

It is possible to imagine *antimatter* made up of antiparticles. The nuclei of antiatoms would be made up of antineutrons and antiprotons, and positrons would circle them. The chemical properties and spectra of such matter would be identical to that of ordinary matter. Scientists have speculated that there may be entire galaxies made of antimatter, perhaps as many as are made of ordinary matter. This hypothesis is difficult to test, because the light given off by such a galaxy would appear identical to the light from a normal galaxy. One common misconception about antimatter is that it has negative mass. The gravitational attraction between a proton and an antiproton would be identical to that between two protons. A satellite made up of antimatter could orbit an ordinary planet as long as it did not get close enough to interact with the atmosphere.

The next elementary particle discovered after the positron and neutron was the *muon* (μ). Its story illustrates the pitfalls that await the unwary in science. In 1935, a Japanese physicist, Hideki Yukawa, published

a theory that accounted for the extremely short range of the strong interaction or nuclear force. He postulated the existence of a particle roughly 200 times the mass of the electron. One year later, a particle fitting this description, the muon, was found in cosmic rays. Its mass was 106 MeV, about 200 times the mass of the electron (0.511 MeV).* The muon was immediately acclaimed as Yukawa's particle, and the investigation of its properties began. This identification was, however, incorrect. The Yukawa particle was supposed to have either spin-0 or spin-1 and to interact strongly with the nucleus. The muon was found to have spin-$\frac{1}{2}$, like the electron. Moreover, it did not interact strongly; that is, the nuclear force did not affect it. Subsequent work has shown that the muon is nothing but a heavy electron. Like the electron-positron pair, there are two muons, a negatively charged one and its positively charged antiparticle. The muon has its own neutrino, which seems to be distinct from the electron's neutrino, although it has the same properties. The muon decays with a half-life of 1.5×10^{-6} sec into an electron and two neutrinos,

$$\mu^{\pm} \longrightarrow e^{\pm} + \bar{\nu} + \nu$$

The electron, its neutrino, the muon, its neutrino, and their respective antiparticles form an eight-member group of elementary particles called *leptons*. The leptons are all light, spin-$\frac{1}{2}$ particles and do not feel the nuclear force. Yukawa's particle, the *pion*, was not discovered until 1947, after new experimental tools became available.

The development of nuclear physics and elementary particles, like most areas of physics, has proceeded hand-in-hand with the development of new experimental techniques. The original observations passively measured the properties of naturally occurring radioactive elements. Later scientists utilized these radiations to stimulate other reactions and to produce artificial radioactivity. They studied cosmic rays; in fact, the positron was first observed in a cosmic-ray experiment. However, researchers realized that radioactivity and cosmic rays were not adequate sources of particles. Radioactivity could not be controlled, and cosmic rays, although having large energy, were infrequent. Scientists needed instead beams of artificially accelerated particles with which to probe nuclear structure. In the early 1930's, the first particle accelerators were built. Then experimental nuclear physics began to rapidly expand.

A typical accelerator takes charged particles (usually protons or electrons) and focuses them into a narrow beam by using magnetic fields.

*In this section, we follow the practice of elementary-particle physicists and refer to mass in energy units. What we are really saying is that (mass)c^2 of a particle has a certain value in joules or MeV.

The beam is then either bent around a circular path or guided down a straight pipe. The individual particles are accelerated with electromagnetic forces. The early accelerators produced beams with particle energies of around 10 MeV. The technology developed rapidly, however, and today protons can be given an energy of up to 500 GeV (1 GeV = 1000 MeV = 10^9 eV). In order to assess the implications of particle beams with such high energy, let us consider the energy associated with various physical phenomena.

In the last chapter we saw that typical binding energies for electrons in atoms were from a few eV to a few hundred eV. If we bombard atoms with a beam of charged particles having energies in this range, we probe the electronic structure of the atom but do not reach the nucleus. Nuclear binding energies, on the other hand, are in the MeV range. Therefore, protons with this energy will reach into the nucleus and reveal its structure. Figure 14.9 shows the spectrum of the single-particle energies met in physics, along with the specific areas and the phenomena associated with each energy range. At the low end of the energy scale are the kinetic energies attributable to thermal motion. At the high end, far above the energies possible with accelerators, are the most energetic cosmic rays, with energies as high as 10^{20} eV (16 joules). The mechanism by which these cosmic rays are accelerated to such high energy is not known, although educated guesses have been made.

Let us consider a proton that has been accelerated to an energy of 20 GeV. The rest energy of a proton is 938 MeV or just under 1 GeV. Thus, this proton has a relativistic mass 20 times its rest mass (see Sec. 12.4). It is moving with a speed of 99.87 per cent of the speed of light and will cross an average nuclear distance of 10^{-15} m in less than 10^{-23} sec. If the proton is accelerated further, both its mass-energy and its momentum will increase, but its speed will be essentially constant. Thus, 10^{-23} sec represents the time scale for its interaction with another proton or nucleus. Many of the elementary particles have half-lives of 10^{-8} or 10^{-10} sec. Although these half-lives may seem very short, they are 10^{10} times longer than the basic interaction time of 10^{-23} sec. A particle with a half-life of 10^{-10} sec moving relativistically (that is, with a speed close to the speed of light) can travel several centimeters before decaying and thus leave a measurable track. Elementary-particle physicists call such particles "stable," even though, in ordinary terms, they are very short-lived.

When a proton with great enough energy strikes a target nucleus, some of its energy may be used in the production of new particles. The energy needed depends on the mass and number of new particles to be produced. For instance, if two particles having kinetic energy of 0.5 GeV were to collide head-on, a particle with a mass of $mc^2 = 1$ GeV could be

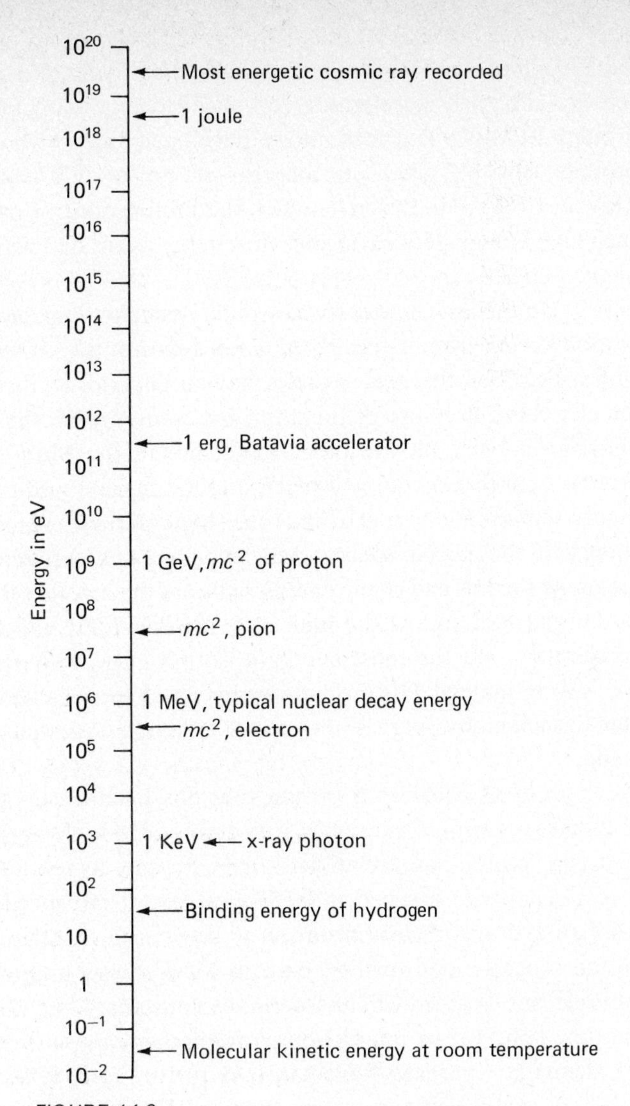

FIGURE 14.9
The range of elementary-particle energies met in physics.

produced. If one of the two particles were initially at rest, as is true in most experiments, more energy would be needed, but the basic process would remain the same. Energy can be converted into new particles.

After the discovery of the muon, no new elementary particles were found until 1947 when the newly developed accelerators were freed from the war effort and could be used for pure physics. From that time to the present, over 200 elementary particles have been observed. Roughly 30 of

these particles are stable. The rest, sometimes called *resonances,* have half-lives of around 10^{-23} seconds, just long enough to have a measurable mass, charge, and spin. These particles are divided into three main families: *leptons, mesons,* and *baryons*. The leptons, which we have already discussed, are distinguished from the other two families, which are known collectively as *hadrons,* by the fact that they do not interact strongly; that is, they do not feel the nuclear force.

The most important members of the meson family are the pions.* These particles were first observed in 1947 in cosmic rays and shortly thereafter were produced in accelerators. There are three pions, two charged, $\pi^{\pm}$, and one neutral, π^0. The two charged pions are a particle-antiparticle pair and have identical masses of 140 MeV or $272m_e$. The mass of the neutral pion is 135 MeV, and it is its own antiparticle. These three particles all have spin-0 and interact strongly in the same way; that is, the nuclear force acts on each of them identically despite their different charges. Electromagnetically, of course, they are easily distinguished. Shortly after their discovery, it became evident that the pions were the particles predicted by Yukawa.

The three pions have the same spin (and the same parity, although we shall not discuss this characteristic), almost the same mass, and are affected by the nuclear force in the same way. Consequently, they are often thought of as one particle having three charge states and are referred to as a *multiplet*. We shall see that most of the mesons and baryons come in multiplets, the most important example being the proton-neutron pair. Pions are produced by nucleon-neucleon collisions. Examples are:

$$p + p \longrightarrow p + p + \pi^0$$
$$n + n \longrightarrow n + n + \pi^0$$
$$p + n \longrightarrow p + n + \pi^0$$
$$p + p \longrightarrow p + n + \pi^+$$
$$n + p \longrightarrow n + n + \pi^+$$
$$n + p \longrightarrow p + p + \pi^-$$
$$n + n \longrightarrow n + p + \pi^-$$

Note that charge and the number of nucleons is conserved in each of these reactions. The charged pions have a half-life of 1.8×10^{-8} sec, decaying into a muon and a neutrino:

$$\pi^+ \longrightarrow \mu^+ + \nu$$
$$\pi^- \longrightarrow \mu^- + \bar{\nu}$$

*Also known as pi-mesons.

The neutral pion decays into two photons:

$$\pi^0 \longrightarrow \gamma + \gamma$$

with the much shorter half-life of 1.2×10^{-16} sec. When a very energetic cosmic-ray proton or neutron strikes the upper atmosphere, it collides with a nucleus and produces many pions in reactions similar to the ones just listed. These pions then decay, forming many of the muons observed at sea level.

Theoretical physicists would have been quite happy if the list of elementary particles had been completed with the discovery of the pions. But between 1947 and 1962, numerous new additions were made to the pion's meson family and to the nucleon's baryon family. Physicists recognized many of the properties of these new particles. They all had charges that were multiples of e. They came in charge multiplets, like the n-p pair and the pion triplet. They could be classed as mesons if they had integral spin (0, 1, 2, . . .) and baryon number zero or as baryons if they had half-integral spin ($\frac{1}{2}, \frac{3}{2}, \frac{5}{2}, \ldots$) and baryon number ± 1.

However, a new quantum number, *strangeness,* had to be introduced to account for "strange" properties in the production and decays of the new particles. Some of these "strange" particles, such as the K-meson or Kaon, should have decayed with a half-life of around 10^{-23} sec. However, they were observed to be comparatively stable, having half-lives around 10^{-10} sec. To explain this behavior, the K^+ and the K^0, a charge doublet, were assigned strangeness $S = +1$; and their antiparticles, the K^- and the $\bar{K}^0$, were assigned strangeness $S = -1$. The pions have strangeness $S = 0$. Strangeness, like baryon number, was an additive quantum number, taking on integral values 0, ± 1, ± 2, and so on. If a decay conserved strangeness (that is, if the net strangeness before equaled the net strangeness after), then the decay would proceed with a half-life of around 10^{-23} sec. Each kaon has a mass of just under 500 MeV and can decay into two or three pions. Such decay, however, does not conserve strangeness. Rather than being forbidden completely, as would be the case for a decay that did not conserve charge or baryon number, the decay is slowed down from 10^{-23} sec to 10^{-10} sec, making the particle almost stable. The strange particles were also observed to be produced only in pairs, one having $S = +1$; the other, $S = -1$, again indicating that strangeness is usually conserved.

In addition, physicists observed strange baryons. The Λ^0, a neutral-charge singlet with a mass of 1115 MeV, and the $\Sigma^0, \Sigma^+, \Sigma^-$, a charge triplet with masses just under 1200 MeV, were observed to have strangeness -1. The Ξ^0, Ξ^-, a charge doublet with mass around 1300 MeV, received strangeness -2. Each of these particles has an associated antibaryon with

baryon number $B = -1$, opposite charge and strangeness, and identical mass. In their decays, all of these particles, except the Σ^0 (which can decay into a Λ^0 and a photon), violate the conservation of strangeness law and thus decay slowly. You should remember that a particle can only decay into lighter particles. Moreover, decays must always conserve charge and baryon number absolutely. Thus, the decay of, say, a Λ^0 into a $\bar{K}^0$ (which has $S = -1$) and a π^0 would be impossible, even though such a decay conserves strangeness and charge and would be rapid if it did not violate conservation of baryon number.

By 1960, elementary-particle physics was in a somewhat chaotic state. With the construction of more powerful particle accelerators, new heavier particles were being discovered monthly. Scientists observed that these particles occurred in charge multiplets, and they assigned quantum numbers to the new particles. The reactions of these particles obeyed the conservation laws, but the relations between them were obscure. In addition, their dynamics, with the exception of electromagnetic properties, were not understood.

In 1961 and 1962, Murray Gell-Mann and Yuval Ne'emann proposed a classification scheme called SU(3). In this scheme the known elementary particles were grouped into supermultiplets of particles. Each member of a supermultiplet had the same spin. From its position in a supermultiplet the properties and reactions of a particle could be predicted. The most immediate success of SU(3) was its prediction of a new "stable" particle, the omega particle Ω^-, whose existence as a baryon singlet with $S = -2$ and mass 1675 MeV was needed to complete the spin-$\frac{3}{2}$ supermultiplet.

This classification scheme served the same purpose for elementary particles that Mendeleev's periodic table served for chemists during the 19th century. It organized the particles into a framework that displayed similarities and enabled predictions to be made. However, as had happened with Mendeleev's periodic table before the development of quantum mechanics, scientists do not understand the basic reasons why SU(3) works. Indeed, at the present time there is an active search for "exotic" particles, such as a baryon with $B = 1$ and $S = 1$, that do not fit into this classification scheme and whose existence would put it into question.

One model was proposed in 1963 by Gell-Mann and also by G. Zweig. It attempts to "explain" SU(3). Called the *quark model* by Gell-Mann, it postulates the existence of three fundamental particles, named *quarks* after a passage in *Finnegan's Wake.* From these three particles and their three antiparticles, all the known hadrons are constructed. Each of the three quarks has baryon number $B = \frac{1}{3}$, and each antiquark has baryon number $B = -\frac{1}{3}$. The mesons, which all have baryon number $B = 0$, are

made of a quark-antiquark pair. The baryons are made of three quarks; and the antibaryons, of three antiquarks. One of the quarks is a charge singlet with strangeness $S = -1$. The other two form a nonstrange charge doublet. Notice that a baryon with $B = 1$ and $S = 1$ could not be formed from three quarks and is thus "exotic," as previously mentioned. The most unusual aspect of the quarks is that they arc required to have fractional charge. The charge singlet has charge $-e/3$, whereas the charge doublet has charge $-e/3$ and $2e/3$. The rules for constructing real particles out of quarks insure that only integral multiples of e occur and that there is no contradiction with observation. The quark model reproduces all of the predictions of SU(3) and thus, in a sense, explains it.

However, with this model there are many fundamental problems. For instance, the force binding together the quarks is not understood. In fact, quarks may well not exist at all. Physicists have searched for fractionally charged particles in accelerators, in cosmic rays, in ocean silt, and in oysters. None has been found. This fact may indicate that quarks are too heavy to be produced by present experimental techniques, or it may indicate that they exist only in the imaginations of theoretical physicists.

Elementary-particle physics is in a state similar to that of atomic physics after the Bohr model but before quantum mechanics. The SU(3) scheme classifies its particles, and the quark model attempts to explain this classification scheme (much as the Bohr model "explained" the periodic table). However, the Bohr model was correct only in certain aspects and then for the wrong reasons. The quark model may have a similar fate. Recent experiments probing the electromagnetic structure of protons and neutrons have indicated that, unlike electrons, these particles do not seem to be true point particles and indeed may be composed of other, perhaps more elementary, objects, called *partons* by R. P. Feynman. Partons may be quarks, or they may be some kind of still-undiscovered particle. Only further experiments and theoretical work will tell. Some physicists feel that physics is approaching a major theoretical breakthrough in which a unifying theory will emerge, explaining elementary particles and providing the cornerstone for all physical science. Others feel that as they probe deeper and deeper into the structure of matter, they will continue to uncover new layers and new systems, which in turn will have to be explained. In any event, physics will remain the branch of science that attempts to understand the workings of nature and the universe at its most fundamental level.

DISCUSSION QUESTIONS

1. Large accelerators can produce many transmutations per second. By selecting an exothermal reaction, each transmutation releases energy. Why are such machines not used as energy sources?

2. Science fiction writers sometimes suggest that in the future cars will be fueled with a pinch of uranium at the factory and never again will have to be refueled. Discuss the difficulties involved.

3. Discuss the difficulties in powering an aircraft with nuclear fuel.

4. The half-life of uranium is about 4.5 billion years and its end product is lead, which is stable. Discuss how measurements on the relative abundances of uranium and lead in an ore sample could be used to estimate the age of the earth.

5. Explain how a patient can be safely given a dose of radioactive iodine to cure a thyroid cancer without danger of producing additional cancer by the radiation emitted by the iodine. (The half-life of this isotope is eight days.)

6. Polycythemia is the overproduction of red blood cells by the bone marrow. Suggest a scheme by which this disease could be arrested by the use of radioactive phosphorus.

7. Discuss a method by which a radioactive isotope could be used to measure the thickness of a steel plate passing through a rolling mill.

8. Discuss the relative virtues of fission and fusion as sources of power for practical use.

9. Would a quark be stable?

10. Devise an experiment where pions are used to test one of the predictions of special relativity.

PROBLEMS

1. Determine the product when ${}_1H^3$ (tritium) decays by the emission of an electron. (*Ans.:* ${}_2He^3$)

2. Determine the product when ${}_{11}Na^{22}$ (a radioisotope of sodium) decays by the emission of a positron.

3. Strontium-89 has a half-life of 55 days. If the initial activity of a sample is 40 disintegrations/sec, compute its activity at the end of 110 days.

4. Observations taken six days in a row on a sample of radon gave the following activities in particles per minute: 1560, 1304, 1090, 908, 760, and 634. From these measurements calculate the approximate half-life of radon.

5. State the numbers and approximate locations of the various particles which make up an atom of ${}_{10}Ne^{22}$. (*Ans.:* Nucleus contains 10 protons and 12 neutrons; 10 electrons move about the nucleus.)

6. Suppose that you were given 15 electrons, 18 protons, and 16 neutrons. What is the most massive neutral atom you could construct using these materials? (*Ans.:* ${}_{15}S^{31}$)

7. If you had 8 protons and 10 neutrons, what other particles would you need to construct a neutral atom?

8. When ${}_6C^{12}$ is bombarded with deuterons (${}_1H^2$), protons are observed. What is the product of this reaction? (*Ans.:* ${}_6C^{13}$)

9. When ${}_7N^{14}$ is bombarded with neutrons, ${}_6C^{14}$ (radioactive) is produced. What is the other particle involved in the reaction?

10. You hope to produce ${}_7N^{13}$ (which is radioactive) by bombarding ${}_8O^{16}$ with protons. What light product-particle would you expect to observe? (*Ans.:* alpha particle)

11. Suppose that the neutrons from each fission process produce on the average two fissions within a time of 10^{-6} sec. Compute approximately the time for a single initial fission to lead to a total of one thousand fissions.

Find the time for a single initial fission to be the forerunner of a million fissions. (*Ans.:* 10^{-4} sec)

12. Compute the change in the sun's mass per second, given that the sun radiates about 1.2×10^{34} joules per year in the form of radiant energy.

13. ${}_2He^6$ is unstable, emitting an electron. What is the (stable) product of the decay? (*Ans.:* ${}_3Li^6$)

14. Suppose that a certain amount of tritium is produced for use in hydrogen bombs. If the half-life of tritium is 12 years, approximately what percentage loss would be incurred if this tritium were not used for 50 years?

15. Wood from the coffin of a pharaoh shows only half the radioactivity per gram that wood today shows. Approximately at what date did the pharaoh die? (*Ans.:* approx. 3760 B.C.; see Sec. 14.4.)

16. Explain why absorbing a small amount of radium (half-life 1600 years) is very dangerous, while doses of iodine-131 (half-life 8 days) are given routinely in hospitals to patients.

17. What is the product when ${}_{92}U^{235}$ emits an alpha particle? (*Ans.:* ${}_{90}Th^{231}$)

18. What is the product when ${}_{11}Na^{24}$ emits an electron?

19. What is the product when ${}_{27}Co^{60}$ emits an electron and a gamma ray? (*Ans.:* ${}_{22}Ni^{60}$)

20. What particle is emitted when ${}_{92}U^{238}$ transforms itself into ${}_{90}Th^{234}$?

21. If a sample of ${}_{14}P^{32}$ (half-life 14 days) emits 200 particles per second initially, how many particles per second will it emit 6 weeks later under the same conditions? (*Ans.:* 25)

22. The half-life of plutonium is about 25,000 years. If this material costs $10,000 per lb to make, what would its value be 100 years from now, if no inflation occurs?

23. The half-life of tritium, which is used in hydrogen bombs, is about 12.5 years. If this material costs $5000 per lb to make, what would its value be 50 years from now, if no inflation occurs? (*Ans.:* $313/lb)

24. State the numbers and approximate locations of the particles which make up an atom of ${}_{11}Na^{23}$.

Repeat for an atom of ${}_{92}U^{234}$. (*Ans.:* 92 protons and 142 neutrons in nucleus and 92 electrons in orbit about the nucleus)

25. Wood from an ancient campfire shows only $\frac{1}{8}$ the radioactivity that wood today shows. Approximately at what date was the fire made?

26. Find the density the sun ($M = 2 \times 10^{30}$ kg) would have if compressed to a radius of 8 km. Compare this with the density of a nucleus. These are typical figures for a neutron star.

27. If 10^{-23} sec is taken to be the nuclear time unit, how long does a charged pion live?

Appendix

Review of Mathematics

A.1 REVIEW OF FRACTIONS AND SIMPLE PROPORTION

Consider the fraction $\frac{6}{8}$. We can simplify this fraction by dividing the numerator (6) and the denominator (8) each by 2, obtaining the equivalent fraction $\frac{3}{4}$. Similarly, the value of the fraction would be unchanged if we were to multiply the numerator and the denominator each by the same factor. If this factor is 3, for instance, the equivalent fraction would be $\frac{18}{24}$. Depending on the situation, we may benefit by changing a fraction to an equivalent form, as described above. In general terms, if we begin with the fraction a/b, then the equivalent fraction is na/nb for any value of n, positive or negative, since $n/n = 1$.

Let us now consider adding $\frac{3}{4}$ to $\frac{1}{8}$. Since fourths and eighths are as different as oranges and apples, we must first arrange things so that both fractions have the same denominator. We do this by multiplying the numerator and the denominator of the first fraction by 2. We find then that

$$\frac{3}{4} + \frac{1}{8} = \frac{6}{8} + \frac{1}{8} = \frac{7}{8}$$

As a more difficult case, let us add $\frac{2}{3}$ and $\frac{3}{7}$. Again we must make the denominators of the fractions the same before we can add them. If we

multiply the first fraction by $\frac{7}{7}$ and the second by $\frac{3}{3}$, we find

$$\frac{2}{3} + \frac{3}{7} = \frac{7}{7}\left(\frac{2}{3}\right) + \frac{3}{3}\left(\frac{3}{7}\right) = \frac{14}{21} + \frac{9}{21} = \frac{23}{21}$$

In the example just above we could just as well have written that

$$\frac{2}{3} + \frac{3}{7} = \frac{2 \times 7 + 3 \times 3}{3 \times 7} = \frac{14 + 9}{21} = \frac{23}{21}$$

In general, if we wish to add the fractions a/b and c/d, we proceed in the same way. Thus we write:

$$\frac{a}{b} + \frac{c}{d} = \frac{d}{d}\frac{a}{b} + \frac{b}{b}\frac{c}{d} = \frac{da + bc}{bd}$$

The final fraction may then possibly be simplified as described in the paragraph above. Subtraction of fractions is basically the same as addition, except that now we must pay attention to the algebraic signs of the numbers. Consider the following example:

$$\frac{2}{3} - \frac{3}{7} = \frac{14}{21} - \frac{9}{21} = \frac{5}{21}$$

In the example given above in terms of letters, one or more of the letters a, b, c, d may have a negative sign which must be taken into account in computing the value of the final fraction.

Suppose now that we must combine three or more fractions. We will treat the case of three fractions, but it is easy to extend the ideas to cover cases involving four or more fractions. As an example, consider adding $\frac{2}{7}$, $\frac{1}{3}$, and $\frac{3}{5}$. The first step is to make all three fractions have the same denominator, which can be done by multiplying the numerator and denominator of each fraction by the product of the denominators of the *other* fractions which are to be added. In this case we have:

$$\frac{2}{7} + \frac{1}{3} + \frac{3}{5} = \left(\frac{3 \times 5}{3 \times 5}\right)\left(\frac{2}{7}\right) + \left(\frac{5 \times 7}{5 \times 7}\right)\left(\frac{1}{3}\right) + \left(\frac{7 \times 3}{7 \times 3}\right)\left(\frac{3}{5}\right)$$

$$= \frac{30}{105} + \frac{35}{105} + \frac{63}{105} = \frac{128}{105}$$

If one or more of the fractions is to be subtracted, a minus sign will appear in place of the plus sign for each such negative fraction. After all fractions

have been changed so as to have the same denominator, they are then added with due attention to the sign of each numerator. Often the resulting fraction can be put in simpler form by dividing its numerator and denominator by some common factor.

Decimal fractions are a special but very common type of fraction for which a different notation is used. For instance, in place of 5/10 it is customary to write 0.5. Similarly 0.012 stands for 0.12/10 which is the equivalent of 1.2/100 or 12/1000. The rule is simple: count the number of digits to the right of the decimal point, including zeros, to find the number of zeros in the denominator. The numerator is then the number to the right of the decimal point, with the decimal point left out. For instance, $0.348 = 348/1000$, $0.0204 = 204/10000$, and $23/1000 = 0.023$. Decimal fractions are particularly useful since it is so easy to multiply and divide by ten or a multiple of ten. The first digit to the right of the decimal point gives the number of tenths, the next digit to the right gives the number of hundredths, the third digit to the right of the decimal point gives the number of thousandths, and so forth. Consider adding 0.135 and 0.203. To get this sum we write

$$\begin{array}{r} 0.135 \\ +0.203 \\ \hline 0.338 \end{array}$$

Suppose that in adding a part of a decimal fraction we get, for instance, 12/100. Since 12/100 is equivalent to $1/10 + 2/100$, we simply carry one unit to the tenths addition. As an example consider adding 0.386 to 0.035. We have then:

$$\begin{array}{r} 0.386 \\ +0.035 \\ \hline 0.421 \end{array}$$

In the example above, we first found 11/1000, so that we carried one into the hundredths column. Then we got 12/100, so that we carried one into the tenths column. The extension of these ideas to the addition or subtraction of more than two decimal fractions is easy and will be left for exercises at the end of this chapter.

Consider the identity $\frac{2}{3} = \frac{4}{6}$. In words we can say that 2 is to 3 as 4 is to 6. This is called a *simple proportion.* If a recipe calls for $\frac{1}{2}$ cup of flour and we wish to triple the recipe, we must use $3 \times (\frac{1}{2}) = \frac{3}{2} = 1\frac{1}{2}$ cups of flour. This is an example of a simple proportion which we can calculate very easily. More generally, a simple proportion can be written in the form

$a/b = c/d$. Suppose that all of the letters had known values except c. Multiplying each side of the equation by d, we obtain $c = da/b$, from which we can compute c. Alternatively, suppose we knew all of the quantities except b. If we multiply both sides of the equation by bd/c, we get $b = da/c$. Thus we see that in any simple proportion we can compute the value of one factor if the values of the other three factors are known. Proportions occur very frequently in science, so you should master the principles used in solving them.

EXAMPLES

1. $2/3 = n/6$. Multiplying both sides of this equation by 6, we find $n = 6 \times (2/3) = 12/3 = 4$.
2. $3/5 = 10/n$. When we multiply each side of this equation by n, we obtain $(3/5) \times n = 10$. If we now multiply each side of the last equation by 5/3, we get $n = (5/3) \times 10 = 50/3 = 16\frac{2}{3}$.
3. $2/7 = (n + 3)/28$. First we multiply each side of this equation by 28, finding $28 \times (2/7) = n + 3$. We now have that $8 = n + 3$. After we subtract 3 from each side of the last equation, we find $n = 5$.

A.2 REVIEW OF ALGEBRA

The three examples at the end of the preceding section are all simple cases of algebraic equations, since in each case we could solve for an unknown quantity symbolized by a letter in terms of known quantities. Consider the simple equation $3n = 8$. If we divide each side of this equation by 3, we find $n = \frac{8}{3} = 2.67$. A slightly more difficult equation is $2n + 4 = 9$. After we subtract 4 from each side of this equation, we get $2n = 9 - 4 = 5$. Instead of saying that we have subtracted 4 from each side of the equation above, we could just as well have said that we can move $+4$ from the left side of the equation if we write it as -4 on the right side of the equation. When we now divide each side of the preceding equation by 2, we find $n = \frac{5}{2} = 2.5$. As a third example, let us consider the equation $10 + 3n = 6$. First we subtract 10 from each side of this equation, so that we obtain $3n = 6 - 10 = -4$. (Subtracting 10 from each side of the equation is the same as moving $+10$ to the right side of the equation and writing it as -10.) Thus, $n = -\frac{4}{3} = -1.33\frac{1}{3}$.

The general scheme in solving equations involving one unknown quantity is first to add or subtract terms on both sides of the equation in such a way that the term involving the unknown appears by itself. Then the value of the unknown is found by dividing both sides of the new equation by the coefficient (multiplier appearing in front of a factor) of the unknown. The resulting value for the unknown may be either positive or negative, depend-

ing on the problem. The general case is the equation $an + b = c$. We first write $an = c - b$. Then we divide by a, obtaining as our solution $n = (c - b)/a$. Note that in this example the letters, a, b, and c may stand for either positive or negative numbers.

In many physical problems two unknown quantities are related by two equations. Suppose that the unknowns n and m are related by the equations $n + 2m = 8$ and $3n + 4m = 18$. In this case we first solve one of the equations for one of the unknowns. For instance, from the first equation we see that $n = 8 - 2m$. When we substitute this value for n into the second equation, we find $3(8 - 2m) + 4m = 18$. After we multiply together the factors involved in the parenthesis, we get $24 - 6m + 4m = 18$. We can move a quantity from one side of an equation to the other side if we change its sign. When we do this in the equation above we have $24 - 18 = 6m - 4m$. This equation now reduces to $6 = 2m$, from which we see that $m = 3$. Returning to the first equation, we find that $n = 8 - (2 \times 3) = 8 - 6 = 2$. Our problem is now solved. To summarize the procedure, in solving for two unknowns which are related by two equations, we first solve one of the two equations for one of the unknowns. We then substitute this value of one unknown into the second equation. This reduces the second equation to one involving only a single unknown, which we can now solve for. After we have determined one of the unknowns, we return to the first equation to calculate the other unknown.

A.3 EXPONENTIAL NOTATION

Let us look at the identities $3 \times 3 = 9$ and $3 \times 3 \times 3 = 27$. In the first case the number 3 occurs twice in the multiplication, while in the second, 3 occurs three times. In order to simplify writing repetitive expressions, it is customary to write $3 \times 3 = 3^2$ and $3 \times 3 \times 3 = 3^3$. These are examples of *exponential notation*. In general, if we wish to multiply the number n by itself m times we write n^m. If m is 1, we mean the number n itself. Thus, $8^1 = 8$ and $n^1 = n$. If m equals zero, this means that the number n is not a factor at all. For this reason, we define 8^0 as 1 and n^0 as 1, regardless of the value of n. Consider now the following identity:

$$\frac{3 \times 3 \times 3}{3 \times 3} = 3$$

In exponential notation the identity above would be written:

$$\frac{3^3}{3^2} = 3^{(3-2)} = 3^1 = 3$$

From this example we see that a number in the denominator of a fraction can be considered as having a negative exponent. Thus, the preceding example could have been written in the form:

$$3^3 \times 3^{-2} = 3^{(3-2)} = 3^1$$

In general we find it useful to mean by n^{-m} the equivalent expression $1/n^m$. In multiplying together powers of the same number we then simply add their exponents, paying due regard to sign if some of the powers are negative, meaning that they occur in the denominator. As an example, $10^{-2} = 1/10^2 = 0.01$. Similarly, $10^3 = 1000$, and $10^3 \times 10^{-2} = 10^1 = 10$.

The foregoing furnishes the basis for the *power-of-ten notation.* Powers of ten are factored from each number, so that the numerical part left over is between 1 and 10. Thus, $3167 = 3.167 \times 10^3$, and $0.0123 = 1.23 \times 10^{-2}$. In either case we count right or left from the decimal point until we reach a number between 1 and 10. If we count to the left, the power of ten is positive (3 in the example above), while if we count to the right, the power of ten is negative (2 in the example above). One useful feature of this notation is the expression of very large or very small numbers. For instance, instead of writing

$$602300000000000000000000$$

we can write 6.023×10^{23}. Similarly, instead of writing

$$0.000000000000000000000000000000911$$

we can write 9.11×10^{-31}. The other advantage of this notation is that in multiplication and division we can combine the powers of ten very easily by addition and subtraction as explained in the previous paragraph. Here are some examples of this procedure.

EXAMPLE Suppose that we would like to multiply 3785 by 163. We can then write:

$$\begin{aligned} 3785 \times 163 &= 3.785 \times 10^3 \times 1.63 \times 10^2 \\ &= (3.785 \times 1.63) \times 10^{3+2} \\ &= 6.17 \times 10^5 \qquad \text{(approximately)} \end{aligned}$$

Next, consider multiplying 450 by 0.12. We can write this in the form:

$$\begin{aligned} 450 \times 0.12 &= 4.50 \times 10^2 \times 1.2 \times 10^{-1} \\ &= 5.40 \times 10^{2-1} \\ &= 5.40 \times 10^1 \\ &= 54 \end{aligned}$$

In the same way we can use the power-of-ten notation where division is involved. Thus, if we wish to divide 4500 by 30,000 we can write:

$$\frac{4500}{30{,}000} = \frac{4.5 \times 10^3}{3 \times 10^4} = 1.5 \times 10^{3-4} = 1.5 \times 10^{-1} = 0.15$$

A somewhat more difficult problem might involve several numbers. Consider the following example:

$$\frac{24 \times 500}{15 \times 80} = \frac{2.4 \times 10^1 \times 5 \times 10^2}{1.5 \times 10^1 \times 8 \times 10^1} = \frac{12 \times 10^3}{12 \times 10^2}\, 10^1 = 10$$

Additional examples are given in the exercises at the end of this chapter and should be worked out by the reader.

While the value of the power-of-ten notation is mainly in its use in expressing large and small numbers and in making arithmetic easier, we can also show most clearly in this notation the estimated accuracy of a number. When we say that we have a dozen oranges, we mean that we have exactly twelve oranges. Mathematically, we might therefore say that we have 12.0000 . . . oranges. In the preceding sentence, the string of dots means that we know the quantity exactly. On the other hand, if we measure the length of a table with a yardstick and say that the length of the table is six feet, we do not claim that the length is exactly six feet. In this case, we might only be able to read the yardstick within a hundredth of a foot, so more carefully we would say that the length of the table is (6 ± 0.01) feet. In the measurement of the size of a cloud, the uncertainty in the measurement would be much larger. In all physical measurements the estimated error can be given as shown above.

If we now consider again the example of the length of the table in the preceding paragraph, another way of showing our estimate of the precision of our measurement is to say that the length is 6.00 feet. By this statement we mean that we think that the length lies between 5.99 feet and 6.01 feet. By convention, unless otherwise stated, we imply an uncertainty of one unit in the last figure quoted. In this example, the value is quoted to three *significant figures,* and the uncertainty occurs in the last significant figure.

Suppose now that I say that the speed of light is 186,000 miles per hour. Do I mean that the speed of light lies between 185,999 and 186,001 miles per hour? Certainly not. However, if I write for the speed of light 1.86×10^5 miles per hour, I now do mean that this value lies between 1.85×10^5 and 1.87×10^5 miles per hour. Thus, by using the power-of-ten notation I can show exactly the number of significant figures to which the value is being quoted.

If I survey a plot for a football field and estimate that my error is probably no more than $\frac{1}{3}$ of an inch (approximately 0.01 yards), I could write that the length of the field was 1.0000×10^2 yards. This statement implies that the uncertainty in the measurement is $0.0001 \times 10^2 = 0.01$ yards. While in this case I could have quoted the length of the field as (100 ± 0.01) yards, when very small or large numbers are involved the use of the power-of-ten notation is clearly much more convenient and clear.

The value of a certain quantity is given as 5.230×10^{10}. In this case the zero is a significant figure. Thus, the probable value of this quantity lies between 5.229×10^{10} and 5.231×10^{10}. Similarly, stating that the value of the quantity is 5.2300×10^{10} implies that the uncertainty is in the second zero, since the value of the quantity was given to five significant figures. In this example we see again that the use of power-of-ten notation gives a precise idea of the accuracy of the measurement, while writing the value as 52,300,000,000 would not.

Index

A

B

C

D

E

F

G

H

I

J

K

L

M

N

O

P

Q

R

S

T

U

V

W

X

Y

Z

T5-DHH-307